EVOLUTIONARY ECOLOGY

Eric R. Pianka
The University of Texas at Austin

HARPER & ROW, PUBLISHERS
New York Evanston San Francisco London

Sponsoring Editor: Alvin A. Abbott
Project Editor: Holly Detgen
Designer: T. R. Funderburk
Production Supervisor: Robert A. Pirrung

EVOLUTIONARY ECOLOGY

Library of Congress Cataloging in Publication Data 73-7476

Pianka, Eric R.
Evolutionary ecology.
Bibliography: p.
1. Ecology. 2. Evolution. I. Title.
QH541.P5 574.5'2 73-7476
ISBN 0-06-045226-9

To Ronald Fisher and Robert MacArthur
whose penetrating insight and inductive genius
laid so much of the foundation of modern ecology.

Contents

Preface

Language forces us to express ourselves in a one-dimensional stream of words. But nature is seldom so simple and tractable; instead, she has many different dimensions. This is particularly true of ecology because its subject matter includes many complexly interrelated concepts and phenomena involving several levels of organization. Thus there is no such thing as an "ideal" outline or a perfect sequential order for the presentation of the subject treated here. To obtain an overview of modern ecology, a student needs to assimilate a great many ideas. Ideally, a reader would know everything in this book even before beginning to read it! Perhaps the only solution is to read it twice. To help a reader follow a train of thought different from that I have chosen to use here, various chapters and sections are cross-referenced.

I have enjoyed writing most of this book, and I hope you will enjoy reading most of it. I claim no great originality for its content, because virtually all of it comes from talking with other people or from reading their writings. Rather it is my own peculiar blend and distillation of what I think are "significant" facts, ideas, and principles; these represent the residue remaining after considerable sifting and sorting of many other facts and ideas. My approach is abstract and conceptual, and I strive to provide an overview of the subject matter. More than anything else, the book is an image of a part of my mind, a part I hope mirrors some sort of reality external to it and common to all of us.

E.R.P.

Acknowledgments

I am very grateful to my students for helping me, indeed, sometimes *forcing* me, to clarify ideas and improve my presentation of them. My own thinking has been indelibly molded by my students, teachers, and colleagues, particularly by Thomas Frazzetta, John Gillespie, Henry Horn, Robert MacArthur, William Neill, Gordon Orians, Daniel Otte, Robert Paine, Christopher Smith, and Mary Willson. I am especially indebted to Gordon Orians for introducing me to "selective thinking" (that is, using the theory of natural selection the way that Ronald Fisher used it). I thank Gordon Orians and my wife, Helen, for carefully reading an early draft of the entire manuscript and for offering many valuable suggestions for its improvement. John Avise and Eric Charnov made useful comments on some chapters. I am also most grateful to the following authorities for reading the chapters indicated and offering suggestions for their improvement: LaMont Cole (Chapter 4), Robert Whittaker (Chapter 7), Edward Wilson (Chapter 8), and Paul Ehrlich (Chapter 9). Amy Kramer provided me with much early encouragement, without which I might never have completed this project. I am also indebted to Anne Reynolds for typing the entire final draft, both promptly and extremely accurately. Finally, I thank Glennis Kaufman and Helen Pianka for proofreading the entire book. Errors remaining, of course, are entirely my own.

1 Introduction

Domain of Ecology: Definitions and Groundwork

Ecologists, along with all scientists, assume that an organized reality exists in nature and that principles can be formulated which will adequately reflect that natural order. A fundamental and important way in which biological phenomena can be ordered is by simple and direct enumeration, as in the classification of organisms or biotic communities. Thus we recognize different ecological systems such as tundra, desert, prairie, savanna, deciduous forest, coniferous forest, and rain forest. Early ecology was primarily descriptive; the forefathers of the science spent most of their time describing, itemizing, and classifying various ecological elements. This process was absolutely necessary before modern ecology could develop. Founded and firmly based upon this older body of descriptive information, modern ecology seeks to develop general theories with predictive powers that can be compared against the real world. Modern ecologists want to understand and to explain, in general terms, the origin and mechanisms of interactions of organisms with one another and with the nonliving world. In order to build such general theories of nature, ecologists construct "models" of reality which generate testable predictions. If the predictive powers of a model fail, it is either discarded or revised. Models and theories that do not conform adequately to reality are gradually replaced by those that better reflect the real world. The great complexity of ecological systems necessitates the use of graphical and mathematical models, so much so that modern ecologists often employ nearly as much mathematics as biology. However, the development of sound ecological principles depends equally as much upon what might be called "biological intuition," and there is certainly no substitute for a firm foundation in natural history. Models based on erroneous biological assumptions, no matter how elegant and elaborate, can hardly be expected to reflect nature accurately! Hence, a good background for comprehension of the field of ecology includes some biology and mathematics as well as a solid basis in general science.

Ecology and environment are words frequently encountered in the news and popular media, almost invariably in conjunction with man and his environment. As often as not they are misused, especially by politicians and other advertisers. Many people use "ecology" to refer primarily to man's ecology. Every year I begin my courses by telling students that the basic science of ecology is not synonymous with a study of the effects of man on his surroundings, himself, and other organisms, but in fact represents a much broader class of subject matter. Yet each year some students complain that there is not enough "ecology" in my courses, presumably because man's particular devastating problems are given short shrift. Some of the problems facing man today illustrate what can happen when ecological systems are not used wisely in accordance with sound ecological principles; as such, human examples (Chapter 9) are very pertinent to the content of this book. However, throughout the book, emphasis is given to the principles of *basic* ecology, particularly as these principles apply to and can be interpreted in terms of the theory of natural selection. Major concepts and principles are stressed more than detail, but references are given at the end of each chapter for those desiring to delve more deeply into particular subjects.

Ecology has been variously defined as "scientific natural history," "the study of the structure and function of nature," "the sociology and economics of animals," "bionomics," "the study of the distribution and abundance of organisms," and "the study of the interrelationships between organisms and their environments." The last of these definitions is probably the best, with "environment" being defined as *the sum total of all physical and biological factors impinging upon a particular organismic unit.* For "organismic unit" one can substitute either "individual," "population," "species," or "community." Thus, we may speak of the environment of an individual or the environment of a population, but to be precise, a particular organismic unit should be understood or specified. The environment of an individual contains fewer elements than the environment of a population, which in turn is a subset of the environment of the species or community.

In order to avoid the apparent circularity in the above definition, ecology might be better defined as *the study of the relations between organisms and the totality of the physical and biological factors affecting them or influenced by them.* Thus, ecologists begin with the organism and seek to understand how the organism affects its surroundings and how these surroundings in turn affect the organism.

Environment includes everything from sunlight and rain to soils and other organisms. An organism's environment consists not only of other plants and animals encountered directly (such as foods, trees used for nesting sites, predators, competitors, and so on), but also of purely physical processes and inorganic substances, such as daily temperature fluctuations and oxygen and carbon dioxide concentrations. Of course the latter may be affected by other organisms which are then indirectly a part of the environment of the first organism. Indeed, any remote connection or interaction between two organismic units means that each is part of the other's environment.

Because there are direct or indirect interactions between almost all the organisms in a given area, the biotic component of the environment of most organisms is extremely complex. Coupling this great complexity with a multifaceted physical environment makes ecology an exceedingly broad subject. It has been said that ecology is "almost boundless" (Kormondy, 1969); indeed, its breadth inspires awe even in an ecologist! No other discipline seeks to explain such a variety of phenomena at so many different levels. As a consequence, ecology takes in aspects of many other fields, including physics, chemistry, mathematics, computer science, geography, climatology, geology, oceanography, economics, sociology, psychology, and anthropology. Ecology is properly classified as a branch of biology; students of ecology attempt to interweave and correlate the subdisciplines of biology such as evolution, genetics, systematics, morphology, physiology, ethology (behavior), as well as various taxonomic subdivisions of biology like algology, entomology, ichthyology, herpetology, mammalogy, and ornithology. Sometimes "plant ecology" is distinguished from "animal ecology"; however, even as basic as this distinction may be, it is most unfortunate. Plants and animals inevitably constitute part of one another's environments, and their ecologies should always be considered together. One of the most promising frontiers in modern ecology is the interface between population genetics and population ecology; as the principles and theories of each of these disciplines gradually fuse into one another, many potent new insights into population biology are emerging.

Obviously, it is impossible for anyone to master all of such an enormous field, and as a result there are many different kinds of ecologists with a wide variety of perspectives on the subject matter of the science. The breadth of ecology, combined with its youth and great relevance to human problems, makes it a fascinating and exciting field, with real potential for growth and refinement. Young sciences, and especially complex biological ones like ecology, can be characterized as "soft" sciences, in that they are not as precise as older and better established "hard" sciences such as chemistry and physics. As every science matures, it becomes more and more abstract and its hypotheses are refined and improved until they eventually attain the status of "laws," as in the familiar laws of chemistry and physics. Ecology at present has very few firm laws, but many hypotheses, and much work and testing of these hypotheses remains. The one concept closest to deserving the status of "law" in ecology, and one which is shared with all of biology, is *natural selection* (see pp. 9–12).

There is a natural sequence to the subject matter of ecology, proceeding from the inorganic to the organic world, which is diagramed in Figure 1.1. The components of ecological systems are here considered in order of generally increasing complexity, or from left to right in the figure. In this book, the organismic world is treated in much greater detail than the non-organismic world.

The climate, soils, bacteria, fungi, plants, and animals at any particular place together constitute an *ecosystem.* Thus, each ecosystem has both abiotic (nonliving) and biotic (living) components. The biotic components of an ecosystem, or all the organisms living in it, taken together, comprise an ecological *community.* The abiotic components can be separated into inorganic and organic, while the biotic components are usually classified as producers, consumers, and decomposers. Producers, sometimes called *autotrophs,* are the green plants that trap solar energy and convert it into chemical energy. Consumers, or *heterotrophs,* are all the animals that either eat the plants or

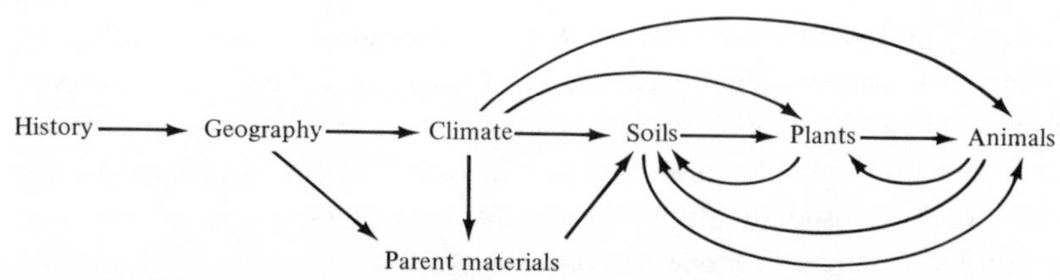

FIGURE 1.1 Diagrammatic representation of the subject matter of ecology showing the natural sequence proceeding from the inorganic to the organic world. Many other arrows and feedback loops could be added, but those depicted are of major importance. Chapter 2 deals with history, geography, and climate; Chapter 3 briefly treats climate, soils, and plants. Remaining chapters examine the interactions between and among plants and animals, especially at the population and community levels.

one another; all heterotrophs are thus directly or indirectly dependent upon plants for energy. Several levels of consumers are recognized (primary, secondary, and tertiary) depending upon whether they eat plants directly or other herbivorous or carnivorous animals. Decomposers, also heterotrophs, are often bacteria and fungi; they function in the ecosystem by breaking down plant and animal material into simpler components and thus returning nutrients to the autotrophs. Decomposers are therefore essential in recycling the matter within an ecosystem.

Plants and animals in ecosystems can be considered at several different levels: individuals, family groups, populations, species, and communities (subsequent chapters deal with each organismic level of organization, but most emphasis is given to populations). None of these levels of organization can be adequately understood in isolation, because each exerts strong influences upon the others. Every individual is simultaneously a member of a population, a species, and a community, and must therefore be adapted to cope with each, and must be considered in that context. An individual's fitness, or its ability to perpetuate itself as measured by its reproductive success, is determined not only by its status within its own population, but also by the various interspecific associations of its species, and especially by the particular community in which it finds itself. Similarly, every community is composed of many populations and numerous individuals, which determine many, but by no means all, of its properties. At each level of organization, important new properties emerge that are not properties of the preceding level. Thus, individuals have a fixed genetic makeup and live or die, whereas populations have gene frequencies, birth rates, and death rates. All these population parameters (and others) can change in time as the composition of the population changes in response to a changing environment.

Basic Mendelian Genetics

Although a background in genetics is not essential for appreciation of many populational and ecological phenomena, it is an extremely useful aid for application to some such phenomena and is required for a full understanding of others. The precise rules of inheritance were unknown when Darwin (1859) developed the theory of natural selection, but were formulated a short time afterward (Mendel, 1865). Darwin accepted the mechanism of inheritance in vogue at the time: *blending inheritance*. Under the blending inheritance hypothesis the genetic makeups of both parents are blended in their progeny, and all offspring produced by sexual reproduction should be genetically intermediate between their parents; genetic variability is thus lost rapidly unless new variation is continually being produced. (Under blending inheritance and random mating, genetic variability is *halved* each generation.) Darwin was forced to postulate extremely high mutation rates in order to maintain the genetic variability observed in most organisms (Fisher, 1930), and he was well aware of the inadequacy of knowledge on inheritance. Mendel's discovery of *particulate inheritance* represents one of the major breakthroughs in biology.

Mendel performed breeding experiments with different varieties of peas, paying particular attention to a single trait at a time. He had two types which bred "true" for yellow and green peas, respectively. When a purebred green pea plant was crossed with a purebred yellow pea plant, all progeny, or individuals of the first filial generation (F_1), had yellow peas. However, when these F_1 plants were crossed with each other or self-fertilized, about one out of every four offspring in the second filial generation (F_2) had green peas. Furthermore, only about one-third of the yellow F_2 pea plants bred true; the other two-thirds, when self-fertilized, produced some offspring with green peas. All green pea plants bred true. Mendel proposed a very simple hypothesis to explain his results and performed many other breeding experiments on a variety of other traits which corroborated and confirmed his interpretations. Subsequent work has only strengthened his hypothesis, although it has also led to certain modifications and improvements.

Mendel postulated that each pea plant had a double dose of the "character" controlling pea color, but that only a single dose was transmitted into each of its sexual cells, or *gametes* (pollen and ovules or sperm and eggs). Purebred plants, with identical doses, produced genetically identical single-dosed gametes; the F_1 plants, on the other hand, with two different doses, produced equal numbers of the two kinds of gametes, half bearing the character for green and half that for yellow. In addition, Mendel proposed that yellow masked green whenever the two occurred together in double dose; thus all F_1 plants had yellow peas, but when self-fertilized produced some F_2 progeny with green peas. All green pea plants, which had a double dose of green, bred true.

Modern terminology for various aspects of Mendelian inheritance is as follows: (1) the "character" or "dose" controlling a particular trait is termed an *allele*; (2) its position on a chromosome (below) is termed its *locus*; (3) a single dose is the *haploid* condition, designated by n, while the double-dosed condition, designated by $2n$, is *diploid* (*polyploids*, such as triploids and tetraploids, are designated by still higher numbers); (4) the set of alternate alleles which may occur at a given locus (there can be only two alleles in a diploid individual, but there may be more than two in any given population) is termed a *gene*; (5) purebred diploid individuals with identical alleles are *homozygotes*, homozygous for the trait concerned; (6) individuals with two different alleles, such as the F_1 plants above, are *heterozygotes*, heterozygous at that locus; (7) an allele which masks the expression of another allele is said to be *dominant* and the one that is masked is *recessive*; (8) unlinked alleles separate, or *segregate*, from each other in the formation of gametes; (9) whenever heterozygotes or two individuals homozygous for different alleles mate, new combinations of alleles arise in the following generation by reassortment and *recombination* of the genetic material; (10) observable traits of an individual (i.e., yellow or green in the above example) are aspects of its *phenotype*, which includes all observable characteristics of an organism; and (11) whether or not an organism breeds true is determined by its *genotype*, which is the sum total of all its genes.

Occasionally, in some organisms, pairs of alleles with so-called *incomplete dominance* occur. In such cases the phenotype of the heterozygote is intermediate between that of the two homozygotes; that is, phenotype accurately reflects genotype and vice versa. Presumably, alleles conferring advantages upon their bearers will usually evolve dominance over time, because such dominance ensures that a maximal number of the organism's progeny and descendants will benefit from possession of that allele. The apparent rarity of incomplete dominance is further evidence that dominance has evolved. Moreover, so-called wild-type alleles, that is, those most prevalent in natural wild populations, are nearly always dominant over other alleles occurring at the same locus. Geneticists have developed numerous theories of "the evolution of dominance," but the exact details of the process have not been completely resolved.

Cytological observations of appropriately prepared cell nuclei confirm Mendel's hypothesis beautifully (Figure 1.2). Microscopic examination of such cells reveals elongated dense bodies in cell nuclei; these are the *chromosomes*, which have been shown to contain the actual genetic material, deoxyribonucleic acid, or DNA. Nuclei of diploid cells, including the zygote (the fertilized ovule or egg) and the somatic (body) cells of most organisms always contain an even number of chromosomes. (The exact number varies widely from species to species, with as few as two in certain arthropods to hundreds in some plants.) Moreover, pairs of distinctly similar *homologous chromosomes* are always present and often easily detected. Gametes, however, contain only half the number of chromosomes found in diploid cells, and, except in polyploids, none of them are homologous. Thus, haploid cells contain only one full set of different chromosomes and alleles, or one *genome*, whereas

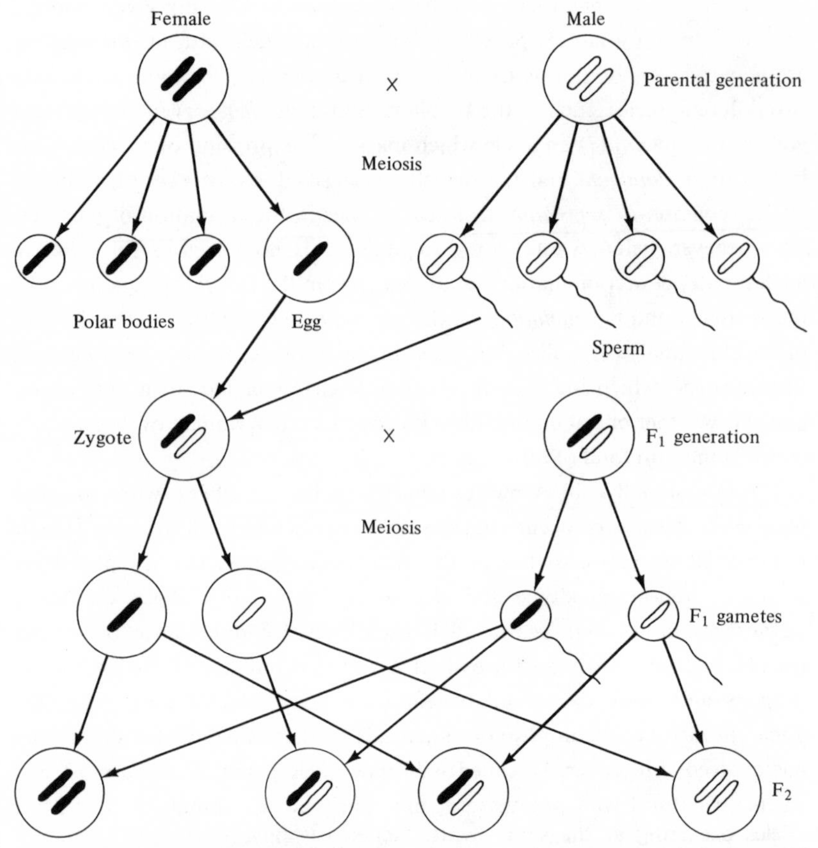

FIGURE 1.2 Diagrammatic representation of the cytological events in cell nuclei showing how the two parental genomes are sorted and recombined in the next generation, or the F_2. For simplicity, only one pair of chromosomes is shown and the complex events of the reduction division (meiosis) are omitted.

diploid cells contain two. During the reduction division (*meiosis*) in which diploid gonadal cells give rise to haploid gametes, homologous chromosomes separate (Figure 1.2). Later, when male and female gametes fuse to form a diploid *zygote* which will develop into a new diploid organism, homologous chromosomes come together again. Hence, one genome in every diploid organism is of paternal origin while the other is of maternal ancestry. Because each member of a pair of homologous chromosomes separates from its homologue independently of other chromosome pairs, the previous generation's chromosomes are reassorted with each reduction division. Thus, the genetic material is regularly rearranged and mixed up by the dual processes of meiosis and the actual fusion of gametes.

Many different loci and allelic systems occur on each chromosome. Had Mendel examined two different traits controlled by alleles located on the same chromosome, he might have discovered the phenomenon of *linkage*, in which two alleles do not segregate truly independently but are statistically associated with or dissociated from one another. During meiosis homologous chromosomes can effectively exchange portions by means of *crossovers*; this process is referred to as *recombination*. Because the frequency of occurrence of crossovers between two loci is a function of the distance between them on the chromosome, geneticists can use crossover frequencies to "map" the effective crossover distance between loci, as well as their positions relative to one another on the chromosome. By means of close linkage whole blocks of statistically associated alleles can be passed on to progeny as a functionally integrated unit of "coadapted" alleles.

In many organisms, a single pair of chromosomes, termed *sex chromosomes*, determine the sex of their bearer (the remaining chromosomes, which are not involved in sex determination, are *autosomes*). Typically one homologue of the sex chromosome pair is smaller. In the diploid state, an individual heterozygous for the sex chromosomes is *heterogametic*. In mammals, males are the heterogametic sex with an XY pair of sex chromosomes, and females are the homogametic sex with an XX. Note that since male–male matings are impossible, the homozygous genotype YY can never occur. In birds and some other organisms, the female is the heterogametic sex.

Natural Selection

The theory of natural selection is a truly fundamental unifying theory of life. A thorough appreciation of it is essential background to understanding modern ecology. Natural selection comes as close to being a "fact" as anything in biology, including Mendelian genetics. While there is no such thing as "proof" in science (except in mathematics where all postulates are taken as given), an enormous body of data has been amassed in support of the theory of natural selection over the last century.

Although natural selection is not a difficult concept, it is frequently misunderstood. A common misconception is that natural selection is synonymous with evolution. It is not. Evolution refers to temporal changes of any kind, whereas natural selection specifies one particular way in which these changes are brought about. There are other possible mechanisms of evolution besides natural selection, such as the inheritance of acquired characteristics and genetic drift. Another frequent misconception is that natural selection occurs mainly through differences between organisms in death rates, or *differential mortality*.

Selection may proceed in a much more subtle and inconspicuous way. Whenever one organism leaves more successful offspring than another, in time its genes will come to dominate the population gene pool. Eventually, the genotype leaving fewer offspring must become extinct in a stable population, unless there are concomitant changes conferring an advantage on it as it becomes rarer. Thus, ultimately, *natural selection operates* ***only by differential reproductive success.*** Differential mortality can be selective *only* to the degree that it creates differences between individuals in the number of reproductive progeny they produce.

Hence, Darwin's choice of words, such as "struggle for existence" and "survival of the fittest," have had a most unfortunate consequence. They have tended to make people think in terms of a dog-eat-dog world and to consider such things as predation and fighting over food as the prevalent means of selection. All too often, natural selection is couched in terms of differential death rates. Thus, the strongest and fastest individuals are often considered to have a selective advantage over weaker and slower individuals. But, if this were the case, every species would continually gain in strength and speed. Because this is not happening, selection against increased strength and speed (counterselection) must be occurring and must limit the process.

Animals are sometimes too aggressive for their own good; an extremely aggressive individual may spend so much time and energy chasing other animals that he spends less than average time and energy on mating and reproduction, and as a result leaves fewer offspring than average. Likewise an individual can be too submissive, and spend too much time and energy running away from other animals. Indeed, it is a basic principle of biology that, under *stable* conditions, the *intermediates* in a population leave more descendants, on the average, than the extremes. We say that they are more "fit." An individual's "fitness" is measured by the proportion of its genes left in the population gene pool. Selection of this sort, which continually crops the extremes and tends to hold constant the intermediate or average phenotype, is termed *stabilizing selection* (Figure 1.3*a*). In a stable environment, genetic recombination increases populational variance each generation, while stabilizing selection reduces it back to approximately what it was in the previous generation.

However, in a *changing* environment, average individuals (modal phenotypes) may not be the most fit members of the population. Under such a situation, *directional selection* occurs, and the population mean shifts

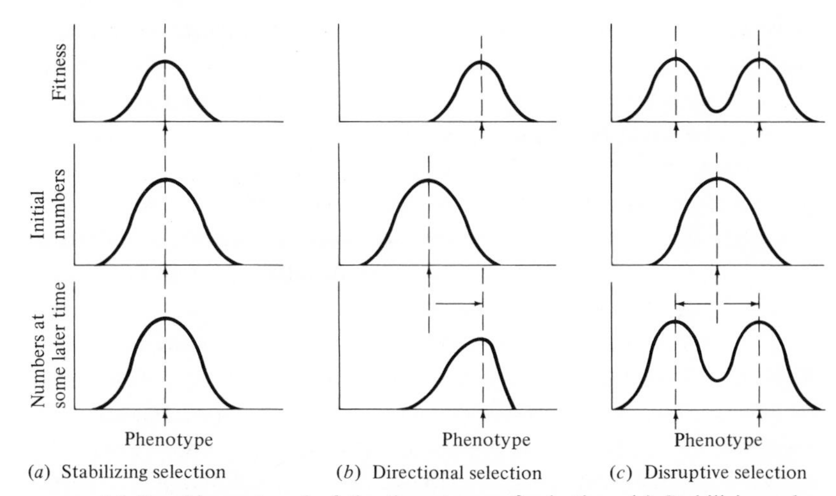

FIGURE 1.3 Graphic portrayal of the three types of selection. (*a*) Stabilizing selection, which occurs in stable environments, holds the modal phenotype constant. (*b*) Directional selection takes place in a changed environment and causes a shift in the modal phenotype. (*c*) Disruptive selection, with two or more modal phenotypes, occurs in patchy environments with two or more discrete phases.

towards a new phenotype (Figure 1.3*b*), better adapted to the altered environment. Eventually, of course, unless the environment continues to change, an equilibrium is reached in which the population is readjusted to the new environment, whereupon stabilizing selection resumes.

A third type of selection, *disruptive selection*, takes place when two or more phenotypes with high fitnesses are separated by intermediate phenotypes of lower fitness (Figure 1.3*c*). This usually occurs in distinctly heterogeneous environments with a discrete number of different "patches." Disruptive selection is one of the mechanisms that produces and maintains polymorphisms, such as the green–brown polymorphisms of some insects (Chapter 5). For instance, some butterflies mimic leaves (commonly called "leaf butterflies"); one population may contain both green and brown animals, with the former matching living leaves, and the latter, dead ones. Through appropriate behavior and selection of resting sites, each phenotype has a relatively high fitness; in contrast a butterfly with a phenotype intermediate between green and brown would match its surroundings less well and have a considerably lower fitness.

Although natural selection actually operates on phenotypes of individuals (that is, an organism's immediate fitness is determined by its total phenotype), the effectiveness of selection in changing the composition of a population depends upon the *heritability* of phenotypic characteristics, or the percentage of variability attributable to genotype. Thus, because non-genetic traits are not inherited, differential reproduction by different phenotypes stemming from such nontransmittable traits cannot alter a population's gene pool. Different genotypes may often have fairly similar phenotypes, and thus similar fitnesses. It has even been suggested that selection favors alleles which are "good mixers," that is, those which work well with a wide variety of other genes in increasing the fitness of their bearer (Mayr, 1959). Conversely, of course, identical genotypes can develop into rather different phenotypes under different environmental conditions.

A word of warning is now appropriate. Overenthusiastic proponents of natural selection have sometimes used it to explain observed biological phenomena in a somewhat after-the-fact manner. Thus, one might say that an animal "does what it does because that particular behavior increases its fitness." Used in this way, natural selection can be misleading; it is so pervasive and powerful that nearly any observable phenomenon can be interpreted as a result of selection, even though some probably are not. Hence, there is a real danger of circularity in such arguments. One should always consider alternative explanations for various biological phenomena.

Levels of Selection

Does natural selection act upon individuals or upon entire groups of individuals, such as families, populations, species, communities, and/or ecosystems? What is the "unit" of selection? These questions are often discussed by ecologists and there is some disagreement as to correct answers.

One frequently reads statements similar to the following: "It is conceivable that reproductive capacity might become so great as to be detrimental *to a species*. The many deleterious effects of overcrowding are well known" (from Cole, 1954b, p. 104, italics mine). Fisher (1958a) recognized the fallacy of such statements and cautioned against arguments invoking "the benefit of the species." In spite of his warning, it has become vogue to consider and interpret many behavioral and ecological attributes of species as having been evolved for the benefit of the group, rather than the individual.

Any individual sacrificing its own reproductive success for the benefit of a group is obviously at a selective disadvantage (within that group) to any other individual not making such a sacrifice. In order for such traits to evolve, natural selection must therefore act upon entire groups of organisms, which in turn must possess differential rates of survivorship and reproduction (that is, differential fitness). Because the course of selection within groups cannot be altered by selection acting between groups (Wright, 1931), group selection requires some very special conditions. In order for group selection to operate, selfish groups must go extinct faster than selfishness arises within altruistic groups, and furthermore, the majority of newly founded groups must be altruistic.

Recently Williams (1966a) reemphasized, restated, and expanded the argument against group selection. He points out that classical Darwinian selection at the level of the individual is adequate to explain the majority of putatively "group-selected" attributes of populations and species, such as those suggested by Wynne-Edwards (1962) and Dunbar (1960, 1968). Williams reminds us that group selection has more conditions and is therefore a more onerous process than classical natural selection; furthermore he urges that it be invoked only after the simpler explanation has clearly failed. Lewontin (1970) makes a fairly convincing argument that group selection may occur, although much less frequently than classical Darwinian selection. We return to this issue from time to time in later chapters.

Self-replicating Molecular Assemblages

Life began with the first self-replicating molecular assemblage; moreover, natural selection begins to operate as soon as any complex of molecules begins to replicate itself. No copying device is perfect, and some variants of the molecular assemblage produced are bound to be better than others in their abilities to survive and replicate themselves under particular environmental conditions. As resources become depleted, competition can occur among various self-replicating units. Furthermore, given enough time, some inferior variants presumably become extinct. Thus, each molecular unit will maximize its own numbers at the expense of all other such units. Even as originally simple self-replicating units become more and more elaborate, eventually attaining the complex form of present-day organisms, the same principles of natural selection are in effect throughout. Thus we can make certain statements about life that are entirely independent of the precise mechanism of replication. For example, should there be self-replicating molecular assemblages on Mars or any other distant planet, the fact that they may not obey the laws of Mendelian genetics would not drastically alter their basic attributes as living material. *Natural selection and competition are inevitable outgrowths of reproduction in a finite environment.* Hence, natural selection exists independently of life on Earth.

Once a self-replicating entity arises, qualitatively new phenomena exist that are not present in an inanimate world. In order to reproduce, living organisms (or replicating molecular assemblages) must actively gather other molecules and energy, that is, they must have some sort of acquisition technique. Direct and indirect disputes over resources place those units best able to acquire materials and energy (and best able to transform them into offspring) at a selective advantage over other such units which are inferior at these processes. Thus natural selection has the same effect as an efficiency expert, optimizing the use of resources in reproduction.

Obviously, the ultimate endpoint of these processes would be for the one best organismic unit to take over all matter and energy and to exclude all others. This has not occurred for a variety of reasons, as discussed earlier, but especially because of the great variability of the Earth's surface, both in space and in time (Chapter 2).

Limiting Factors, Tolerance Limits, and the Principle of Allocation

Ecological events and their outcomes, such as growth, reproduction, photosynthesis, primary production, and population size, are often regulated by the availability of one or a few factors or requisites in short supply, while other resources and raw materials present in excess may go partially unused. This principle has become known as the "law of the minimum" (Liebig, 1840). For instance, in arid climates, primary production (the amount of solar energy trapped by green plants) is strongly correlated with precipitation (Figure 1.4); here water is a "master limiting factor." Of the many different factors that can be limiting, frequently among the most important are various nutrients, water, and temperature.

When considering populations, we often speak of those which are food-limited, predator-limited, or climate-limited. Populations may be limited by other factors as well; for example, the density of breeding pairs of blue tits (*Parus caeruleus*) in an English woods was doubled by addition of many new nesting boxes (Lack, 1954, 1966), an indication that nest sites were limiting. However, limiting factors are not always so clear-cut, but may usually interact so that a process is limited simultaneously by several factors, with a change in any one of them resulting in a new equilibrium. For instance, both increased food availability and decreased predation pressures might result in a larger population size.

A related concept, developed by Shelford (1913b), is now known as the "law of tolerance." Too much or too little of anything can be detrimental

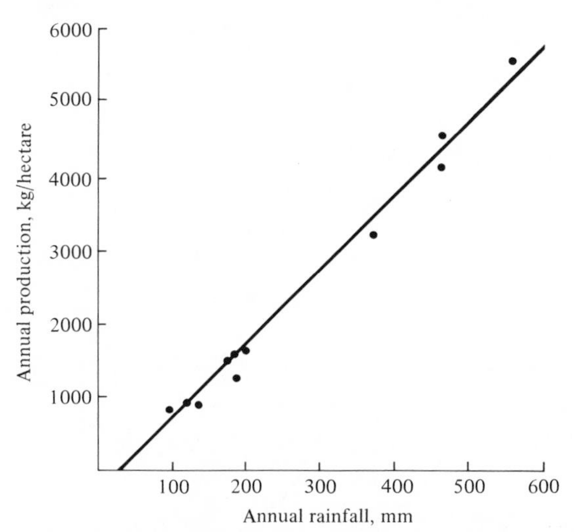

FIGURE 1.4 An example of the strong correlation between annual rainfall and primary production along a precipitation gradient in a desert region of South West Africa. [Adapted from Odum (1959) after Walter (1939).]

to an organism. In the early morning, a desert lizard finds itself in an environment which is largely too cold, whereas later in the day its environment is too hot. The lizard compensates somewhat for this by spending most of its time during the early morning in sunny places, while later on most of its activities take place in the shade. Each lizard has a definite optimal range of temperature, with both upper and lower *limits of tolerance*. More precisely, when measures of performance (such as fitness, survivorship, or foraging efficiency), are plotted against important environmental variables, bell-shaped curves, such as those shown in Figure 1.5, usually result.

Similar considerations apply to the performance of organisms along all sorts of environmental gradients. For instance, the fitness of some hypothetical organism in various microhabitats might be a function of relative humidity, somewhat as shown in Figure 1.5*a*. Moreover, assume that fitness varies similarly along a temperature gradient (Figure 1.5*b*). Figure 1.5*c* combines humidity and temperature gradients to show the variation in fitness with respect to both variables simultaneously. The range of thermal conditions tolerated is less at very low and very high humidities than it is at intermediate (more optimal) humidities. Similarly, an organism's *tolerance range* for relative humidity is less at extreme temperatures than it is at more optimal ones. Fitness is maximal when both temperature and humidity are optimal. There is thus an interaction between temperature tolerance and tolerance of relative humidities.

Levins (1968) suggested that such performance curves must obey

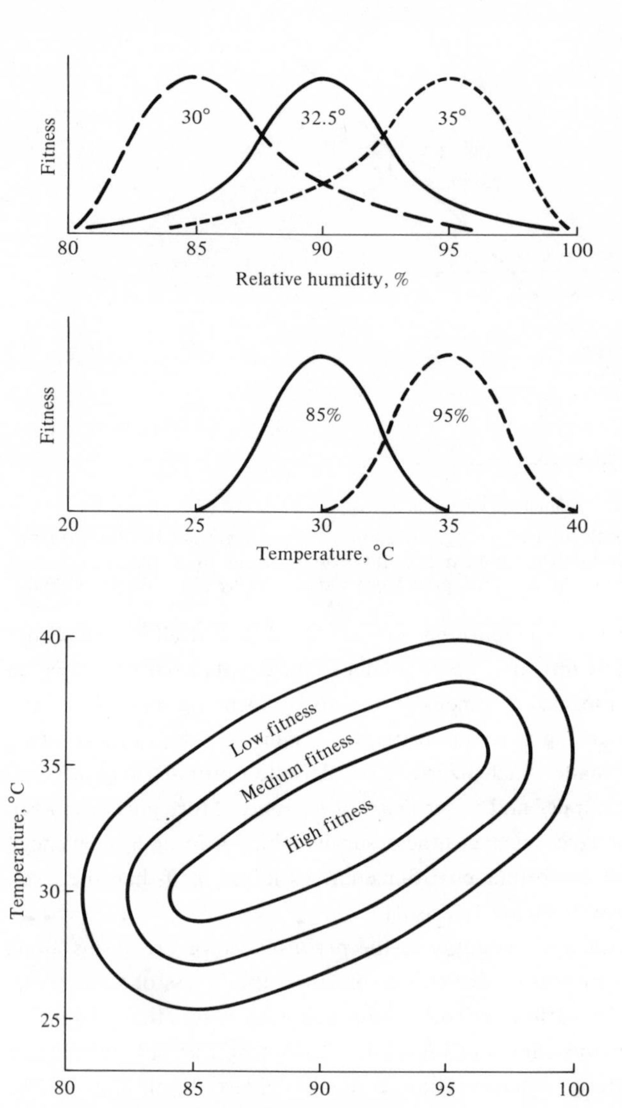

FIGURE 1.5 Hypothetical response curves showing how two variables can interact to determine an organism's fitness. Note that fitness is reduced at extremes of either temperature or humidity. Furthermore, notice that the range of humidities tolerated is less at extreme temperatures than it is at intermediate ones.

certain constraints: namely, that their breadth (or variance) cannot be increased without decreasing their height and vice versa (he coined this "the principle of allocation"). Because any individual organism has a finite energy budget, it must also have a limited capacity for regulation or homeostasis (which is the ability to reestablish a normal state after being disturbed). Organisms stressed along any one environmental variable are thus able to tolerate a lesser range of conditions along all other environmental variables.

Exactly analogous considerations apply to an organism's use of any resource. For example, in a given animal, there is an optimal size of prey items and other prey are less than optimal either because they are too large or too small. Any given animal has its own "utilization curve" which indicates the actual numbers of prey of different sizes taken per unit time under particular environmental conditions. In an idealized, perfectly stable, and infinitely productive environment, a utilization curve could become a spike with no variance, with the organism using only its most optimal resource. However, in actuality, limited and changing availabilities of resources, both in time and in space, result in utilization curves with breadth as well as height. In terms of the principle of allocation, an individual with a generalized diet adapted to eat a wide range of prey sizes presumably is not as effective at exploiting prey of intermediate size as another, more specialized, feeder. In other words, the jack-of-all-trades is a master-of-none. We take this subject up in detail in Chapter 6.

Levels of Approach to Science

Why do migratory birds fly south in the autumn? A physiologist might tell us that decreasing daylength (photoperiod) stimulates hormonal changes which in turn alter bird behavior so that they become restless. Eventually their wanderlust gets the upper hand and they head south. In contrast, an evolutionist would most likely explain that, by virtue of reduced winter mortality, those birds which flew south lived longer and therefore left more offspring than their nonmigratory relatives. Over a long period of time, natural selection resulted in intricate patterns of migratory behavior by means of differential reproductive success.

The physiologist's answer concerns the *mechanism* by which avian migratory behavior is influenced by *immediate* environmental factors, whereas the evolutionist's response is couched in terms of what might be called the "*strategy*" by which individual birds have left the most offspring in response to *long-term* consistent patterns of environmental change (i.e., high winter mortality). The difference between them is in outlook, between thinking in an "ecological" time scale (now time) or in an "evolutionary" time scale (geological time). At the physiologist's level of approach to science his answer is complete, as is the evolutionist's answer at his own level. Mayr

(1961) has termed these the "how?" and "why?" approaches to biology. They have also been called the "functional" and "evolutionary" explanations and the "proximate" and "ultimate" factors influencing an event (Baker, 1938). Neither is more correct; a really thorough answer to any question must include both, although often only the first can be tested directly. Nor are these two ways of looking at biological phenomena mutually exclusive; ecological events can always be profitably considered within an evolutionary framework and vice versa.

The evolutionary approach to biological questions is relatively new, and has resulted in a minor revolution in biology during the last 50 years. Before then, most biologists merely accepted as immutable a broad range of facts, such as the fact that sex ratios are often near equality (50:50), without considering why such facts might be so or how they could have evolved. Although we may not fully understand the causes and consequences of many populational phenomena, we can be certain that all have an evolutionary explanation. This is true of a broad spectrum of observations and facts, such as the following: (1) some genes are dominant, others are recessive; (2) some organisms live longer than others; (3) some organisms produce many more offspring than others; (4) some organisms are common, others are rare; (5) some organisms are generalists, others are specialists; and (6) some species are promiscuous, some polygamous, and some monogamous. All these variables are subject to the effects of natural selection. Population biologists are now thinking in an evolutionary time scale, and we have made substantial progress toward a theoretical understanding of why many of the above differences occur. Ronald Fisher was one of the first to recognize the power of rigorous application of the genetical theory of natural selection to population biology (Fisher, 1930), and his book has become a classic. Numerous other biologists have expanded, experimented with, and built upon Fisher's groundwork. For instance, Lack (1954, 1966, 1968) showed that reproductive rates are subject to natural selection. Others have worked on the evolution of dominance, mating systems, sex ratio, old age, life history phenomena, foraging strategies, and so on. Many of these subjects are taken up in Chapters 4, 5, and 6.

Selected References

Domain of Ecology: Definitions and Groundwork

Allee (1951); Allee *et al.* (1949); Andrewartha (1961); Andrewartha and Birch (1954); Billings (1964); Clarke (1954); Collier *et al.* (1973); Connell, Mertz, and Murdoch (1970); Daubenmire (1947, 1968); Dawson and King (1971); Elton (1927, 1958); Greig-Smith (1964); Harper (1967); Hazen (1964, 1970); Kendeigh (1961); Kershaw (1964); Knight (1965); Kormondy (1969); Krebs (1972); MacArthur (1972); MacArthur and Connell (1966); MacFadyen (1963); Maynard Smith (1968); Odum (1959, 1963, 1971); Oosting (1958); Pielou (1969); Platt (1964); Ricklefs (1973); Shelford (1963); Smith (1966); Watt (1973); Whittaker (1970); Wilson and Bossert (1971).

Basic Mendelian Genetics

Darlington and Mather (1949); Ehrlich and Holm (1963); Ford (1931, 1964); Maynard Smith (1958); Mendel (1865).

Natural Selection

Birch and Ehrlich (1967); Darwin (1859); Ehrlich and Holm (1963); Emlen (1973); Fisher (1930, 1958a, 1958b); Ford (1964); Haldane (1932); Kettlewell (1956, 1958); MacArthur (1962); Maynard Smith (1958); Mayr (1959); Mettler and Gregg (1969); Orians (1962); Salthe (1972); Williams (1966a); Wilson and Bossert (1971); Wright (1931).

Levels of Selection

Brown (1966); Cole (1954b); Darlington (1971); Dunbar (1960, 1968); Emlen (1973); Fisher (1958a); Lewontin (1970); Maynard Smith (1964); Van Valen (1971); Wiens (1966); Williams (1966a, 1971); Wright (1931); Wynne-Edwards (1962, 1964, 1965a, 1965b).

Self-replicating Molecular Assemblages

Bernal (1967); Blum (1968); Calvin (1969); Ehrlich and Holm (1963); Fox and Dose (1972); Jukes (1966); Oparin (1957); Ponnamperuma (1972); Salthe (1972); Wald (1964).

Limiting Factors, Tolerance Limits, and the Principle of Allocation

Errington (1956); Lack (1954, 1966); Levins (1968); Liebig (1840); Odum (1959, 1963, 1971); Shelford (1913b); Terborgh (1971); Walter (1939).

Levels of Approach to Science

Baker (1938); Fisher (1930); Lack (1954, 1966, 1968); MacArthur (1959); Mayr (1961); Orians (1962).

2 | *The Physical Environment*

Earth supports an enormous variety of organisms; for example, plants range from microscopic short-lived aquatic phytoplankton to small annual flowering plants to larger perennials to gigantic ancient Sequoia trees. Animals, while they never attain the massive size of a redwood tree, include forms as diverse as marine zooplankton, jellyfish, sea stars, barnacles, clams, snails, fish, whales, beetles, butterflies, worms, frogs, lizards, sparrows, hawks, bats, elephants, and lions. Different species have evolved and live under different environmental conditions. Some organisms are relatively specialized either in the variety of foods they eat or in the microhabitats they exploit, whereas others are more generalized; some are widespread, occurring in many different habitats, whereas still others have more restricted habitat requirements and geographic ranges. Temporal and spatial variation in the physical conditions for life often make possible or even actually necessitate variety among organisms, both directly and indirectly. Of course, interactions among organisms also contribute to the maintenance of this great diversity of life. But, before considering such biological interactions, we first examine briefly the nonliving world, which sets the background for all life and which often strongly influences the ecology of any particular organismic unit. A major factor in the physical environment is climate, which, in turn, is the ultimate determinant of water availability and the thermal environment; moreover, as we will see in Chapter 3, the latter two interact to determine the actual amount of solar energy that can be captured by plants (primary productivity) at any given time and place. Finally, because climate is a major determinant of both soils and vegetation, there is a close correspondence between particular climates and the types of natural biological communities that exist under those climatic conditions. The interface between climate and vegetation is considered in Chapter 3.

Major global and local patterns of climate are described briefly in this chapter. Entire books have been devoted to some of these subjects, and the reader interested in greater detail is referred to the references at the end of the chapter.

Major Determinants of Climate

The elements of climate (sun, wind, and water) are complexly interrelated. Incident solar energy produces thermal patterns, which, coupled with earth's rotation and movements around its sun, generate the prevailing winds and ocean currents. These currents of air and water, in turn, strongly influence the distribution of precipitation, both in time and in space. We first develop an oversimplified static and global view of the major factors determining climate, and then consider temporal variation and local perturbations.

The amount of solar energy intercepting a unit area of earth's surface varies markedly with latitude for two reasons. First, at high latitudes a beam of light hits the surface at an angle, and its light energy is spread out

TABLE 2.1 Average Annual Temperature (°F) at Different Latitudes

Latitude	Year	January	July	Range
90°N	−8.9	−42.0	30.0	72.0
80°N	−1.0	−26.0	35.6	61.6
70°N	12.7	−15.3	45.1	60.4
60°N	30.0	3.0	57.4	54.4
50°N	42.5	19.2	64.6	45.4
40°N	57.4	41.0	75.2	34.2
30°N	68.7	58.1	81.1	23.0
20°N	77.5	71.2	82.4	11.2
10°N	80.1	78.4	80.9	2.5
Equator	79.2	79.5	78.1	1.4
10°S	77.5	79.3	75.0	4.3
20°S	73.2	77.7	68.0	9.7
30°S	61.9	71.4	58.5	12.9
40°S	53.4	60.1	48.2	11.9
50°S	42.4	46.6	38.1	8.5
60°S	25.9	35.8	15.6	20.2
70°S	7.5	25.7	− 9.4	35.1
80°S	−16.6	12.6	−39.1	51.7
90°S	−27.6	7.7	−54.0	61.7

Source: Adapted from Haurwitz and Austin (1944).

over a large surface area. Secondly, a beam that intercepts the atmosphere at an angle must penetrate a deeper blanket of air, and hence more solar energy is reflected by particles in the atmosphere and radiated back into space. (Local cloud cover also affects the amount of the sun's energy that reaches the ground.) A familiar result of both these effects is that average annual temperatures tend to decrease with increasing latitude (Table 2.1). The poles are cold and the tropics are generally warm (seasons are discussed later).

Water in the atmosphere is warmed by heat radiating from the surface of the earth; much of this heat is radiated back to earth again. The net result is the so-called greenhouse effect, which leads to the retention of heat, keeping earth relatively warm even at night when there is temporarily no influx of solar energy. Without this effect, earth's surface would cool down to many degrees below zero much as does the dark side of the moon's surface. Thus the atmosphere buffers day–night thermal change.

Hot air rises. The ground, and air masses above it, receive more solar energy at low latitudes than at higher ones (Figure 2.1). Thus, tropical air masses, especially those near the equator, are warmed relatively more than temperate air masses, and an equatorial zone of rising air is created. These equatorial air masses cool as they rise and eventually move northward and southward high in the atmosphere above earth's surface (Figure 2.2*a*). As this cold air moves toward higher latitudes it sinks slowly at first, and then

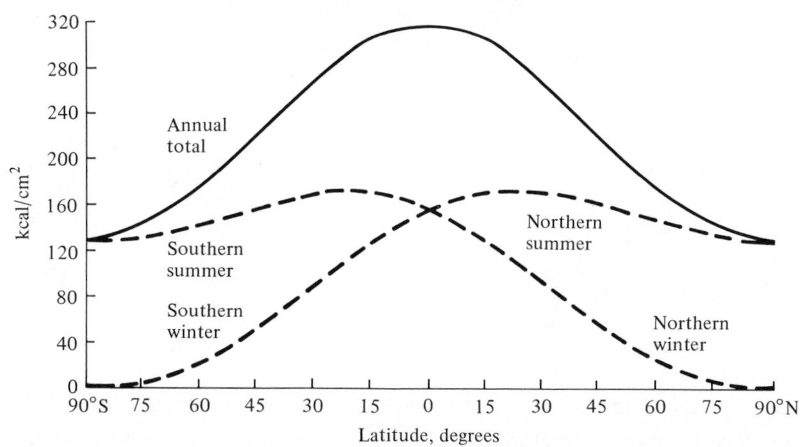

FIGURE 2.1 Estimated amount of incoming solar radiation that would intercept earth's surface in the absence of an atmosphere as a function of latitude. The six-month period from the spring equinox to the fall equinox (see Figure 2.7) is labeled "northern summer" and "southern winter" while the six months from the fall equinox to the spring equinox represent the "northern winter" and the "southern summer." [After Haurwitz and Austin (1944).]

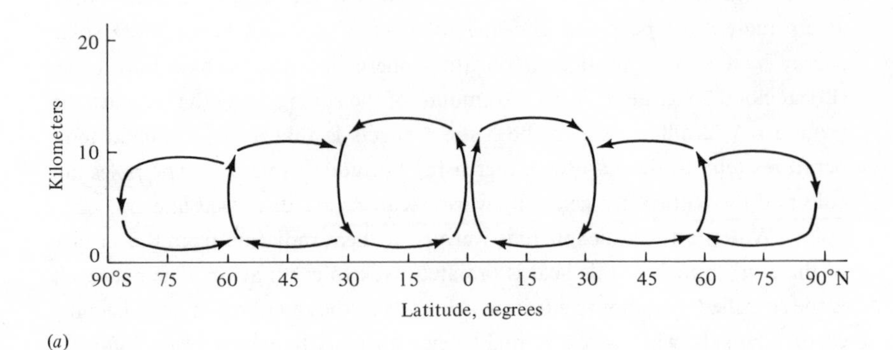

(*a*)

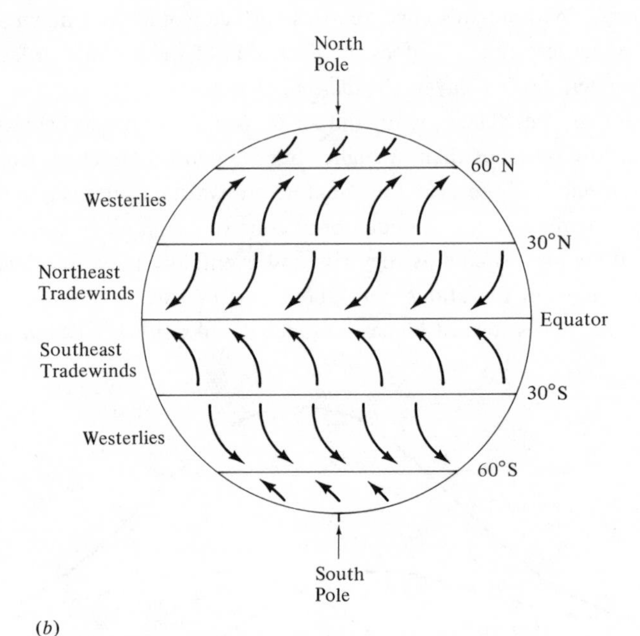

(*b*)

FIGURE 2.2 Idealized atmospheric circulation patterns. (*a*) Vertical profile against latitude. (*b*) Prevailing wind currents on earth's surface. These belts of moving air move north and south with the seasons. [After many sources, especially MacArthur and Connell (1966) and MacArthur (1972).]

descends rapidly to the surface at the so-called horse latitudes of about 30°N and 30°S. At ground level at these latitudes, some of the air moves toward the equator again and some of it moves toward the poles. (The amount of air in the atmosphere is finite, so that air masses leaving one place must always be replaced by air coming from somewhere else; thus a closed system of circulating air masses is set up.) An idealized diagram of the typical vertical and horizontal movements of atmospheric currents is shown in Figure 2.2*a*.

At the surface, the equator is a zone of convergence of air masses, while they are diverging at the horse latitudes. Between latitudes 0° and 30°, surface air generally moves toward the equator, while between latitudes 30° and 60°, it generally moves away from the equator. As air masses move along the surface, they are slowly warmed, and eventually rise again.

Movements of air masses are not strictly north–south as suggested by Figure 2.2*a*, but instead they acquire an east–west component due to the rotation of earth about its axis (Figure 2.2*b*). Earth rotates from west to east. A man standing on either pole would rotate slowly around and do a full "about face" each 24 hours. (Near the North Pole, the ground moves counter-clockwise under one's feet, whereas near the South Pole it moves clockwise; other related important differences between the hemispheres are considered in the following paragraphs.) Someone located near the equator, however, travels much farther during a 24-hour period; indeed, he would traverse a distance equal to the circumference of the earth, or about 25,000 miles, during each rotation of the globe. Hence the velocity of a body near the equator is approximately 1000 miles/hour (relative to earth's axis), while a body at either pole is, relatively speaking, at a comparative standstill.

Above considerations, plus the law of conservation of momentum, dictate that objects moving north in the Northern Hemisphere must speed up, relative to the earth's surface, and thus veer toward the right. Similarly, objects moving south in the Northern Hemisphere are going slower than earth's surface; thus they slow down relative to the surface, which means that they also veer to the right. In contrast, moving objects in the Southern Hemisphere always veer to the left; that is, northward-moving objects slow down and southward-moving ones speed up. These forces, known collectively as the "Coriolis force," act on north–south wind and water currents to give them an east–west component. The Coriolis force is maximal at the poles, where a slight latitudinal displacement is accompanied by a large change in velocity, and minimal at the equator, where a slight latitudinal change has little effect upon the velocity of an object.

Equator-bound surface air between latitudes 0° and 30° slows down relative to the surface and veers toward the west in both hemispheres, producing winds from the east (i.e., "easterlies"); these constitute the familiar trade winds, known as the northeast trades between 0° and 30°N and the southeast trades between 0° and 30°S. Between latitudes 30° and 60°, surface air moving toward the poles speeds up (again, relative to earth's surface), and veers toward the east, producing the familiar prevailing "westerly" winds at these latitudes in both hemispheres (Figure 2.2*b*).

These wind patterns, in turn, coupled with the action of the Coriolis force on water masses moving north to south, drive the world's ocean currents; in the Northern Hemisphere the ocean waters rotate generally clockwise, while they rotate counterclockwise in the Southern Hemisphere (Figure 2.3). During their movement westward along the equator, oceanic waters are warmed by solar irradiation. (These waters also "pile up" on the western sides of oceanic basins; in Central America the sea level of the Atlantic is several feet higher than that of the Pacific.) As this warm equatorial water approaches the eastern sides of land masses, it is diverted northward and/or southward to higher latitudes, carrying equatorial heat toward the poles along the eastern coasts of continents. Cold polar waters flow toward the equator on west coasts (this is the main reason the Pacific Ocean off southern California is cold but at the same latitude the Atlantic Ocean off Georgia is quite warm).

Heat is molecular movement. Compressing a volume of air results in more collisions between molecules and increased molecular movement; hence, compression causes an air mass to heat up. Exactly analogous considerations apply in the reverse case: allowing compressed air to expand decreases the number of molecular collisions and the air mass cools off. As warm air rises, atmospheric pressure decreases, and the air expands and is cooled "adiabatically," or without change in total heat content. Some of the air's own heat is used in its expansion. As descending cold air is compressed it is adiabatically warmed in a similar manner.

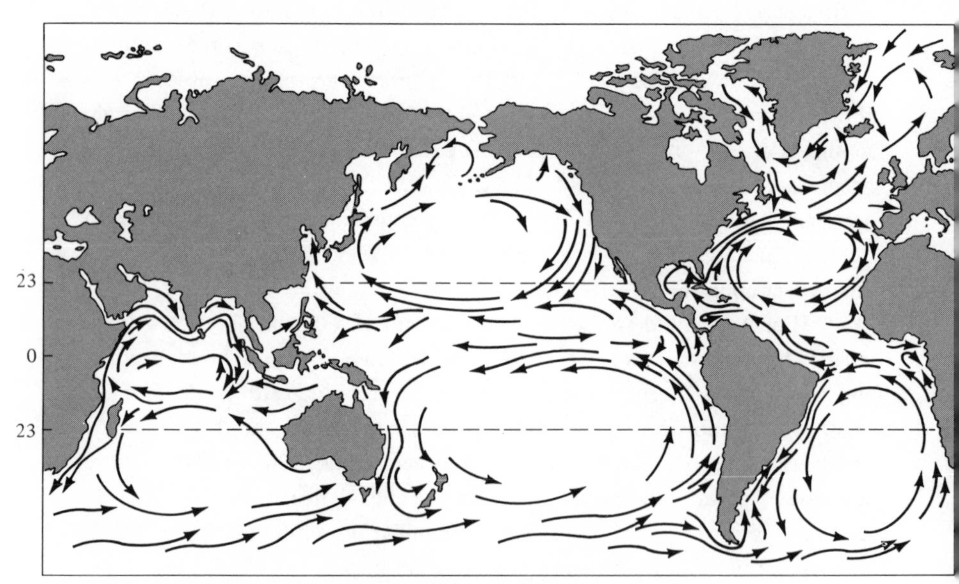

FIGURE 2.3 Major surface currents of the world's oceans.

Warm air can carry more water vapor than cold air. As warm, water-laden, equatorial air rises and cools adiabatically, it first becomes saturated with water at its dew point, and then its water vapors condense to form precipitation. As a result, regions of heavy rainfall tend to occur near the equator (Figure 2.4). In contrast, at the horse latitudes, as cold dry air masses descend and warm adiabatically, they take up water; such desiccating effects help to produce earth's major deserts at these latitudes (see Figure 2.4 and 2.5). Also, notice that deserts occur mainly on the western sides of continents, where cold offshore waters are associated with a blanket of cold, and therefore dry, air; westerly winds coming in off these oceans do not contain much water to give up to the descending dry air, and precipitation must therefore be scanty. The global pattern of annual precipitation shown in Figure 2.5 conforms in general outline to the expected, but there are many local anomalies and exceptions, some of which we discuss next.

Local Perturbations

The major climatic trends presented above are modified locally by a variety of factors, most notably by the size(s) and position(s) of nearby water bodies and land masses, including topography, especially mountains. Ascending a mountain is, in many ways, comparable to moving toward a higher latitude; thus mountains are usually cooler and windier than adjacent valleys and generally support communities of plants and animals characteristic of lower elevations at higher latitudes (1000 feet of elevation corresponds roughly to

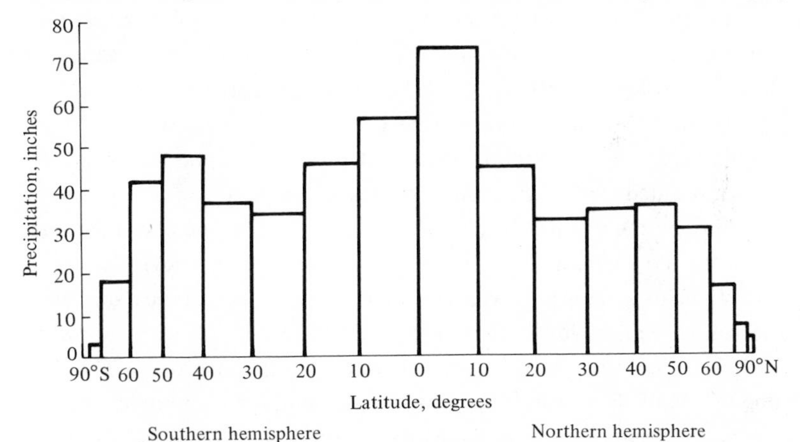

FIGURE 2.4 Histogram of average annual precipitation versus latitudinal zones. Note earth's major deserts in the horse latitudes. [From Haurwitz and Austin (1944).]

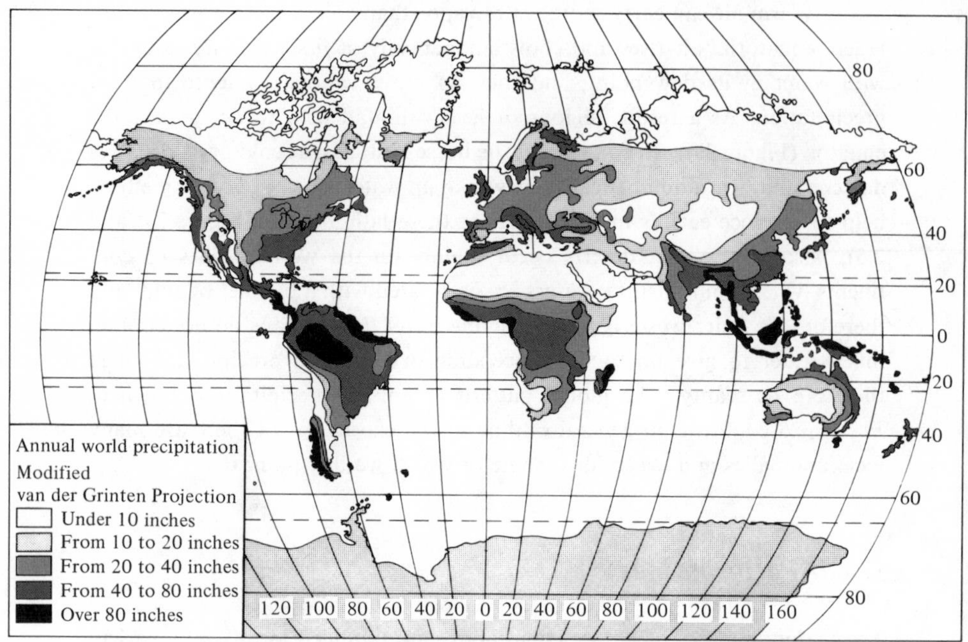

FIGURE 2.5 The geographic distribution of average annual precipitation. [After MacArthur and Connell (1966) after Koeppe.]

100 miles of latitude). In addition to this thermal effect, mountains also markedly modify water availability and precipitation patterns. Water rapidly runs off a slope but sits around longer on a flatter place and soaks into the ground. For this reason, precipitation falling on a mountainside is generally less effective than an equivalent amount falling on a relatively flat valley floor. Precipitation patterns themselves are also directly affected by the presence of mountains. Consider a north–south mountain range receiving westerly winds, such as the Sierra Nevada of the western United States. Air is forced upward as it approaches the mountains, and, as it ascends, it cools adiabatically, becomes saturated with water, and releases some of its water content as precipitation on the windward side of the Sierra. After going over the ridge, this same air, now cold and dry, descends, and as it warms adiabatically, it takes up much of the moisture available on the leeward side of the mountains. The so-called "rainshadow" effect (Figure 2.6) is produced, with windward slopes being relatively wet and leeward slopes being much drier. The desiccating effects of these warm dry air masses extend for many miles beyond the Sierra, and help to produce the Mojave Desert in southern California.

Water has a high specific heat; that is, a considerable amount of heat energy is needed to change its temperature. Conversely, a body of water can

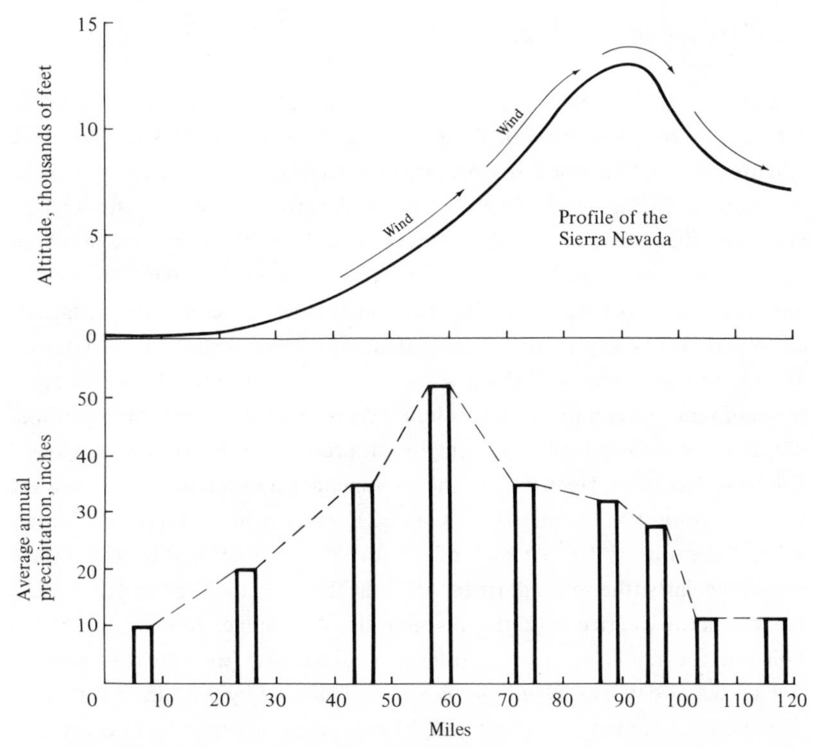

FIGURE 2.6 Illustration of the "rainshadow" effect of the Sierra Nevada mountains in central California.

give up a relatively large amount of heat without cooling down very much. A result of these heat "sink" attributes is that large bodies of water, particularly oceans, effectively reduce temperature changes of nearby land masses. Thus, coastal "maritime" climates are distinguished from inland "continental" climates, with the former being much milder and less variable. Large lakes, such as the Great Lakes, also decrease thermal changes on adjacent land masses, and produce a more constant local temperature. Still another effect of the differential rates of heating and heat-holding capacities of land and water concerns the direction of local breezes along shorelines. During the day, a landmass and the air above it heat up more rapidly than air over an adjacent water body; as this air expands, breezes are produced from land to water. At dusk, air over the land cools off faster than that over the water, contracts, and breezes blow from water to land.

Variations in Time and Space

The seasons are produced by the annual elliptical orbit of earth around its sun and by the inclination of the planet's axis relative to this orbital plane (Figure 2.7). For historical reasons, these movements and patterns have been described from the point of view of the Northern Hemisphere, although by symmetry the same events occur some six months out of phase in the Southern Hemisphere. Twice each year, at the vernal equinox (March 21) and the autumnal equinox (September 23), solar light beams intercept the surface of earth perpendicularly on the equator (that is, the sun is "directly overhead" at its zenith and equatorial shadows point exactly east to west). At two other times of year, the summer solstice (June 22) and the winter solstice (December 22), the axis of earth is tilted maximally with respect to the sun's rays. Viewed from the Northern Hemisphere, the axis inclines approximately 23° toward the sun during the summer solstice and 23° away from it during the winter solstice (see Figure 2.7). At each of the solstices, rays of light hit the surface perpendicularly (the sun is at its zenith) near the Tropic of Cancer (23°N) and the Tropic of Capricorn (23°S), respectively. At summer solstice, the North Pole is in the middle of its six-month period of sunlight (the "polar summer"). The excess of daylight in summer is exactly balanced by the winter deficit, so that the total annual period of daylight is precisely six months at every latitude; the equator has invariate days of exactly 12-hours duration, whereas the poles receive their sunshine all at once over a six-month interval and then

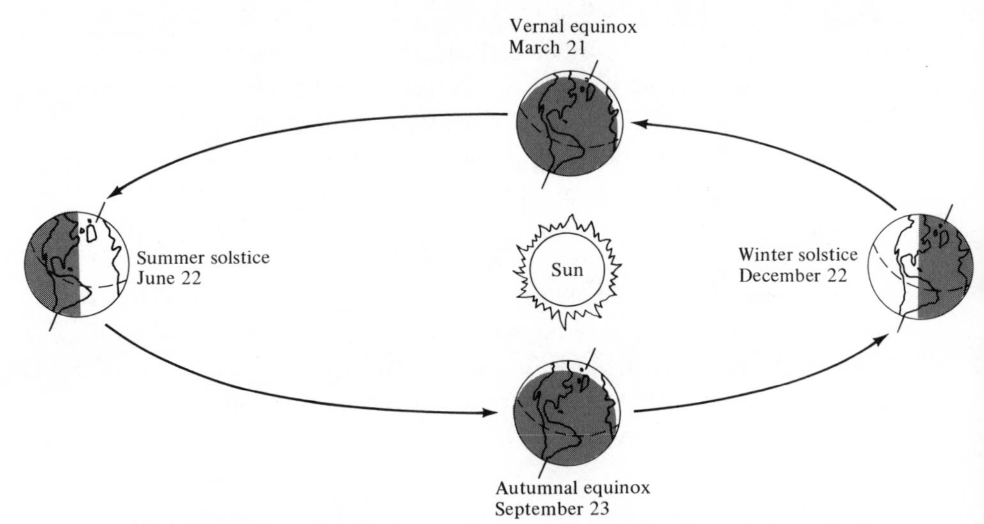

FIGURE 2.7 Diagram of earth's annual elliptical orbit around the sun, which produces the seasons. [After MacArthur and Connell (1966).]

face six months of twilight and total darkness. Because earth is closest to its sun shortly after the winter solstice and farthest from it about the summer solstice, events in the Northern and Southern Hemispheres are not exactly symmetrical. Thus, summer in the Northern Hemisphere takes place when earth is farthest from the sun and northern winters occur when earth is nearest the sun, whereas the reverse is true of summers and winters in the Southern Hemisphere.

Of course, prevailing winds and ocean currents are not static as suggested by Figures 2.2 and 2.3, but in fact they vary seasonally with earth's movement about its sun. The latitudinal belt receiving the most solar radiation (the "thermal equator") gradually shifts northward and southward between latitudes 23°S and 23°N; moreover, the latitudinal belts of easterlies and westerlies also move northward and southward with the seasons, producing seasonal weather changes at higher latitudes. Because of the spherical shape of earth, seasonal changes in insolation increase markedly with increasing latitude.

Although temperatures are modified by prevailing winds, topography, altitude, proximity to bodies of water, cloud cover (Figure 2.8), and other factors, annual marches of average daily temperatures at any given place nevertheless closely reflect earth's movement around the sun. Thus, average daily temperatures on the equator change very little seasonally, while those at higher latitudes usually fluctuate considerably more (Figure 2.9); moreover,

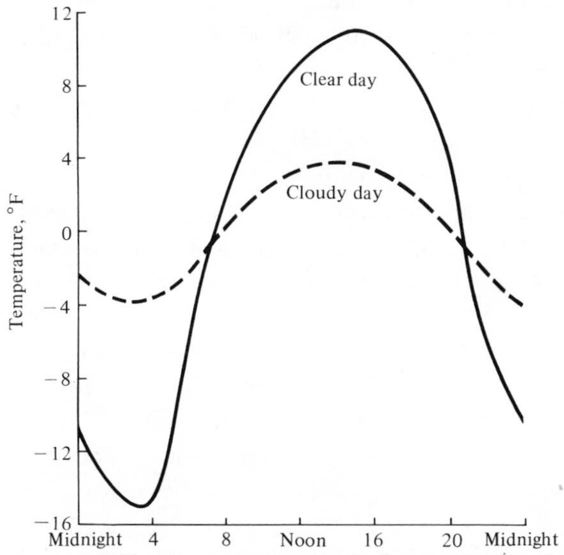

FIGURE 2.8 Cloud cover reduces both the amplitude and the extremes of the daily march of temperature. [After Haurwitz and Austin (1944).]

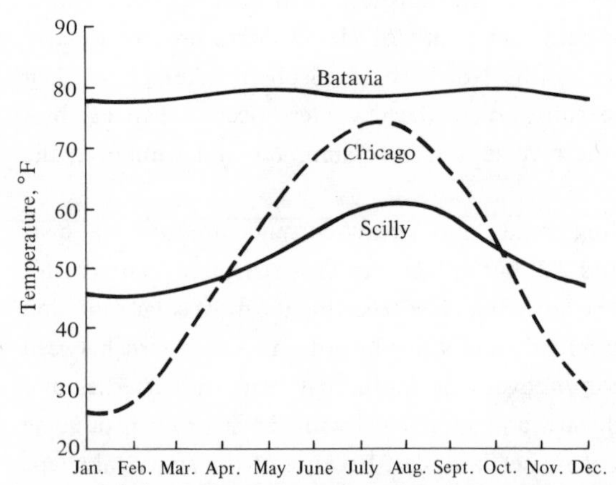

FIGURE 2.9 Annual marches of average daily temperature at an equatorial locality (Batavia) and two temperate areas, one coastal (Scilly) and one inland (Chicago). [After Haurwitz and Austin (1944).]

the annual range in temperature is also much greater in the temperate zones (Table 2.1).

Annual patterns of precipitation also reflect earth's orbital movements (Figure 2.10), although precipitation patterns are perhaps modified locally more easily than thermal patterns. We now examine briefly some of these more subtle determinants of precipitation patterns.

At very low latitudes, say 10°S to 10°N, there are often two rainfall maxima each year (Figure 2.10*a*), one following each equinox as the region of rising air (the thermal equator) passes over the area. Thus the fact that the sun passes over these equatorial areas twice each year produces a bimodal annual pattern of precipitation. Bimodal annual precipitation patterns also occur in other regions at higher latitudes, such as the Sonoran Desert (Figure 2.10*b*), but for different reasons. However, not all equatorial areas have such a rainfall pattern, and some have only one rainy season each year (Figure 2.10*c*).

During the summer, air masses in the central parts of continents at high latitudes tend to warm faster than those around the periphery, which are cooled somewhat by the nearby oceans. As the hot air rises from the center of continents, a "low" pressure area is formed, and cooler, but still warm and water-laden, coastal air is drawn in off the oceans and coastlines. When this moist air is warmed over the land, it rises, cools adiabatically, and releases much of its water content on the interiors of continents. Thus continental, in contrast to coastal, climates are often characterized by summer rains

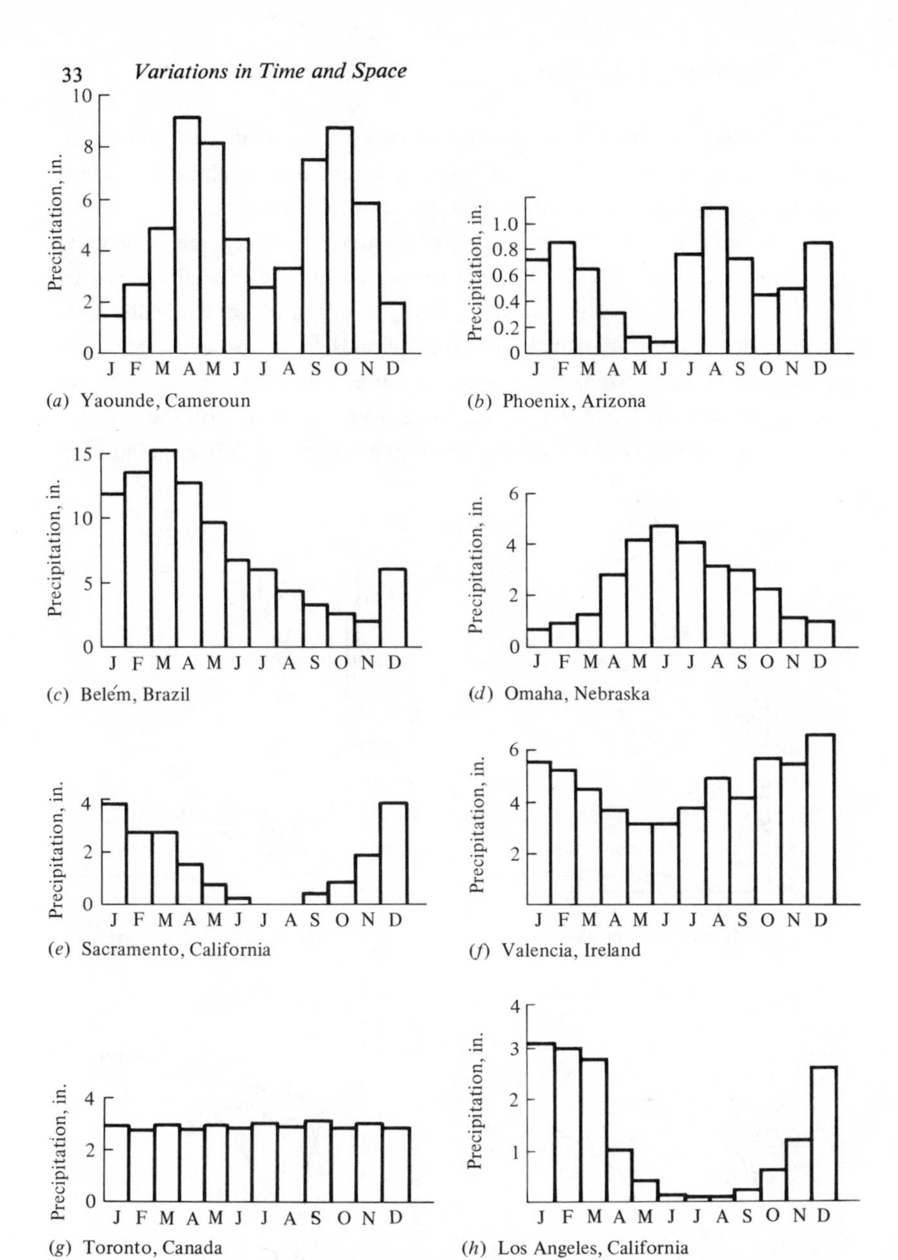

FIGURE 2.10 Annual marches of average yearly precipitation from eight selected localities (see text). (*a*) Bimodal equatorial rainfall, (*b*) the bimodal annual precipitation pattern of the Sonoran Desert, (*c*) unimodal annual rainfall pattern of an equatorial area, (*d*) typical continental summer rainfall regime, (*e*) a coastal area with winter precipitation, (*f*) a typical maritime rainfall regime with more winter precipitation than summer rain, (*g*) the annual march of precipitation in eastern North America is distributed evenly over the year, and (*h*) an area with a pronounced summer drought that would support a chaparral vegetation. [After Haurwitz and Austin (1944).]

(Figure 2.10*d*). In winter, as the central regions of continents cool down relative to their water-warmed edges, a "high" pressure area develops, and winds reverse, with cold dry air pouring outward toward the coasts.

A weather "front" is produced when a cold (usually polar) air mass and a warm air mass collide. The warmer lighter air is displaced upward by the heavy cold air; as this warm air rises, it is cooled adiabatically, and, provided it contains enough water vapor, clouds form and eventually precipitation falls. Such fronts typify the boundaries between the polar easterlies and the midlatitude westerlies, which move north to south with the seasons.

The seasonality of coastal rainfall is affected by the differential heating

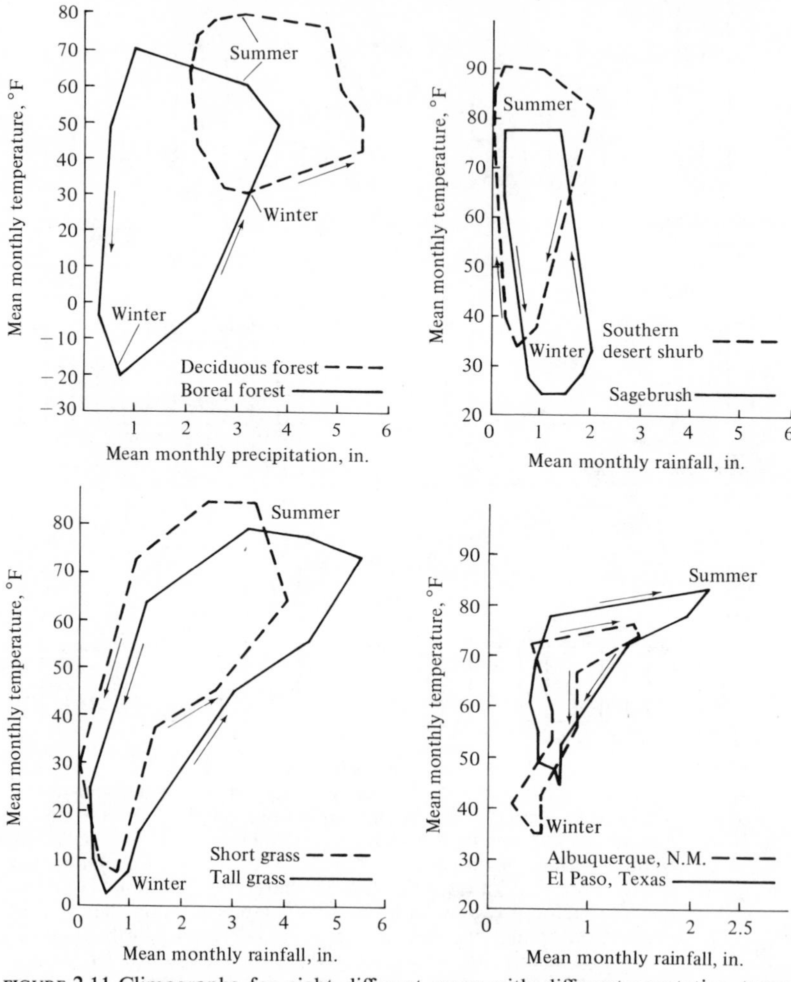

FIGURE 2.11 Climographs for eight different areas with different vegetation types. [After Smith (1940).]

of land and water in another, almost opposite, way. At high latitudes, say 40° to 60°, landmasses cool down in winter and become much colder than the oceans; along the west coasts of continents, water-rich westerly winds coming in off such a relatively warm ocean are rapidly cooled when they meet a cold landmass and hence deposit much of their moisture as winter rain and/or snow along the coast (Figure 2.10*e*). Maritime climates typically have precipitation through the entire year, with somewhat more during the winter months (Figure 2.10*f*). In eastern North America, precipitation is spread out fairly evenly over the year (Figure 2.10*g*). Areas with a long summer drought and winter rains (Figure 2.10*h*) typically support a vegetation of chaparral (see also next chapter).

A convenient means of graphically depicting seasonal climates is the "climograph," which is simply a plot of average monthly temperature against average monthly precipitation (Figure 2.11). Although such graphs do not reflect year to year variability in climate, they do show at a glance the changes in both temperature and precipitation within an average year, as well as the season(s) during which the precipitation usually falls. (Without actually identifying points by months, however, it is not possible to distinguish spring

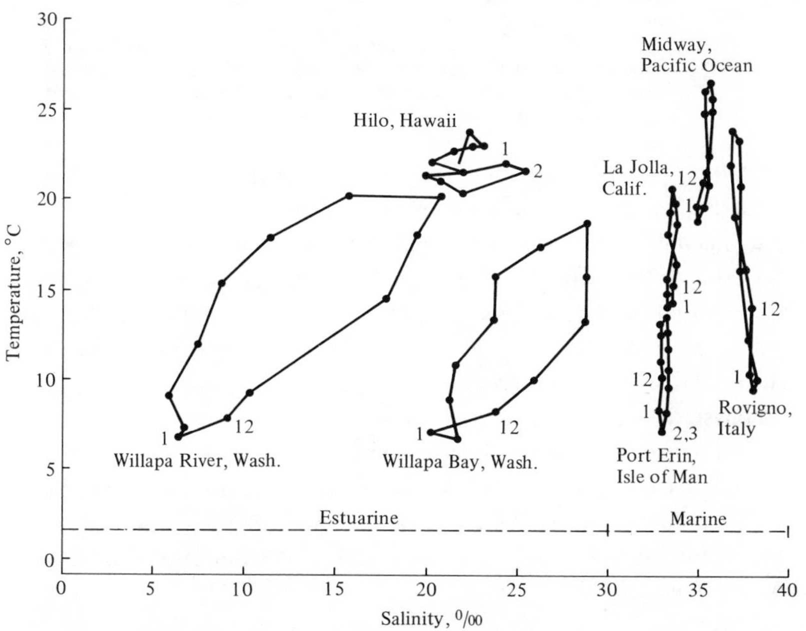

FIGURE 2.12 Graphs of average monthly temperature versus salinity of some estuarine and marine waters; seasonal variation is great in the brackish waters, whereas salinity varies little in true marine waters. [After Odum (1971) after Hedgepeth.]

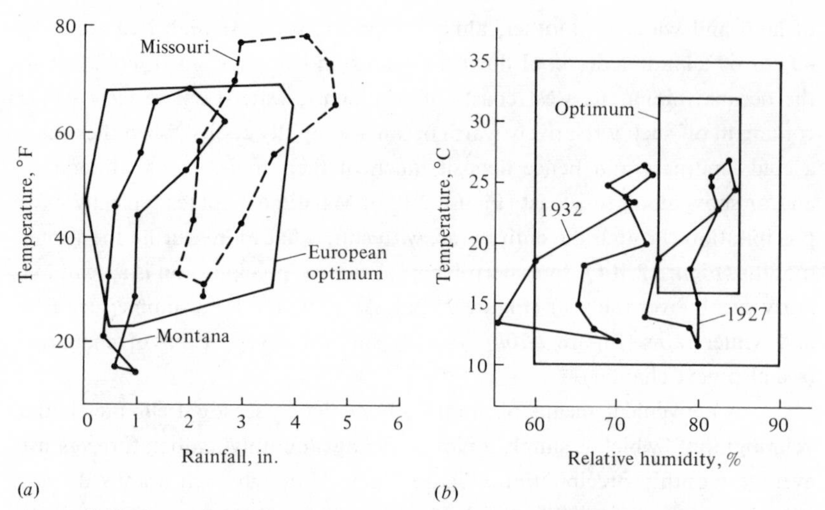

FIGURE 2.13 Two plots of temperature against moisture. (*a*) Climographs for an area in Montana, where Hungarian partridge were introduced successfully, and a Missouri locality where its introduction failed, compared to the average climatic conditions of its European geographic range. Apparently Missouri was too hot and too wet for these birds. (*b*) Plots of temperature versus relative humidity in 1927 and 1932 in Israel superimposed upon optimal (inner rectangle) and favorable (outer rectangle) conditions for the Mediterranean fruit fly. [Note that, as drawn, these rectangles assume no interaction between temperature and humidity; in actuality the curves would probably be rounded due to the principle of allocation—see also Chapter 1.] Damage to fruit crops by these flies was much greater in 1927. [After Odum (1959).]

from fall since both are seasons of moderate temperatures.) Exactly analogous plots are often made for any two physical variables, such as temperature versus humidity, or temperature versus salinity in aquatic systems (Figure 2.12). Many other variables of biological importance, such as pH and dissolved phosphorus or nitrogen, can be treated similarly. When coupled with information on the tolerance limits of organisms, climographs and their analogs can be useful in predicting the responses of organisms to changes in their physical environments (Figure 2.13).

Although there are an infinite number of different types of climates, attempts have been made to classify them. One scheme (Köppen's) recognizes five major climatic types (as well as many minor ones): tropical rainy, dry, warm temperate rainy, cool snow forest, and polar. Another classification is shown in Figure 2.14.

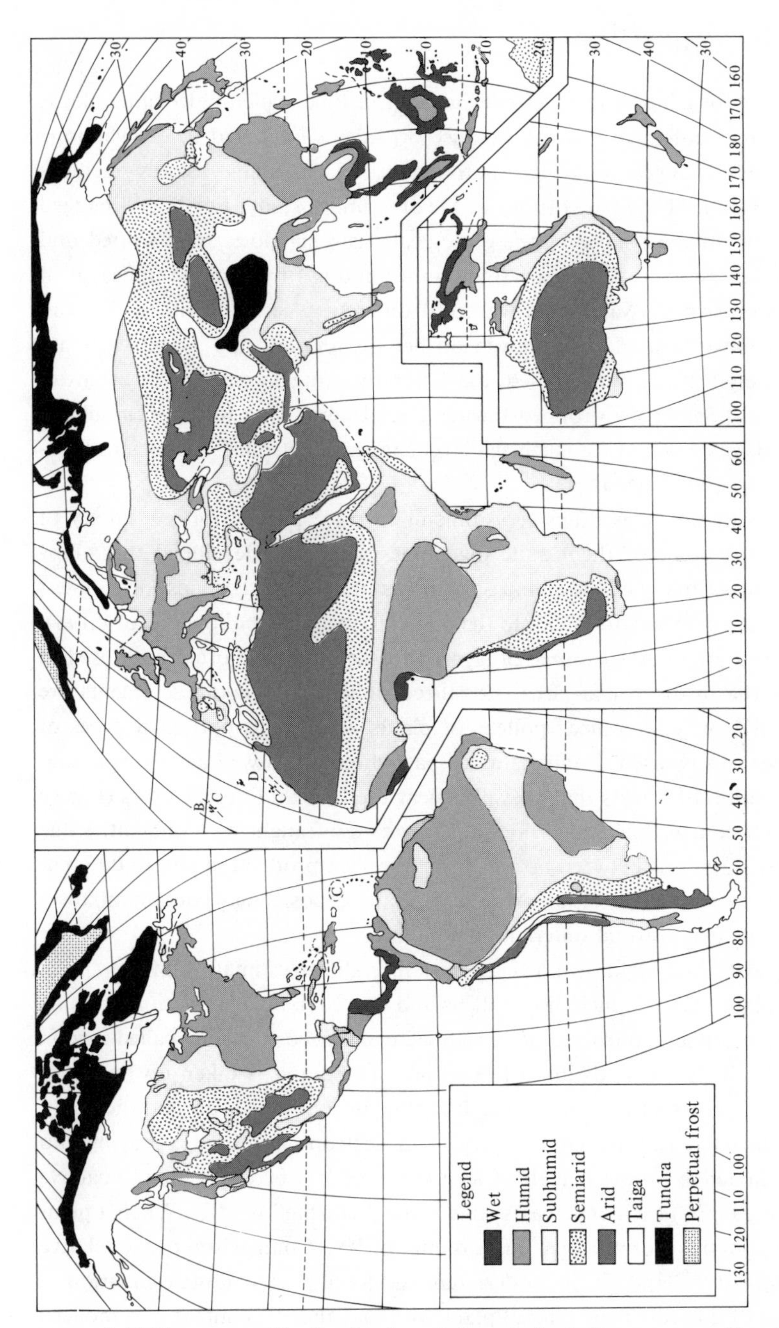

FIGURE 2.14 The geographic distribution of the principal climates, according to the Thornthwaite classification. [After Blumenstock and Thornthwaite (1941).]

The Geological Past

The study of climatic changes over geological time, called paleoclimatology, is of some ecological interest, because organisms have had to evolve along with such changes. A really thorough ecological study usually includes consideration of the past history of the area under study. Earth has changed in innumerable ways during the geological past; her poles have shifted and wandered, periods of orogeny (mountain building) by tectonic upheaval of earth's crust have waxed and waned at different places and times, the continents have "drifted" and moved on her mantle, and the planet itself has alternately warmed and cooled, the latter resulting in periods of extensive glaciation. Sea levels dropped during glacial periods as water accumulated on land as snow and ice (such sea level changes controlled by glacier alterations are called "eustatic").

Although it is often very difficult to trace past history, a variety of techniques, some of them quite ingenious, have been developed that allow us to deduce many of the changes earth has undergone. Perhaps the simplest and most direct way to look into the past is to examine the fossil record. Lake sediments are an ideal source of layered fossils, and have often been used to follow the history of an area. Fossilized pollens in a lake's sediments are relatively easily identified; pollens of plants adapted to particular types of climates can be used as indicators of past climates, as well as the types and composition of forests that prevailed near the lake at different times (Figure 2.15). Of course, such palynological analyses are fraught with difficulties due both to variations between taxa in rates of pollen production and to differential transport and deposition. Moreover, many deposits may contain mixtures of pollens from several different communities.

A technique known as carbon dating allows estimation of the age of fossil plant remains, including pollens and charcoal. Solar radiation converts some atmospheric nitrogen into a radioactive isotope of carbon called carbon 14 (^{14}C). This ^{14}C is oxidized to carbon dioxide and is taken up by plants in photosynthesis in proportion to its abundance in the air around the plant. All radioactive isotopes emit neutrons and electrons, eventually decaying into nonradioactive isotopes. Half of a quantity of ^{14}C becomes nonradioactive carbon 12 (^{12}C) each 5600 years (this is the "half-life" of ^{14}C). When a plant dies it contains a certain maximal amount of ^{14}C. Comparison of the relative amounts of ^{14}C and ^{12}C in modern day and fossil plants allows estimation of the age of a fossil. Thus a fossil plant with half the ^{14}C content of a modern plant is about 5600 years old; one with one-quarter as much ^{14}C is 11,200 years old; etc. The carbon dating method has been checked against ancient

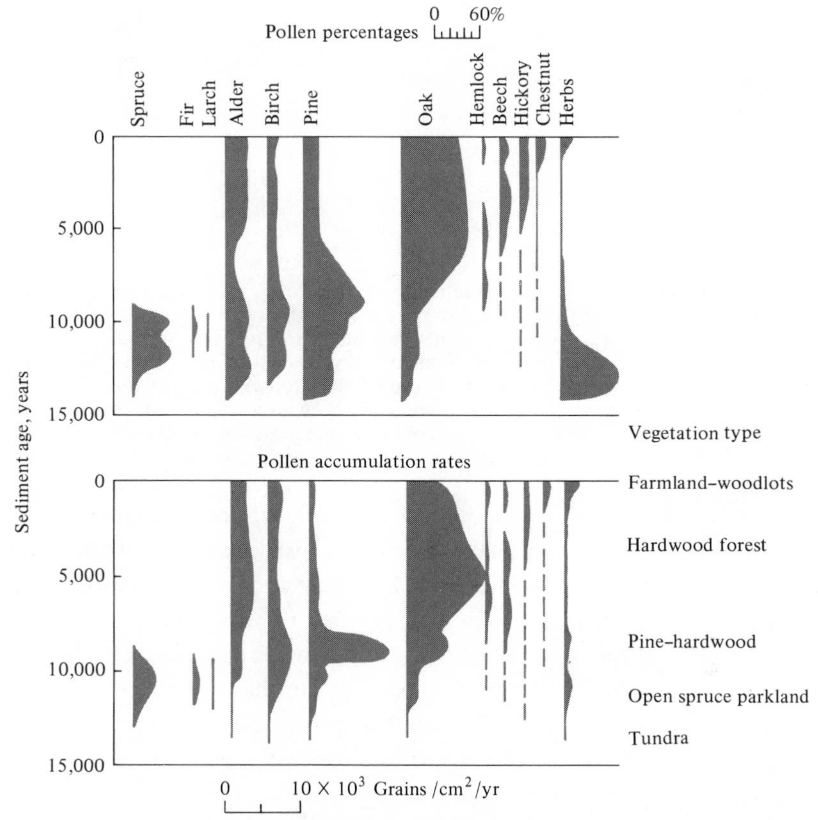

FIGURE 2.15 Fossil pollen profiles from dated layers of lake sediments in the northeastern United States for the period following the last ice age. Upper plot shows the number of pollen grains of each species group as a percentage of the total sample. Lower plot gives the estimated rates of deposition of each type of pollen and, at the right, the type of vegetation that probably prevailed in the area. [Adapted from Odum (1971) after Davis.]

Egyptian relics of known age made from plant materials; it accurately estimates their ages, confirming that the rate of production of ^{14}C and the proportion of ^{14}C to ^{12}C have not changed much over the last 5000 years. The technique allows the assignment of very accurate ages to all sorts of fossil plant materials.

A similar technique makes use of the fact that the uptake of two oxygen isotopes, ^{16}O and ^{18}O, into carbonates is temperature dependent. Thus the proportion of these two isotopes in a fossil seashell presumably reflects the temperature of an ancient ocean in which that particular mollusk lived.

There is now a massive body of evidence that the continents have

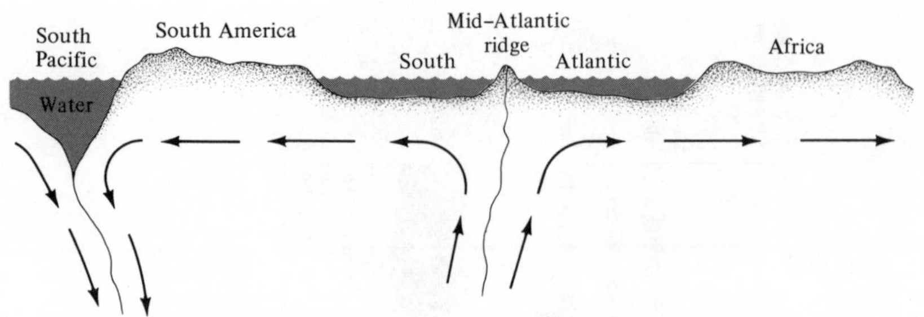

FIGURE 2.16 Diagrammatic cross-sectional view of the probable movements of earth's mantle and crust that lead to sea-floor spreading and continental drift. Upwelling of deep mantle materials in the mid-Atlantic is accompanied by a surface movement away from the mid-oceanic ridge. The continents, which float on top of these moving denser materials, are carried along. An oceanic trench is shown on the far left, which would correspond to an area where these materials sink back down into the mantle, forming a closed system of circulating materials.

actually moved or "drifted" away from one another during geological time (Wilson, 1971, 1973). Certain types of rocks, particularly basalts, retain a magnetic "memory" of the latitude in which they were solidified. Such paleomagnetic evidence strongly indicates that the continents have moved with respect to one another. The continents are formed of light "plates" of siliceous, largely granitic, rocks, about 30 km thick which in turn float on denser mantlelike basaltic blocks. The ocean floors are composed of a relatively thin altered top of the earth's mantle. A mountain range on the sea floor in the mid-Atlantic has recently been mapped and this ridge is now thought to represent a region of upwelling of the mantle. Under this interpretation, as the upwelling proceeds, sea floors spread and continents move apart (Figure 2.16). Thus modern theory holds that, except for the Pacific, which is shrinking, the oceans are growing with very young ocean floors in mid-ocean and progressively older floors toward the continents. Other evidence, such as the apparent ages of islands and the depths of sediments, nicely corroborates this conclusion.

The reasons for many of these past changes, such as polar movements and the alternate warming and cooling of the planet, are little known and may well involve solar changes. Piecing together all this diverse, sometimes conflicting, evidence on earth's past history is a most difficult and very challenging task, and one which occupies many of our finest scientists.

Selected References

Major Determinants of Climate; Local Perturbations; Variations in Time and Space

Blair and Fite (1965); Blumenstock and Thornthwaite (1941); Byers (1954); Chorley and Kennedy (1971); Collier *et al.* (1973); Finch and Trewartha (1949); Flohn (1969); Gates (1962); Haurwitz and Austin (1944); Lowry (1969); MacArthur (1972); MacArthur and Connell (1966); Taylor (1920); Thornthwaithe (1948); Trewartha (1943); U.S. Department of Agriculture (1941).

The Geological Past

Birch and Ehrlich (1967); Dansereau (1957); Darlington (1957, 1965); Hesse, Allee, and Schmidt (1951); Jelgersma (1966); Martin and Mehringer (1965); Sawyer (1966); Udvardy (1969); Wilson (1971, 1973); Wiseman (1966); Wright and Frey (1965).

3 *The Interface Between Climate and Vegetation*

Climate is the major determinant of vegetation. Plants, in turn, exert some degree of influence on climate. Both climate and vegetation profoundly affect soil development and the animals that live in an area. Here we examine some of the ways in which climate and vegetation interact. Much more emphasis is given to terrestrial ecosystems than to aquatic ones, although some aquatic analogs are noted. Topics presented rather briefly in this chapter are treated in much greater detail in other texts, such as Allee *et al.* (1949), Andrewartha and Birch (1954), Clarke (1954), Colinvaux (1973), Collier *et al.* (1973), Daubenmire (1947, 1968), Kendeigh (1961), Knight (1965), Kormondy (1969), Krebs (1972), Lowry (1969), Odum (1959, 1971), Oosting (1958), Ricklefs (1973), Smith (1966), and Whittaker (1970).

Plant Life Forms and Biomes

Terrestrial plants adapted to a particular climatic regime often have similar morphologies, or plant growth forms. Thus climbing vines, epiphytes, and broad-leafed species characterize tropical rain forests. Evergreen conifers dominate very cold areas at high latitudes and/or altitudes, while small frost-resistant tundra species occupy still higher latitudes and altitudes. Seasonal temperate zone areas with moderate precipitation usually support broad-leafed, deciduous trees, while tough-leafed (sclerophyllous) evergreen shrubs, or so-called chaparral-type vegetation, occur in regions with winter rains and a pronounced long water deficit during spring, summer, and fall (see also

Figure 3.8*b*). Chaparral vegetation is found wherever this type of climate prevails, including southern California, Chile, Spain, Italy, southwestern Australia, as well as the northern and southern tips of Africa (see Figure 3.1), although the actual plant species comprising the flora usually differ. In general, there is a very close correspondence between climate and vegetation (compare Figures 2.5 and 2.14 with Figure 3.1); indeed, climatologists have sometimes used vegetation as the best indicator of climate! Thus rain forests occur in rainy tropical and rainy warm-temperate climates, forests exist under more moderate mesic climates, savannas and grasslands prevail in semiarid climates, and deserts characterize still drier climates. Topography and soils, of course, also play a part in the determination of vegetation types, which are sometimes termed "plant formations." Such major communities of characteristic plants and animals are known as biomes. Classification of natural communities is discussed later in this chapter.

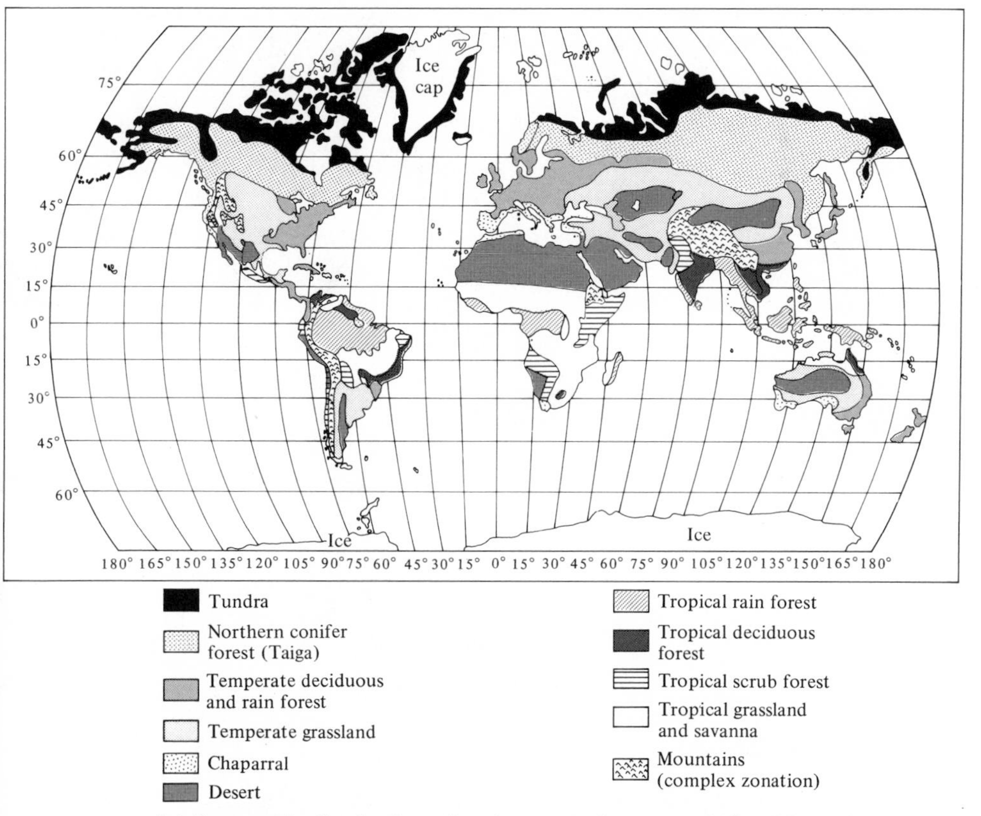

FIGURE 3.1 Geographic distribution of major vegetation types. [After MacArthur and Connell (1966) after Odum.]

Microclimate

Even in the complete absence of vegetation, major climatic forces, or macroclimates, are expressed differently at a very local spatial level, which has resulted in the recognition of so-called microclimates. Thus, the surface of the ground undergoes the greatest daily variation in temperature, and daily thermal flux is progressively reduced with both increasing distance above and below ground level (Figure 3.2). During the daylight hours, the surface intercepts most of the incident solar energy and rapidly heats up, whereas at night this same surface cools down more than its surroundings. Such plots of temperature versus height above and below ground are called thermal profiles. An analogous type of graph, called a bathythermograph, is often made for aquatic ecosystems, by plotting temperature against depth (Figure 3.16).

Daily temperature patterns are also modified by topography, again, even in the absence of any vegetation. A slope facing the sun intercepts light beams more perpendicularly than does a slope facing away from the sun; as a result a south-facing slope in the Northern Hemisphere receives more solar energy than a north-facing slope, and the former heats up faster

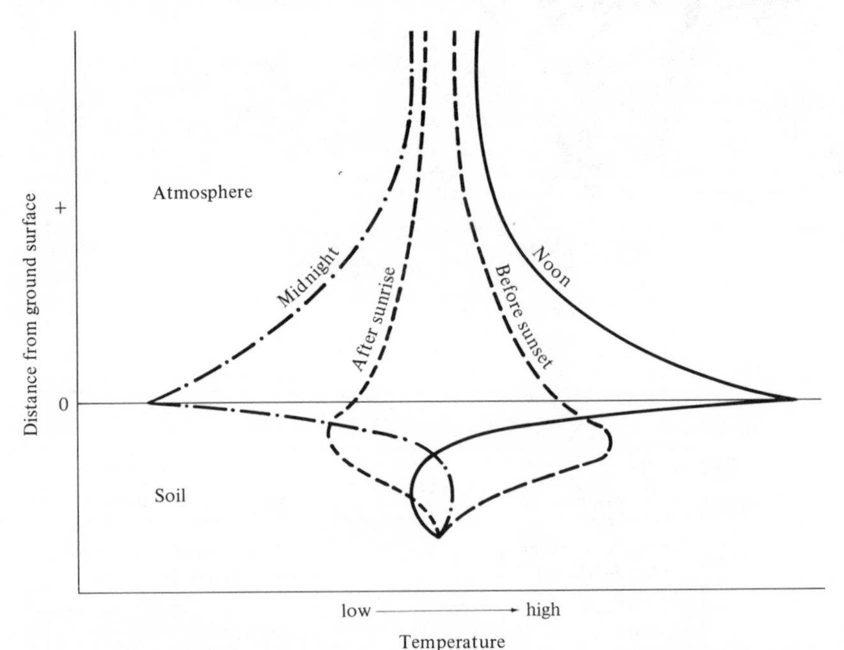

FIGURE 3.2 An idealized thermal profile showing temperatures at various distances above and below ground at four different times of day. [After Gates (1962).]

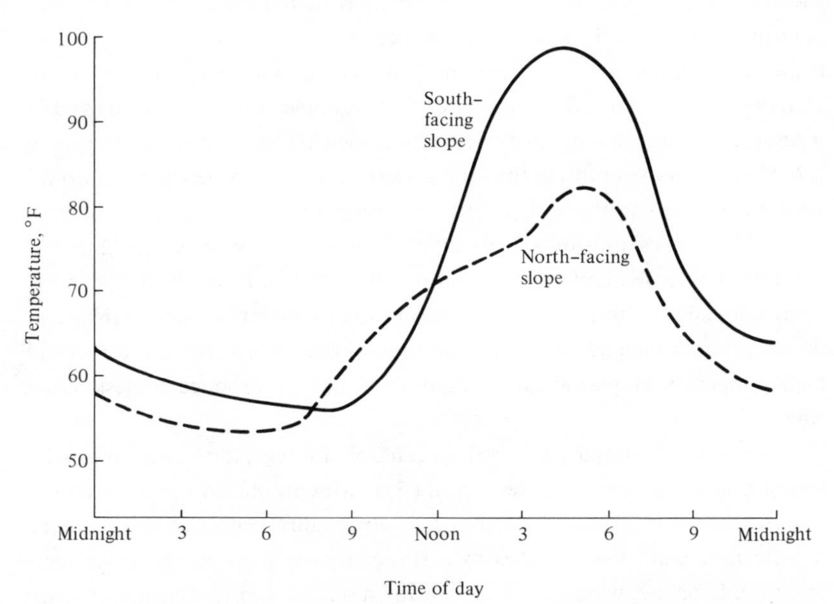

FIGURE 3.3 Daily marches of temperature on an exposed south-facing slope (solid line), and a north-facing slope (dashed line) during late summer in the northern hemisphere. [After Smith (1966) after van Eck.]

and gets hotter during the day (Figure 3.3). Moreover, such a south-facing slope is typically drier than a north-facing one because it receives more energy and therefore more water is evaporated.

By orienting themselves either parallel to or at right angles to the sun's rays, organisms (and parts of organisms such as leaves) may either decrease or increase the total amount of solar energy they actually intercept. Leaves in the brightly illuminated canopy often droop during midday, whereas those in the shaded understory typically present their full surface to incoming beams of solar radiation. Similarly, many desert lizards position themselves on the ground perpendicular to the sun's rays in the early morning when environmental temperatures are low, but during the high temperatures of midday these same animals may climb up off the ground into cooler air temperatures and orient themselves parallel to the sun's rays by facing into the sun.

The major effect of a blanket of vegetation is to moderate most daily climatic changes, such as changes in temperature, humidity, and wind. (Plants, however, do generate daily variations in concentrations of oxygen and carbon dioxide through their photosynthetic and respiratory activities.) Thermal profiles at midday in cornfields at various stages of growth are shown in Figure 3.4, demonstrating the marked reduction in ground temperature due especially to shading. Notice that, in the mature field, the air is warmest at about 4 feet above ground. Similar vegetational effects on microclimates occur in natural communities. For example, a patch of open sand in a desert might have a daily thermal profile somewhat like that shown in Figure 3.2, whereas temperatures in the litter underneath a nearby dense shrub would vary much less with the daily march of temperature.

Humidities are similarly modified by vegetation, with relative humidities within a dense plant being somewhat greater than those of the air in the open adjacent to the plant. Likewise, moisture content is more stable, and therefore more dependable, deeper in the soil than it is at the surface, where high temperatures periodically evaporate water to produce a desiccating effect.

Wind velocities are also reduced sharply by vegetation and are usually lowest near the ground (Figures 3.5 and 3.6). Moving currents of air promote rapid exchange of heat and water; hence an organism cools or warms more rapidly in a wind than it does in a stationary air mass at the same temperature. Likewise, winds often carry away moist air and replace it with drier air, thus promoting evaporation and water loss; the desiccating effects of such dry winds can be extremely important to an organism's water balance.

In aquatic systems, water turbulence parallels wind in many ways; rooted vegetation around the edges of a pond or stream reduces water turbulence. At a more microscopic level, algae and other organisms that attach themselves to underwater surfaces (so-called periphyton) create a thin film of distinctly modified microenvironment in which water turbulence, among

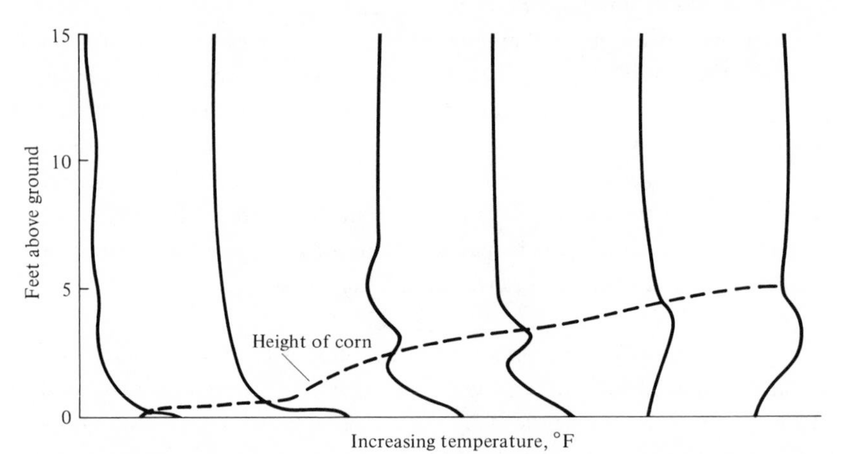

FIGURE 3.4 Temperature profiles in a growing cornfield at midday, showing the effect of vegetation on thermal microclimate. [After Smith (1966).]

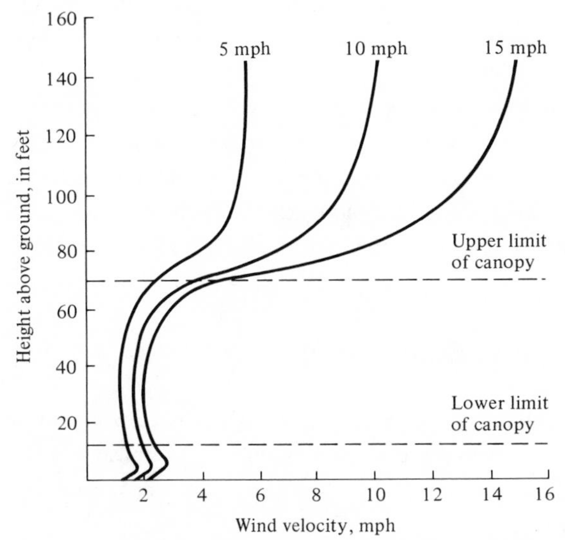

FIGURE 3.5 Wind velocities within a forest vary relatively little with changes in the wind velocity above the canopy. [After Fons (1940).]

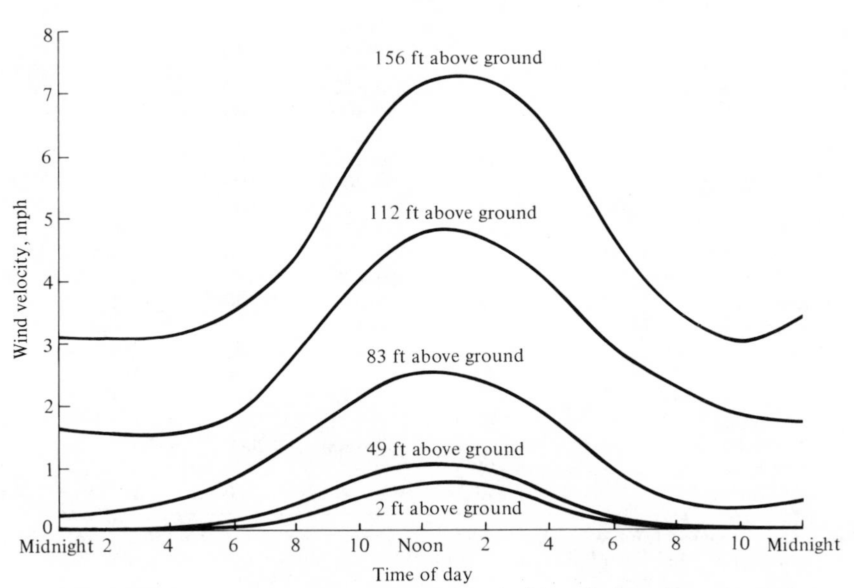

FIGURE 3.6 Daily march of average wind velocities during June at various heights inside a coniferous forest in Idaho. [After Smith (1966) after Gisborne.]

other things, is reduced. Localized spatial patches with particular concentrations of hydrogen ions (pH), salinities, dissolved nitrogen and phosphorus, etc., form similar aquatic microhabitats.

By actively or passively selecting such microhabitats, organisms can effectively reduce the overall environmental variation they encounter and enjoy more optimal conditions than they could without microhabitat selection. Innumerable other microclimatic effects could be cited, but these should serve to illustrate their existence and their significance to plants and animals.

Primary Production and Evapotranspiration

In terrestrial ecosystems, climate is by far the most important determinant of the amount of solar energy plants are able to capture as chemical energy, or the gross primary productivity (Table 3.1). In warm arid regions, water is a

TABLE 3.1 Net Primary Productivity and World Net Primary Production for the Major Ecosystems

	Area[a] (10^6km^2)	Net Primary Productivity, Per Unit Area[b] (dry g/m^2/yr)		World Net Primary Production[c] (10^9 dry tons/yr)
		normal range	mean	
Lake and stream	2	100–1500	500	1.0
Swamp and marsh	2	800–4000	2000	4.0
Tropical forest	20	1000–5000	2000	40.0
Temperate forest	18	600–2500	1300	23.4
Boreal forest	12	400–2000	800	9.6
Woodland and shrubland	7	200–1200	600	4.2
Savanna	15	200–2000	700	10.5
Temperate grassland	9	150–1500	500	4.5
Tundra and alpine	8	10–400	140	1.1
Desert scrub	18	10–250	70	1.3
Extreme desert, rock, and ice	24	0–10	3	0.07
Agricultural land	14	100–4000	650	9.1
Total land	149		730	109.0
Open ocean	332	2–400	125	41.5
Continental shelf	27	200–600	350	9.5
Attached algae and estuaries	2	500–4000	2000	4.0
Total ocean	361		155	55.0
Total for earth	510		320	164.0

Source: Adapted from Whittaker (1970), *Communities and Ecosystems.* Reprinted with permission of Macmillan Publishing Co., Inc. © Copyright Robert H. Whittaker, 1970.

[a] Square kilometers × 0.3861 = square miles.
[b] Grams per square meter × 0.01 = t/ha, × 0.1 = dz/ha or m centn/ha (metric centers, 100 kg, per hectare, 10^4 square meters), × 10 = kg/ha, × 8.92 = lbs/acre.
[c] Metric tons (10^6 g) × 1.1023 = English short tons.

master limiting factor, and, in the absence of runoff, primary production is strongly positively correlated with rainfall in a linear fashion (see Chapter 1, pp. 14–15 and Figure 1.4). Above about 80 cm of precipitation/year, primary production slowly decreases with increasing precipitation and then levels off (asymptotes) (Figure 3.7). Notice that some points fall below the line in this figure, presumably because there is some water loss by runoff and seepage into groundwater supplies.

Evapotranspiration refers to the release of water into the atmosphere as water vapor, both by the physical process of evaporation and by the biological processes of transpiration and respiration. The amount of water vapor thus returned to the atmosphere is strongly dependent on temperature, with greater evapotranspiration at higher temperatures. The *theoretical* temperature-dependent amount of water that *could* be "cooked out" of an ecological system, given its input of solar energy and provided that much water fell on the area, is called its *potential evapotranspiration* (PET). In many ecosystems, water is frequently in short supply, so that *actual evapotranspiration* (AET) is somewhat less than potential (clearly AET can never exceed PET and is equal to PET only in completely water-saturated habitats). Actual evapotranspiration can be thought of as the reverse of rain, for it is the amount of water that actually goes back into the atmosphere at a given spot.

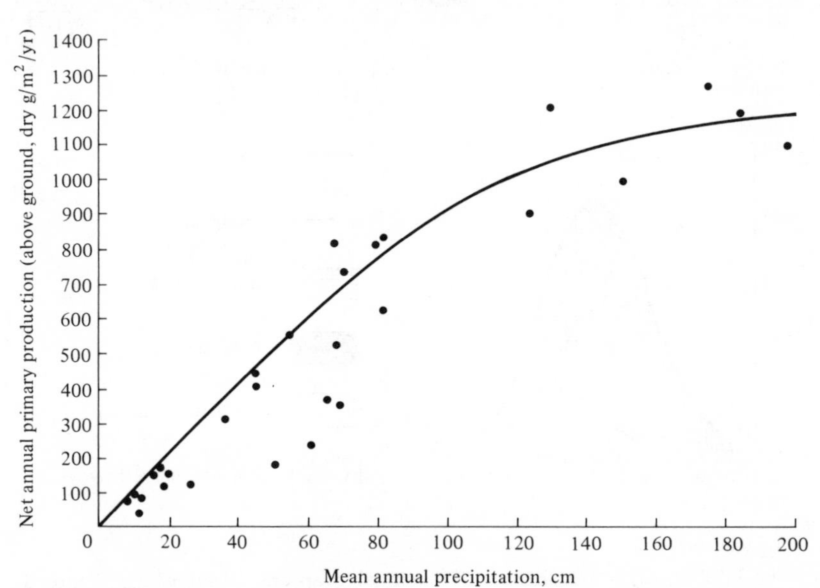

FIGURE 3.7 Net primary productivity (above ground) plotted against average annual precipitation. [From Whittaker (1970). Reprinted with permission of Macmillan Publishing Co., Inc., from *Communities and Ecosystems* by Robert H. Whittaker. © Copyright Robert H. Whittaker, 1970.]

The potential evapotranspiration for any spot on earth is determined by the same factors that regulate temperature, most notably by latitude, altitude, cloud cover, and slope (topography). There is a nearly one-to-one correspondence between PET and temperature, and an annual march of PET can be plotted in centimeters of water. By superimposing the annual march of precipitation on these plots (Figure 3.8), seasonal changes in water availability are depicted graphically. A water deficit occurs when PET exceeds precipitation, whereas a water surplus exists when the situation is reversed. During a period of water surplus, some water may be stored by plants and some may accumulate in the soil as soil moisture depending upon

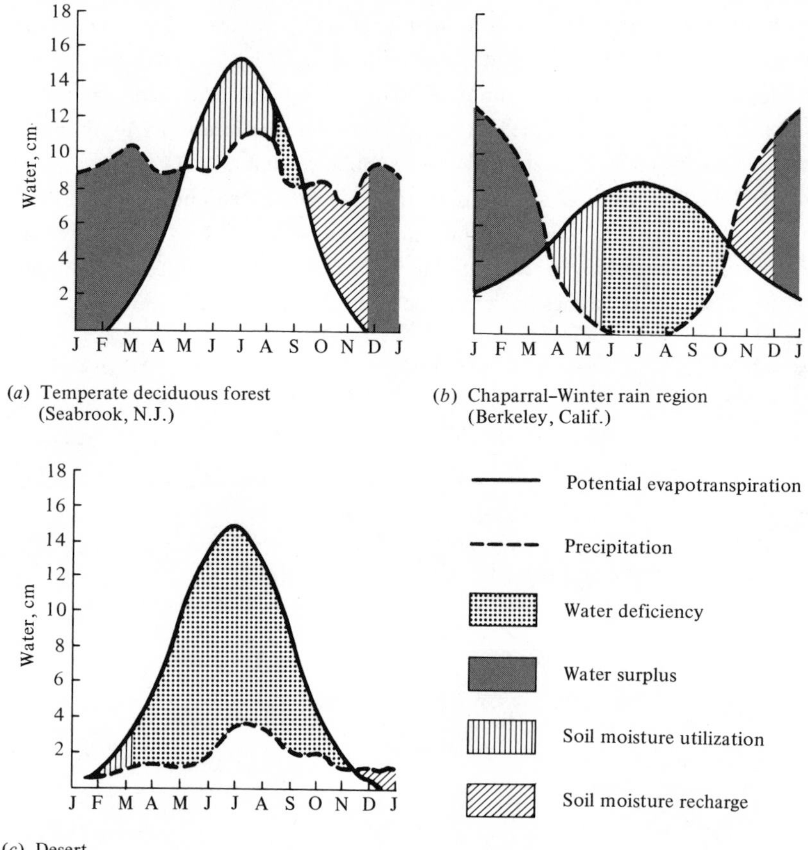

FIGURE 3.8 Plots of annual march of potential evapotranspiration superimposed upon the annual march of precipitation for three ecologically distinct regions showing the water relations of each. (*a*) Temperate deciduous forest, (*b*) chaparral with winter rain, (*c*) desert. [From Odum (1959) after Thornthwaite.]

runoff and the capacity of soils to hold water; during a later water deficit, such stored water can then be used by plants and released back into the atmosphere. Winter rain is generally much less effective than summer rain because of the reduced activity (or complete inactivity) of plants in winter; indeed two areas with the same annual march of temperature and total annual precipitation may differ greatly in the types of plants they support and in their productivity depending upon their seasonal patterns of precipitation. For example, an area receiving about 50 cm of precipitation annually might support either a grassland vegetation or chaparral, depending upon whether the precipitation falls in the summer or the winter, respectively.

Rosenzweig (1968) found that net annual primary production above ground is strongly correlated with actual evapotranspiration, or AET (Figure 3.9). This correlation is remarkable in that AET was crudely estimated using only monthly macroclimatic statistics with no allowance for either runoff and water-holding capacities of soils or for groundwater usage. Rosenzweig suggested that the reason for the observed correlation is that AET measured simultaneously two of the most important factors limiting primary production

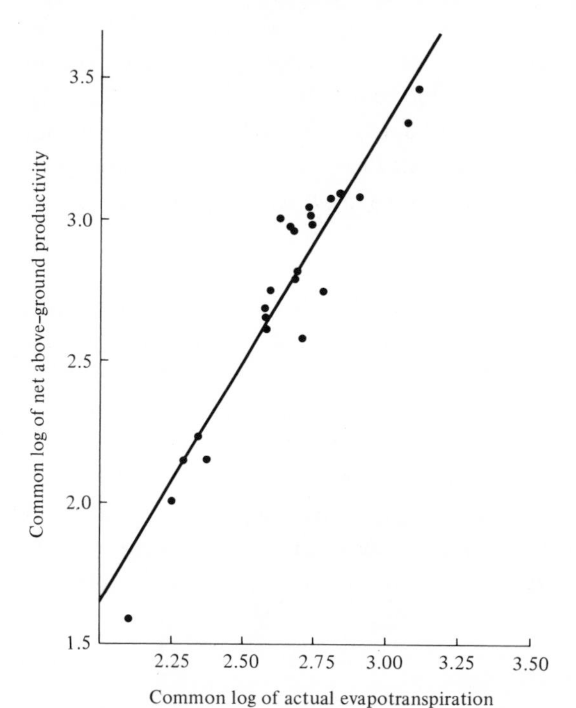

FIGURE 3.9 Log–log plot of net primary productivity above ground against estimated actual evapotranspiration for 24 areas ranging from barren desert to luxurious tropical rainforest. [After Rosenzweig (1968).]

on land: namely water and solar energy. Photosynthesis, that fundamental process upon which nearly all life depends for energy, is represented by the following chemical equation:

$$6CO_2 + 12H_2O \xrightarrow[\text{energy}]{\text{solar}} C_6H_{12}O_6 + 6O_2 + 6H_2O$$

where $C_6H_{12}O_6$ is the energy-rich glucose molecule. Carbon dioxide concentration in the atmosphere is fairly constant at about 0.03 percent and does not strongly influence the rate of photosynthesis, except under most unusual conditions of high water availability and full or nearly full sunlight (Meyer, Anderson, and Bohning, 1960). Rosenzweig notes that, on a geographical scale, each of the other two requisites for photosynthesis, water and solar energy, are much more variable in their availabilities; moreover, AET measures the availability of both. Temperature is often a rate-limiting factor and markedly affects photosynthesis; it, too, is presumably incorporated into an estimated AET value. Primary production may be influenced by nutrient availability as well, but in many terrestrial ecosystems these effects are often relatively minor. In aquatic ecosystems, however, nutrient availability is often a major determinant of the rate of photosynthesis (see later section on aquatic ecosystems, pp. 61–64). Primary production in aquatic systems of fairly constant temperature, such as the oceans, is often strongly affected by light. Because light intensity decreases rapidly with depth in lakes and oceans, most primary productivity is concentrated near the surface. However, short wavelengths (blue light) penetrate deeper into water than longer ones, and some benthic (bottom-dwelling) marine "red" algae have evolved unique photosynthetic adaptations to utilize these wavelengths.

Similarly, within forests, light intensity varies markedly with height above the ground. Tall trees in the canopy receive the full incident solar radiation, while shorter trees and shrubs receive progressively less and less light. Indeed, in really dense forests, less than 1 percent of the incident solar energy impinging on the canopy actually penetrates to the forest floor. Although a tree in the canopy therefore has more solar energy available to it than a fern on the forest floor, the canopy tree must also expend much more energy on vegetative supporting tissues (wood) than the fern. Hence, each life form and each growth strategy has its own associated costs and profits.

Soil Formation and Succession

Soils are a key part of the terrestrial ecosystem because many processes critical to the functioning of ecosystems occur in the soil. This is where dead organisms are decomposed and where their nutrients are retained until used by plants, and, indirectly, returned to the remainder of the community. Soils are essentially a meeting ground of the inorganic and the organic worlds. Many organisms live in the soil, and the vast majority of insects spend at least a part of their life cycle there. Certain soil-dwelling organisms, such as earthworms, often play a major role in breaking down organic particles into smaller pieces which present a larger surface area for microbial action, thus facilitating decomposition. Indeed, the activities of such soil organisms sometimes constitute a "bottleneck" for the rate of nutrient cycling, and, as such, they can regulate nutrient availability and turnover rates in the entire community. In aquatic ecosystems, bottom sediments like mud and ooze are closely analogous to the soils of terrestrial systems.

Much of modern soil science, or pedology, was anticipated in the late 1800s by a prominent Russian pedologist named Dokuchaev. He devised a theory of soil formation, or pedogenesis, based largely on climate, although he also recognized the importance of time, topography, organisms (especially vegetation), and parent materials (the underlying rocks from which the soil is derived). The relative importance of each of these five major soil-forming factors varies from situation to situation. Figure 3.10 shows how markedly the soil changes along the transition from prairie to forest in the midwestern United States, where the only other conspicuous major variable is vegetation type. Jenny (1941) gives many other examples of how each of the five major soil forming factors influence particular soils.

The marked effects some soil types can have on plants are well illustrated by so-called serpentine soils (Whittaker, Walker, and Kruckeberg, 1954), which are formed over a parent material of serpentine rock. These soils often occur in localized patches, surrounded by other soil types; typically the vegetation changes abruptly from nonserpentine to serpentine soils. Serpentine soils are rich in magnesium, chromium, and nickel, but contain

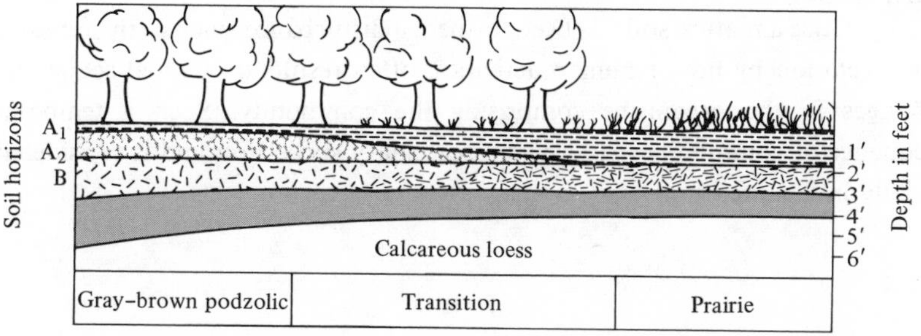

FIGURE 3.10 Diagram showing the typical soil changes along the transition from prairie (deep black topsoil) to forest (shallow topsoil) at the edge of the North American great plains. [After Crocker (1952).]

very little calcium, molybdenum, nitrogen, and phosphorus. They usually support a stunted vegetation and are relatively less productive than adjacent areas with different, richer soils; indeed, entire floras of specialized plant species have evolved that are tolerant of the conditions of serpentine soils, particularly their low calcium levels.

Soil development from bare rock, or primary succession, is a very slow process that often requires centuries. Rock is fragmented by temperature changes, by the action of windblown particles, and, in colder regions, by the alternate freezing and thawing of water. Chemical reactions, such as the formation of carbonic acid (H_2CO_3) from water and carbon dioxide, may also help to dissolve and break down certain rock types. Such weathering of rocks releases inorganic nutrients that can be used by plants. Eventually, lichens may establish themselves, and as other plants root and grow, the rock is further broken up into still smaller fragments. As these plants photosynthesize, they convert inorganic materials into organic matter. Such organic material, mixed with inorganic rock fragments, accumulates and soil is slowly formed. Early in primary succession, production of new organic material exceeds its consumption and organic matter accumulates; as soil "maturity" is approached, soil eventually ceases to accumulate (see also Figure 9.1).

There are fairly close parallels between concepts of soil development and those for the development of ecological communities; pedologists speak of "mature" soils at the steady state, while ecologists recognize the "climax" communities that grow on and live in these same soils. Indeed, these two components of the ecosystem (soils and vegetation) are intricately interrelated and interdependent, with each strongly influencing the other. Except in forests and rain forests, there is usually a one-to-one correspondence between them (Figure 3.11). [Compare also the geographic distribution of soil types (Figure 3.12) with the distribution of vegetation types shown in Figure 3.1.]

Once a mature soil has been formed, a disturbance, such as the removal of vegetation by fire or human activities, often results in gradual sequential changes in the organisms comprising the community. Such a temporal sequence of communities is termed a secondary succession and is considered briefly in Chapter 7.

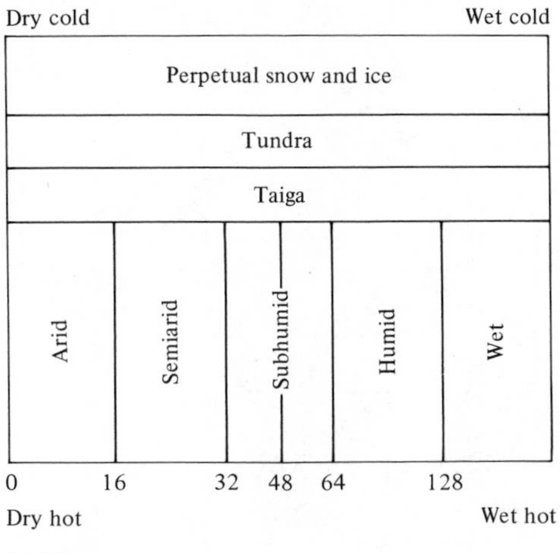

(*a*) Climate

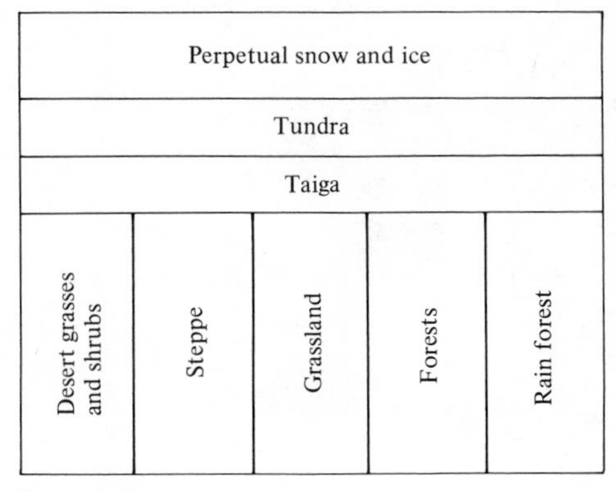

(*b*) Vegetation

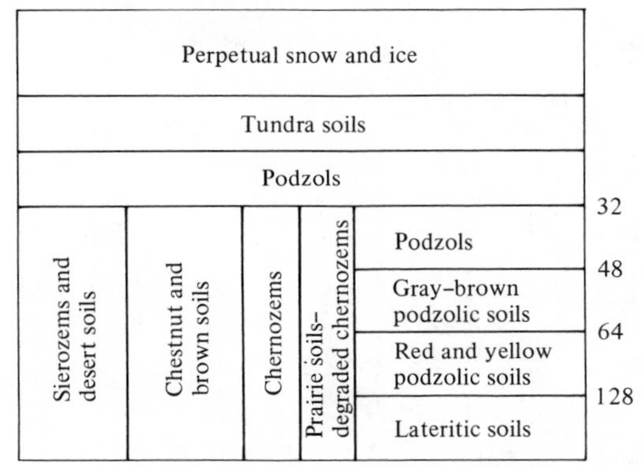

(*c*) Soils

FIGURE 3.11 Relationships between temperature and precipitation and climatic types (*a*), vegetational formations (*b*), and major zonal soil groups (*c*). [From Blumenstock and Thornthwaite (1941).]

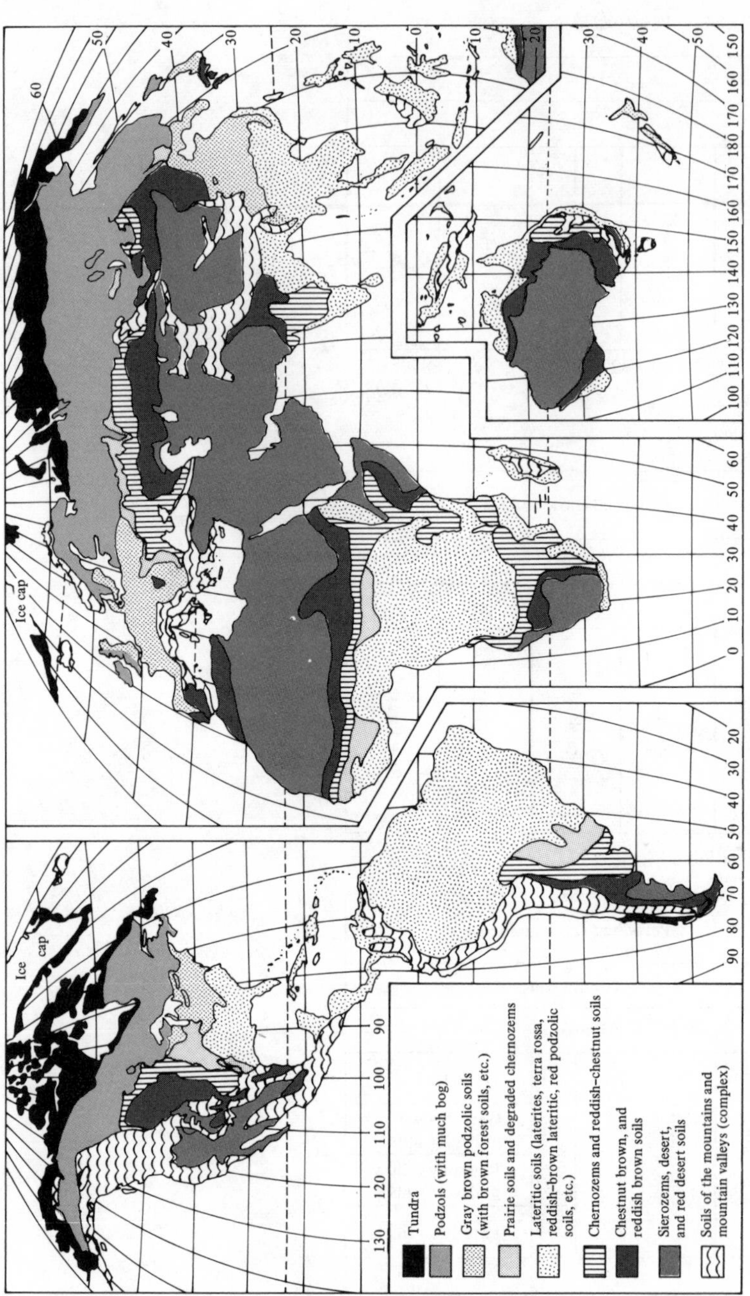

FIGURE 3.12 Geographic distribution of the primary soil types. Compare with Figure 3.1. [After Blumenstock and Thornthwaite (1941).]

Classification of Natural Communities

Biotic communities have been classified in various ways. An early attempt to classify communities was that of Merriam (1890), who recognized a number of different "life zones" which were defined solely in terms of temperature and ignored precipitation. His somewhat simplistic scheme is no longer used, but his approach did link climate with vegetation in a more or less predictive manner.

Shelford (1913a, 1963) and his students have taken a somewhat different approach to the classification of natural communities that does not attempt to correlate climate with the plants and animals occurring in an area. Rather, they classify different natural communities into a large number of so-called biomes and associations, relying largely upon the characteristic plant and animal species that compose a particular community. As such, this scheme is descriptive, rather than predictive. Such massive descriptions of different communities (see, for example, Dice, 1952, and Shelford, 1963) can often be quite useful in that they allow one to become familiar with a particular community with relative ease.

Workers involved in such attempts at classification typically envision communities as discrete entities with relatively little or no intergradation between them; thus the Shelford school considers biomes to be distinct and real entities in nature rather than artificial and arbitrary human constructs. Another school of ecologists, represented by McIntosh (1967) and Whittaker (1970), takes an opposing view, emphasizing that communities grade gradually into one another, and form so-called continua or ecoclines (Figures 3.13 and 7.15).

Two recent attempts to relate climate and vegetation are shown in Figures 3.14 and 3.15 (compare these with Figure 3.11). Whittaker (Figure 3.14) simply plots the average annual temperature against average annual precipitation and fills in the vegetational formation typically occurring under such a climatic regime; thus he demonstrates how macroclimate determines the vegetation of an area, although he also emphasizes that these correlations are not hard and fast but that local vegetation type depends upon other factors such as soil types, rainfall regime, and the frequency of disturbance by fires.

At first glance, Holdridge's scheme (Figure 3.15) appears to be much more complex and more sophisticated than Figure 3.14. But closer inspection shows that there is really very little difference between these two plots. Figure 3.15 plots mean annual "biotemperature" (defined as simply mean annual temperature excluding temperatures below 0°C and above 30°C)

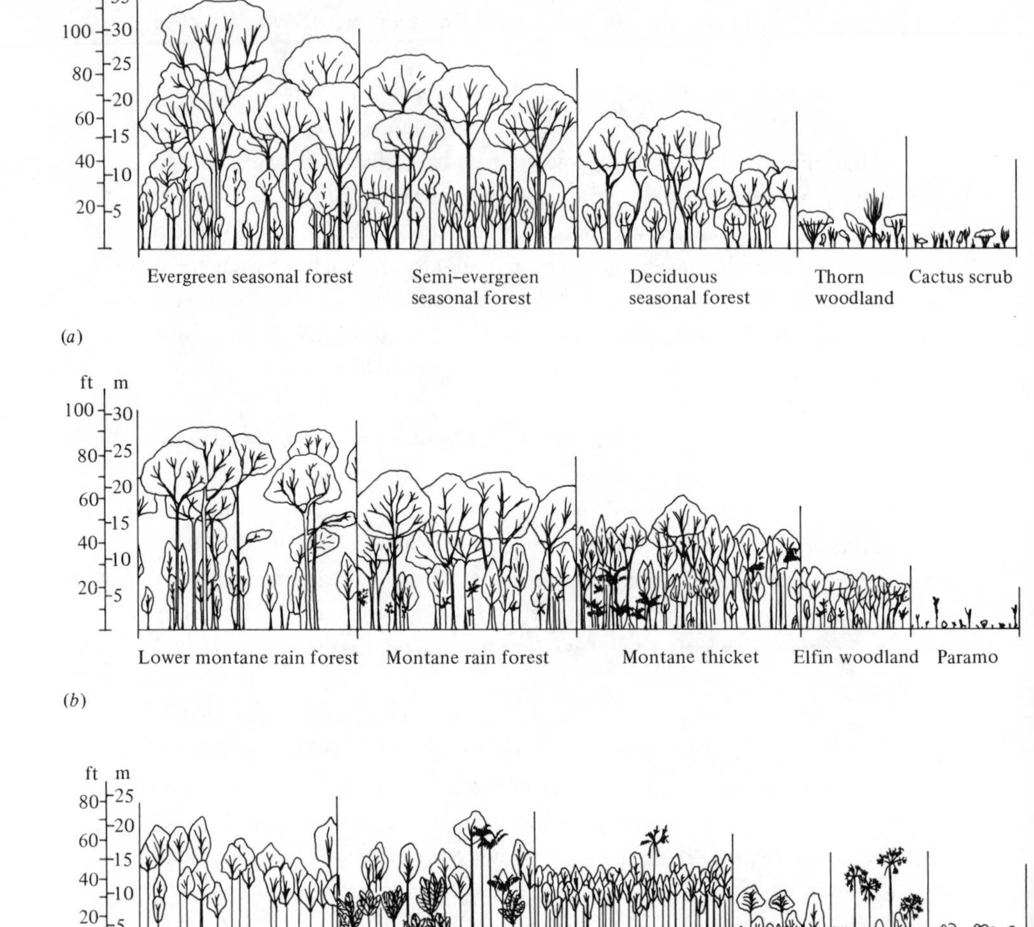

FIGURE 3.13 Vegetation profiles along three ecoclines. (*a*) A gradient of increasing aridity from seasonal rainforest to desert. (*b*) An elevation gradient up a tropical mountainside from tropical rainforest to alpine meadow (paramo). (*c*) A humidity gradient from swamp forest to savanna. [From Beard (1955).]

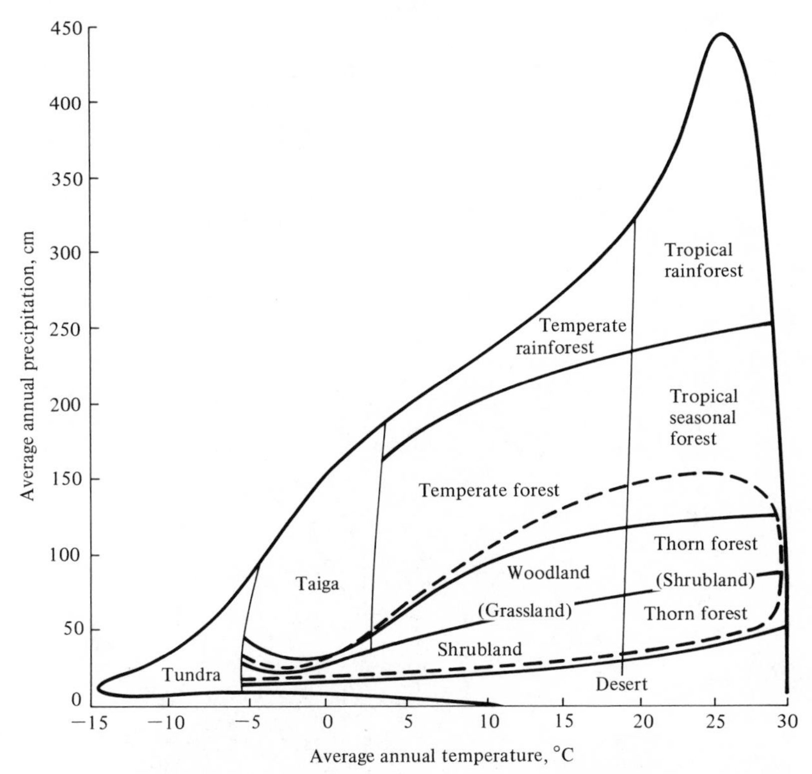

FIGURE 3.14 Diagrammatic representation of the correlation between climate, as reflected by average annual temperature and precipitation, and vegetational formation types. Boundaries between types are approximate and are influenced locally by soil type, seasonality of rainfall, and disturbances such as fire. The dashed line encloses a range of climates in which either grasslands or woody plants may constitute the prevailing vegetation of an area. Compare this figure with Figure 3.15. [After Whittaker (1970). Reprinted with permission of Macmillan Publishing Co., Inc., from *Communities and Ecosystems* by Robert H. Whittaker. © Copyright Robert H. Whittaker, 1970.]

against average annual precipitation (these two axes are labeled in boldface letters in the figure). There is a direct positive relationship between mean annual biotemperature and mean annual potential evapotranspiration (Holdridge, 1959), as indicated on the far right of Figure 3.15. Holdridge's "latitudinal regions" and "altitudinal belts" are also entirely dependent upon mean annual biotemperature. The interaction of mean annual biotemperature and mean annual precipitation, in turn, completely determines the position on the other two axes, "potential evapotranspiration ratio" and "humidity provinces." Hence, this elaborate figure actually plots only two truly *independent* variables.

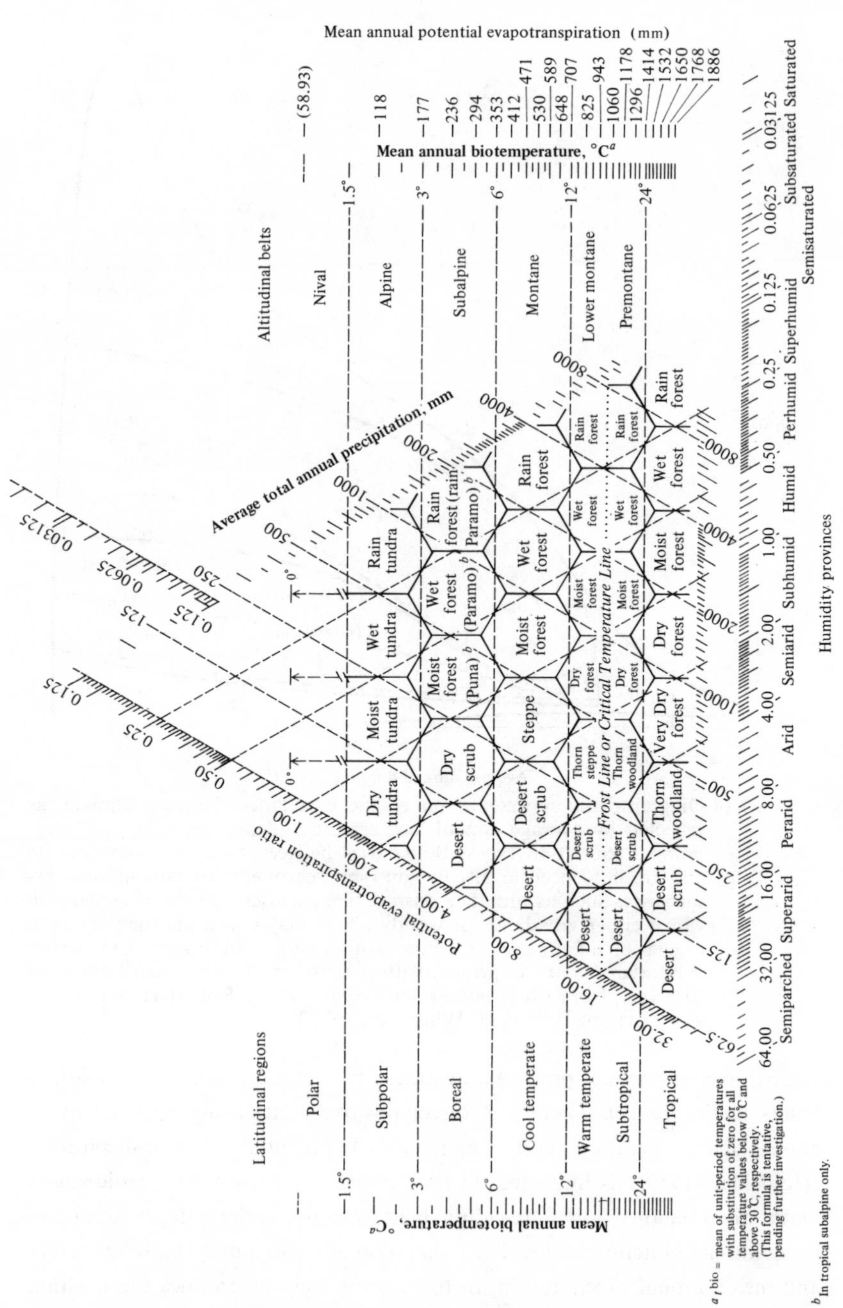

FIGURE 3.15 Another diagrammatic representation of the correlation between climate and vegetation type. Although this plot seems very complicated, and, at first glance, it may appear to plot three (or more) different variables, this scheme is really not markedly different from the one shown in Figure 3.14. See text. [After Holdridge (1967).]

These two figures neatly underscore the strong correlation between climate and vegetation mentioned at the beginning of this chapter.

Some Considerations of Aquatic Ecosystems

Although the same ecological principles presumably operate in both aquatic and terrestrial ecosystems, there are some striking and interesting fundamental differences between these two sorts of ecological systems. For example, primary producers on land are sessile and tend to be large and relatively long-lived (air does not provide much support and woody tissues are needed), while those in planktonic aquatic communities are typically free-floating, microscopic, and very short-lived (the buoyancy of water may make supportive plant tissues unnecessary; indeed a large planktonic plant might be easily broken by water turbulence). Most ecologists study either aquatic or terrestrial systems, and various aquatic subdisciplines of ecology are recognized, such as aquatic ecology and marine ecology. Limnology is the study of freshwater ecosystems (ponds, lakes, and streams), whereas oceanography is concerned with bodies of salt water. Because the preceding part of this chapter and most of the remainder of the book emphasize terrestrial ecosystems, certain salient properties of aquatic ecosystems, especially lakes, are briefly considered in this section. Lakes are particularly appealing subjects for ecological study in that they are self-contained ecosystems, discrete and largely isolated from other ecosystems. Nutrient flow into and out of a lake can often be estimated with relative ease. The study of lakes is fascinating and the interested reader is referred to Ruttner (1953) and Hutchinson (1957b, 1967).

Water has some peculiar physical and chemical properties that strongly influence the organisms that live in it. As indicated earlier, water has a high specific heat; moreover, in the solid (frozen) state, its density is less than it is in the liquid state (that is, ice floats). Water is densest at 4°C and water at this temperature "sinks." Furthermore, water is nearly a "universal solvent," in that many important substances go into aqueous solution. The significance of these properties becomes apparent in the following discussion.

A typical, relatively deep lake in the temperate zones undergoes marked and very predictable seasonal changes in temperature. During the warm summer months its surface waters are heated up, and, because warm water is less dense than colder water, a distinct upper layer of warm water termed the *epilimnion* is formed (Figure 3.16*a*). (Movement of heat within a lake is due to water currents produced primarily by wind—see below.) Deeper waters, termed the *hypolimnion*, remain relatively cold during summer, often

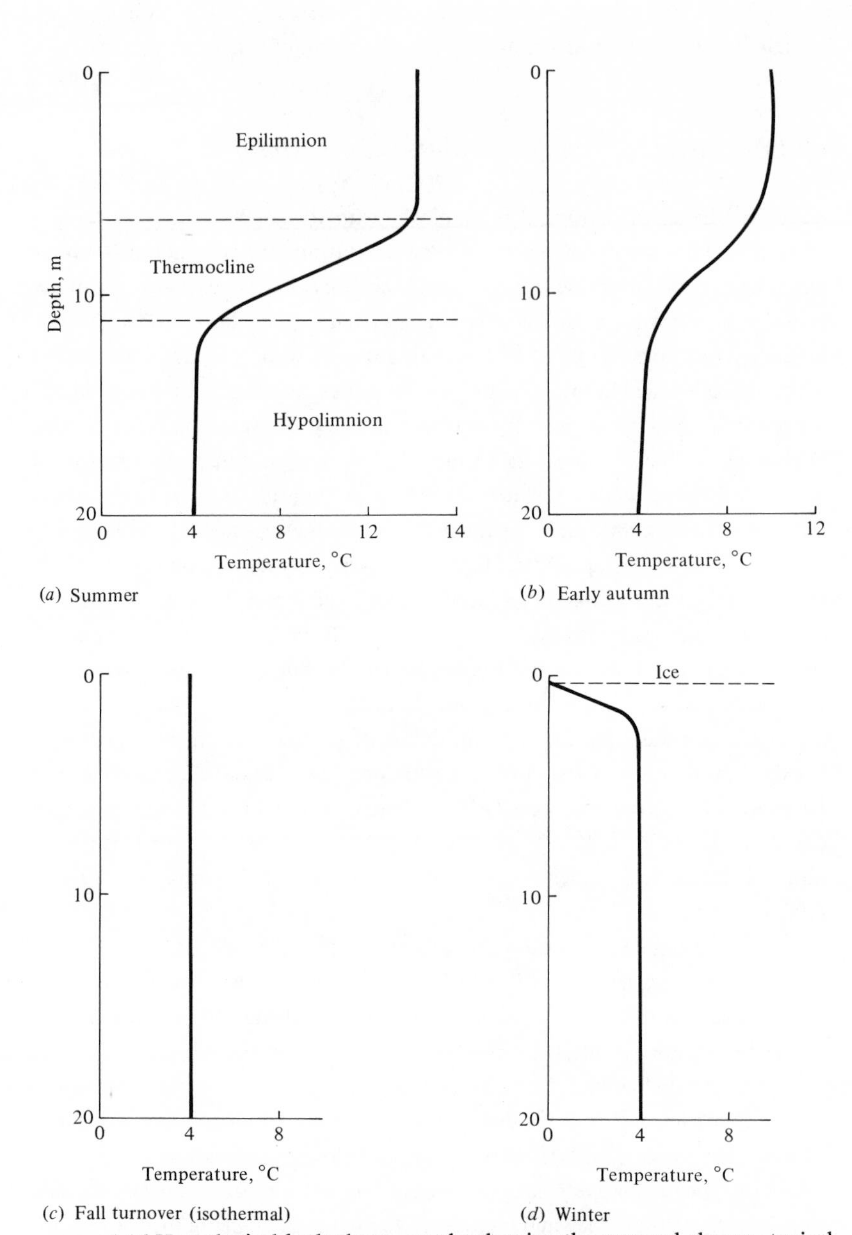

FIGURE 3.16 Hypothetical bathythermographs showing the seasonal changes typical of a deep temperate zone lake. (*a*) A stratified lake during summer. (*b*) In early autumn the upper waters cool down somewhat. (*c*) In late autumn or early winter the lake's waters are all at exactly the same temperature, here 4°C (i.e., the lake is "isothermal"). (*d*) During the freezing winter months, a layer of surface ice chills the uppermost water.

at about 4°C; also, an intermediate layer of rapid temperature change, termed the *thermocline*, separates the epilimnion from the hypolimnion (see Figure 3.16*a*). (A swimmer sometimes experiences these water layers of different temperatures when he dives into deep water or when he treads water and his feet drop down into the cold hypolimnion.) A lake with a thermal profile, or bathythermograph, like that shown in Figure 3.16*a* is said to be "stratified," because of its layering of warm water over cold water.

With the decrease in incident solar energy in autumn, surface waters cool off and give up their heat to adjacent landmasses and the atmosphere (Figure 3.16*b*). Eventually the epilimnion cools down to the same temperature as the hypolimnion and the lake becomes isothermal (Figure 3.16*c*). This is the time of the "fall turnover." With winter's freezing temperatures, the lake's surface turns to ice and its temperature versus depth profile looks something like that in Figure 3.16*d*. Finally, in spring the ice melts, the lake is briefly isothermal once again until its surface waters are rapidly warmed, when it again becomes stratified and the annual cycle repeats itself.

Because prevailing winds produce surface water currents, a lake's waters circulate. In stratified lakes, the epilimnion constitutes a more or less closed cell of circulating water, while the deep cold water scarcely moves or mixes with the warmer water above it. During this period, as dead organisms and particulate organic matter sink into the noncirculating hypolimnion, the lake undergoes what is known as summer stagnation. When a lake becomes isothermal, its entire water mass can be circulated and nutrient-rich bottom waters brought up to the surface; limnologists thus speak of the spring and the fall "turnover." Meteorological conditions, particularly wind velocity and duration, strongly influence such turnovers; indeed, if there is little wind during the period a lake is isothermal, its waters might not be thoroughly mixed and many nutrients may remain locked up in its depths. After a thorough turnover, the entire water mass of a lake is equalized, and concentrations of various substances, such as oxygen and carbon dioxide, are similar throughout the lake.

Because lakes differ in their nutrient content and degree of productivity, they can be arranged along a continuum ranging from those with low nutrient levels and low productivity (so-called *oligotrophic* lakes) to those with high nutrient content and high productivity (*eutrophic* lakes). Clear, cold, and deep lakes high in the mountains are usually relatively oligotrophic, whereas shallower, warmer, and more turbid lakes such as those in low-lying areas are generally more eutrophic. Oligotrophic lakes typically support game fish such as trout, whereas eutrophic lakes contain "trash" fish such as carp. As they age and fill up with sediments, many lakes gradually undergo a natural process of eutrophication, steadily becoming more and more productive. Man accelerates this process by enriching lakes with his wastes, and many oligotrophic lakes have rapidly become eutrophic under his influence. A good indicator of the degree of eutrophication is the oxygen content of deep water during summer. In a relatively unproductive lake, oxygen content varies little with depth and there is ample oxygen at the bottom of the lake. In contrast, oxygen content diminishes rapidly with depth in productive lakes, and anaerobic processes sometimes characterize their depths during the summer months. With the autumn turnover, oxygen-rich waters again reach the bottom sediments and aerobic processes become possible. However, once such a lake becomes stratified, the oxygen in its deep water is quickly used up by benthic organisms (in the deep water there is little or no photosynthesis to replenish the oxygen).

These seasonal physical changes profoundly influence the community of organisms living in a lake. During the early spring and after the fall turnover, surface waters are rich in dissolved nutrients and temperate lakes are very productive, whereas during the middle of summer, many nutrients are unavailable to phytoplankton in the upper waters and primary production is much reduced.

Organisms within a lake community are usually distributed quite predictably in time and space. Thus, there is typically a regular seasonal progression of planktonic algae, with diatoms most abundant in the winter, changing to desmids and green algae in the spring, which flora gradually gives way to blue-green algae during the summer months. The composition of the zooplankton also varies seasonally. Although plankton are moved about by water currents, many are strong enough swimmers to select a particular depth. Such species often actually "migrate" vertically during the

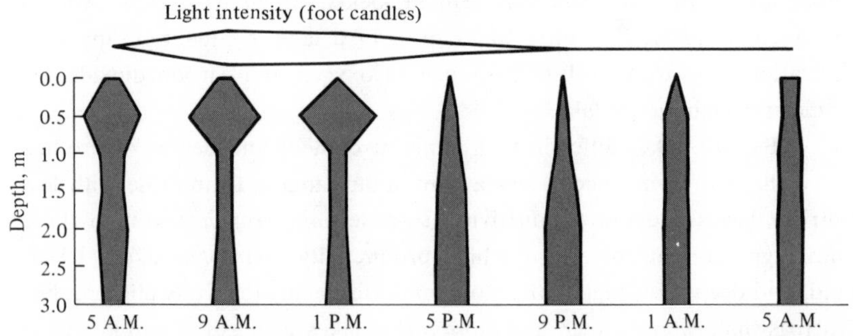

FIGURE 3.17 Many freshwater planktonic animals move vertically during the daily cycle of illumination somewhat as shown here (the widths of the bands represent the density of animals at a given depth at a particular time). [After Hutchinson (1967) after Cowles and Brambel.]

day and/or with the seasons, being found at characteristic depths at any given time (Figure 3.17).

Selected References

Allee *et al.* (1949); Andrewartha and Birch (1954); Clapham (1973); Clarke (1954); Colinvaux (1973); Collier *et al.* (1973); Daubenmire (1947, 1956, 1968); Kendeigh (1961); Knight (1965); Kormondy (1969); Krebs (1972); Lowry (1969); Odum (1959, 1971); Oosting (1958); Ricklefs (1973); Smith (1966); Watt (1973); Weaver and Clements (1938); Whittaker (1970).

Plant Life Forms and Biomes

Cain (1950); Clapham (1973); Horn (1971); Raunkaier (1934); Whittaker (1970).

Microclimate

Collier *et al.* (1973); Gates (1962); Geiger (1966); Lowry (1969); Schmidt-Nielson (1964); Smith (1966).

Primary Production and Evapotranspiration

Collier *et al.* (1973); Gates (1965); Horn (1971); Meyer, Anderson, and Bohning (1960); Odum (1959, 1971); Rosenzweig (1968); Whittaker (1970); Woodwell and Whittaker (1968).

Soil Formation and Succession

Black (1968); Burges and Raw (1967); Crocker (1952); Crocker and Major (1955); Doeksen and van der Drift (1963); Eyre (1963); Fried and Broeshart (1967); Jenny (1941); Joffe (1949); Oosting (1958); Schaller (1968); Waksman (1952); Whittaker, Walker, and Kruckeberg (1954).

Classification of Natural Communities

Braun-Blanquet (1932); Clapham (1973); Dice (1952); Gleason and Cronquist (1964); Holdridge (1947, 1959, 1967); McIntosh (1967); Merriam (1890); Shelford (1913a, 1963); Tosi (1964); Whittaker (1962, 1967, 1970).

Some Considerations of Aquatic Ecosystems

Clapham (1973); Ford and Hazen (1972); Frank (1968); Frey (1963); Grice and Hart (1962); Hutchinson (1951, 1957b, 1961, 1967); Mann (1969); National Academy of Sciences (1969); Ruttner (1953); Watt (1973); Welch (1952); Weyl (1970).

4 Principles of Population Ecology

Introduction

Each generation, sexually reproducing organisms mix their genetic materials. Such shared genetic material is called a *gene pool*, and all the organisms involved in a gene pool are collectively termed a Mendelian *population*. Populations are more abstract conceptual entities than cells or organisms and are somewhat more elusive, but they are nonetheless real. A gene pool has continuity in both space and time, and organisms belonging to a given population either have a common immediate ancestry or are potentially able to interbreed. Alternatively, a population may be defined as a cluster of individuals with a high probability of mating with each other compared to their probability of mating with a member of some other population. As such, Mendelian populations are groups of organisms with a substantial amount of genetic exchange. Such populations are also called *demes*, and the study of their vital statistics is termed *demography*.

In practice, it is extremely difficult to draw boundaries between populations except in most unusual circumstances. Certainly the English sparrows introduced into Australia are no longer exchanging genes with those introduced into North America, and so each represents a functionally distinct population. Although they might well be potentially able to interbreed, the probability of their doing so is remote because of geographical separation. Such differences also exist at a much finer and more local level, both between and within habitats. For example, English sparrows in eastern

Australia are separated from those in western Australia by a desert that is uninhabitable to these birds and so they form different populations.

By the above definition, organisms reproducing asexually, as for example a plant that buds off another individual, strictly speaking do not form true populations. There is no gene pool, no interbreeding, and all offspring are essentially identical genetically. However, even such non-interbreeding plants and animals often form collections of organisms with many of the populational attributes of true sexual populations. In any case, most of the animal and many of the plant species that have been studied resort to sexual reproduction at least periodically, and therefore mix up their genes to some extent and form true Mendelian populations. Populations vary in size from very small (for example, a few individuals on a newly colonized island) to very large, such as some wide ranging and common small insects with populations in the millions. Populations are more usually in the thousands. Consideration of both asexual and sexual organisms at the population level often allows us to extend our insight into the activities of individuals in remarkable ways.

Each member of a population has its own relative fitness within that population, which determines in part the fitness of the other members of the population; likewise, every individual's fitness is influenced by all the other members of its population. Fitness can be defined and understood only in the context of an organism's total environment.

If we consider any continuously varying measurable character, such as height or weight, a population under consideration has a *mean* (average) and a *variance* (a statistical measure of dispersion based on the average of the squared deviations from the mean). Any one individual has only a single value, but the population of individuals has both a mean and a variance. These are *population parameters*, characteristic of the population concerned, and impossible to define unless we consider a population. Populations also have birth rates, death rates, age structures, sex ratios, gene frequencies, genetic variability, growth rates and growth forms, densities, and so on. Here we examine a variety of such populational characteristics.

Life Tables and Tables of Reproduction

Insurance companies employ actuaries to calculate insurance risks. An actuary obtains a large sample of data on some past event and uses them to estimate the average rate of occurrence of a phenomenon; the company then allows itself a suitable profit and margin of safety and sells insurance on the

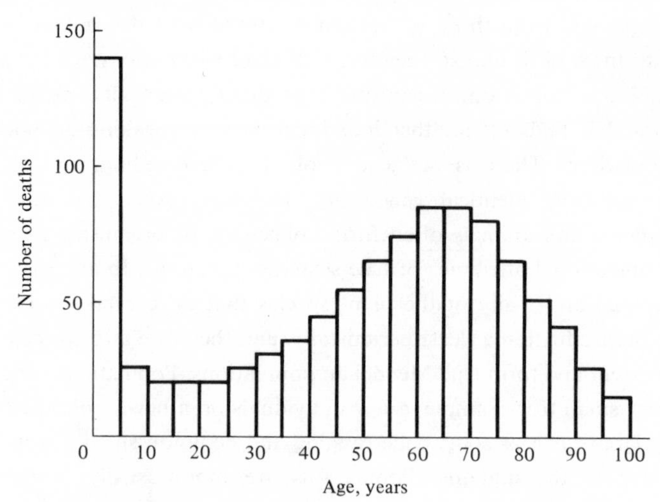

(*a*) Frequency distribution of age at death

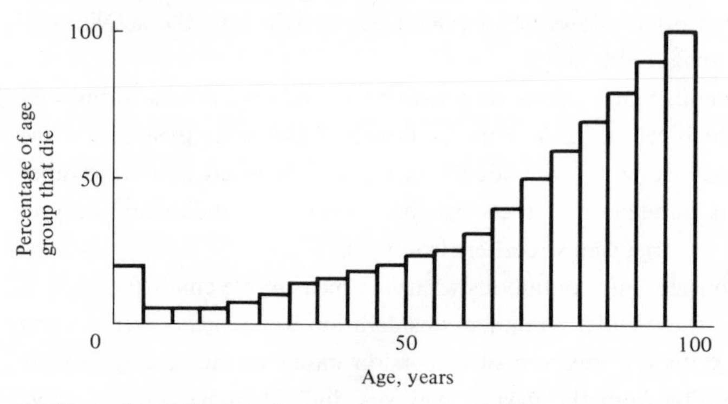

(*b*) Force of mortality in various age groups

(*c*) Survivorship of various age groups

FIGURE 4.1 Hypothetical death data of a life insurance actuary, who would treat the sexes separately. (*a*) The "raw" data, which consist of a frequency distribution of the age of death of, say, 1000 individuals. For convenience,

event concerned. Let us consider how an actuary calculates life insurance risks. His raw data consist simply of the average number of deaths at every age in a population: that is, a frequency distribution of deaths by ages (Figure 4.1*a*). From these values and the age distribution of the population, he calculates age-specific death rates, which are simply the percentages of individuals of any age group who die during that age period (Figure 4.1*b*). Death rate at age x is designated by q_x, which is sometimes called the "force of mortality" or the *age-specific death rate*. In Figure 4.1, deaths are combined into age classes covering a five-year period. If the population is large and age groups are fine, consisting of, for instance, only individuals born on a given day, these curves would be much smoother or more nearly continuous (the distinction between *discrete* and *continuous* events or characteristics will be made repeatedly in this chapter). Demographers begin with discrete age intervals, and use the methods of calculus to fit continuous functions to these for estimates at various points within age intervals.

Still another useful way of manipulating life tables is to calculate age-specific percentage survival (Figure 4.1*c*). Starting with an initial number or *cohort* of newborn individuals, one calculates the percentage of the population alive at every age by sequentially subtracting the percentage of deaths at each age. A smoothed continuous version of survivorship (as in Figures 4.2 and 4.3) is called a *survivorship curve*. The percent surviving at age x divided by 100 gives the probability that an average newborn will survive to that age, which is usually designated by l_x.

Finally, our actuary is interested in estimating the expectation of further life. How long, on the average, will someone of age x live? For newborn individuals (age 0) the average life expectancy is equal to the mean length of life of the cohort. In general, expectation of life at any age x is simply the mean life span remaining to those individuals attaining age x. In symbols,

$$E_x = \frac{\sum_{y=x}^{\infty} l_y}{l_x} \quad \text{or} \quad E_x = \frac{\int_x^{\infty} l_y dy}{l_x} \tag{1}$$

---

data are lumped into 5-year age intervals. (*b*) The death rate at age x, or q_x, expressed as the percentage of each age group that die during that age interval. (The age distribution of the population at large is needed to compute such age-specific mortality rates.) Values are high in older age classes because these age groups contain relatively few individuals, many of which die during that age interval. (*c*) Percentage surviving from an initial cohort of individuals with the above age-specific death rates. When divided by 100, these values give the probability that an average newborn will survive to age x.

where E_x is the expectation of life at age x, and y subscripts age. The left-hand equation is the discrete version, the right-hand one a continuous version using the symbolism of integral calculus. Calculation of E_x is illustrated in Table 4.1 (p. 74).

Life insurance premiums for men are higher than they are for women because males have a steeper survivorship curve, and therefore, at a given age, a shorter life expectancy than females. Figures 4.2 and 4.3 show a variety of survivorship curves, representing the great range they take in

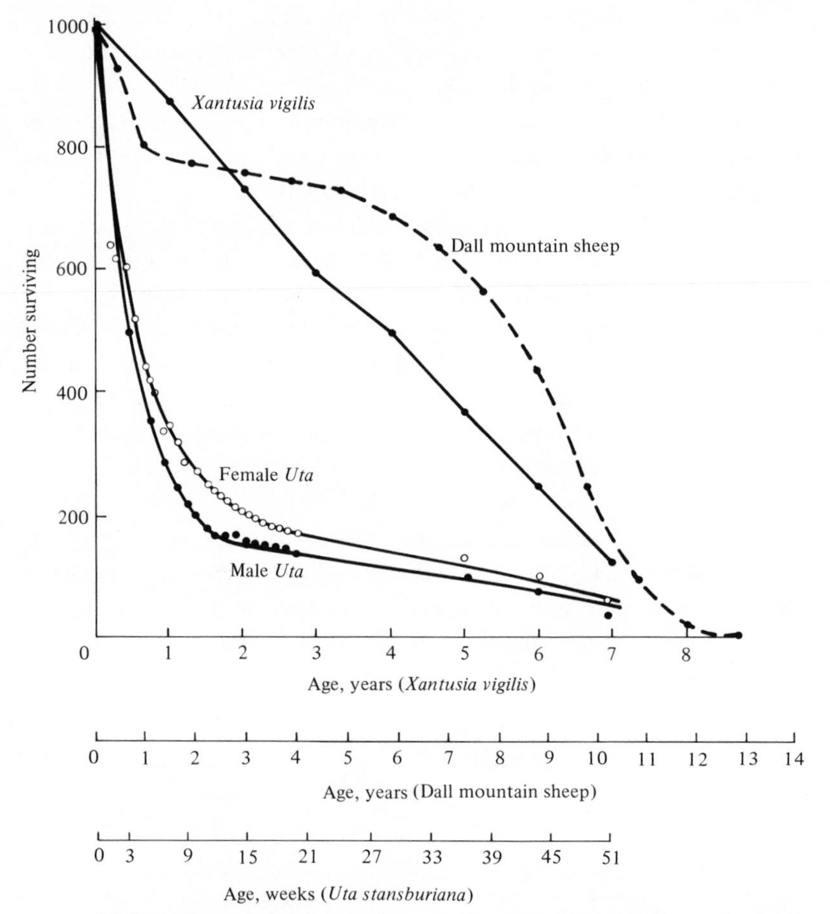

FIGURE 4.2 Several survivorship schedules plotted with an arithmetic vertical axis. Compare with the more rectangular semilogarithmic plots of Figure 4.3. Although both types of plots are in common use, logarithmic ones are preferable. In a logarithmic plot, survivorship of the lizard *Xantusia* becomes rectangular (rather than diagonal), whereas that of another lizard, *Uta*, is diagonal (rather than inversely hyperbolic). [After Deevey (1947), Tinkle (1967), and Zweifel and Lowe (1966).]

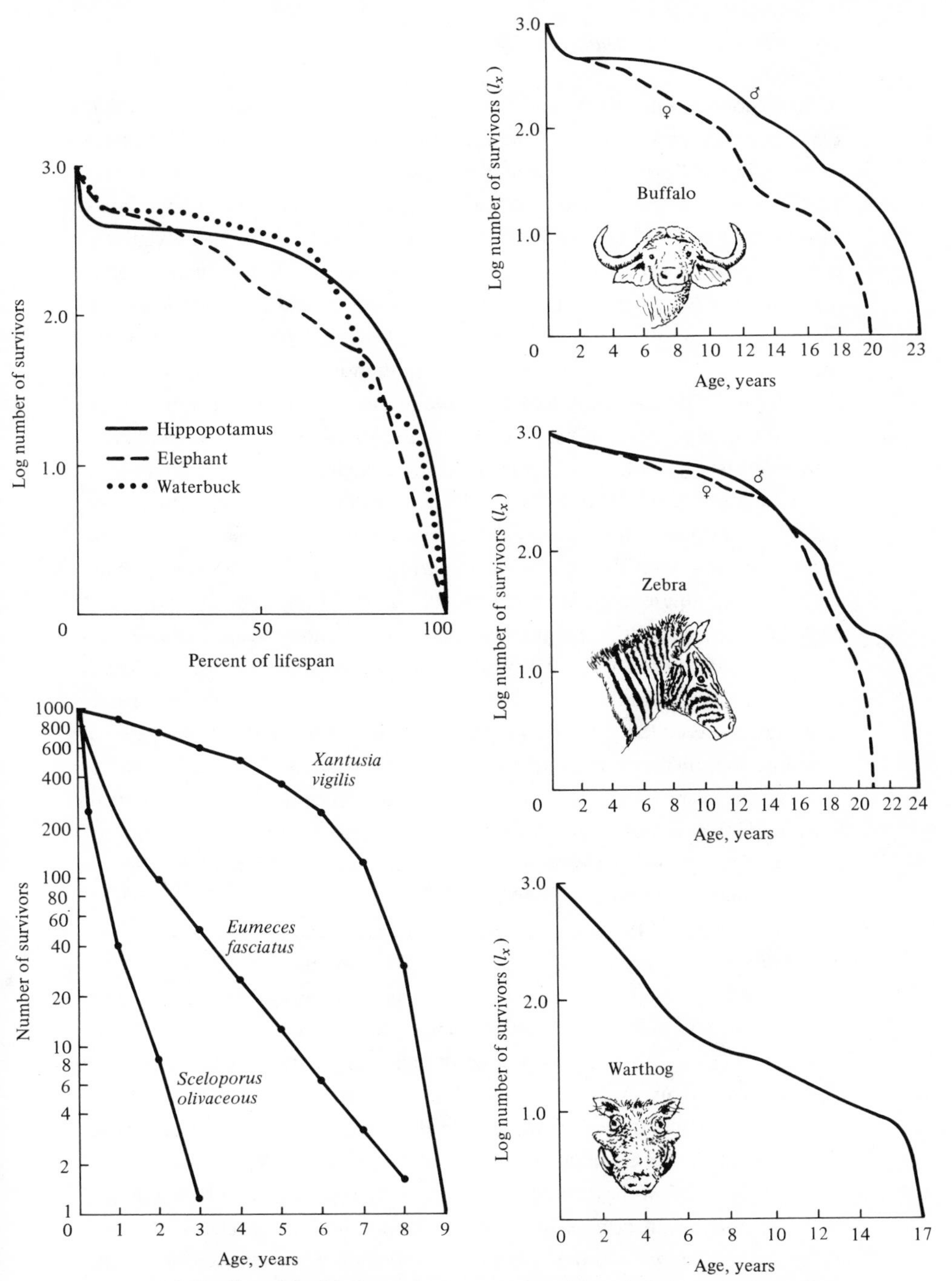

FIGURE 4.3 Semilogarithmic plots of some survivorship schedules. Compare *Xantusia* survivorship in Figures 4.2 and 4.3, plotted from the same data. The virtue of a semilogarithmic plot is that a straight line implies equal mortality rates with respect to age. [After Zweifel and Lowe (1966) and Spinage (1972). Spinage by permission of Duke University Press.]

natural populations. Rectangular survivorship on a semilogarithmic plot, that is, little mortality until some age and then fairly steep mortality thereafter, as in the lizards *Xantusia vigilis* and *Lygosoma laterale*, dall mountain sheep, most African ungulates, humans, and perhaps most mammals (Caughley, 1966), has been called Type I survivorship (Pearl, 1928). Relatively constant death rates with age produce diagonal survivorship curves on semilogarithmic plots as in the lizards *Uta stansburiana* and *Eumeces fasciatus*, the warthog, and most birds; these are classified as Type II curves [actually there are two kinds of Type II curves, representing, respectively, constant risk of death per unit time and constant numbers of deaths per unit time (Slobodkin, 1962)]. Many fish, marine invertebrates, and most insects have extremely steep juvenile mortality and relatively high survivorship afterward; that is, inverse hyperbolic or Type III survivorship. Of course, nature does not fall into three or four convenient categories, and many real survivorship curves are intermediate between the various "types" categorized above. Later we examine evolution of death rates and old age, but first we must consider the other important populational phenomenon—reproduction.

The number of offspring produced by an average organism of age x during that age period is designated m_x; only those progeny that enter age class zero are counted. Males and females are each credited with one-half of one reproduction for every such offspring produced, so that in order to replace itself an organism must have two progeny. (This procedure makes biological sense in that a sexually reproducing organism passes only half its genome to each of its progeny.) The sum of m_x over all ages, or the total number of offspring that would be produced by an average organism in the absence of mortality, is termed the *gross reproductive rate* (GRR). Patterns of reproduction and fecundities, or m_x schedules, vary widely among organisms. Some, such as annual plants and many insects, breed only once during their lifetime. Others, such as perennial plants and many vertebrates, breed repeatedly. The number of eggs produced, and their size relative to the parent, also vary over many orders of magnitude. Litter size (usually designated by B) refers to the number of young produced during each act of reproduction, and is discussed further under Evolution of Reproductive Rates (pp. 92–98).

Reproduction may be delayed until fairly late in life, or reproductive activities may begin almost immediately after hatching or birth. The age of first reproduction is usually termed α, and the age of last reproduction, ω. For an organism that breeds only once, the average time from egg to egg, or the time between generations, termed *generation time* (T), is simply equal to α. But generation time in animals that breed repeatedly is somewhat more complicated. The average time between generations of repeated reproducers can be roughly estimated as $\bar{T} = (\alpha + \omega)/2$. A more accurate calculation of T is possible by weighting each age by its total realized fecundity, using the following equations (see also Table 4.1):

$$T = \sum_{x=\alpha}^{\omega} x l_x m_x \quad \text{or} \quad T = \int_{\alpha}^{\omega} x l_x m_x \, dx \tag{2}$$

[These equations apply only in a nongrowing population; if a population is expanding or contracting the right-hand side must be divided by the net reproductive rate, R_0 (below), in order to standardize for the average number of successful offspring per individual.]

Net Reproductive Rate and Reproductive Value

Clearly not many organisms live to realize their full potential for reproduction, and we need an estimate of the number of offspring produced by an organism suffering average mortality. Thus, the *net reproductive rate* (R_0) is defined as *the average number of age class zero offspring produced by an average organism during its lifetime*. Mathematically, R_0 is simply the product of the age-specific survivorship and fecundity schedules, over all ages at which reproduction occurs, as follows:

$$R_0 = \sum_{x=0}^{\infty} l_x m_x \quad \text{or} \quad R_0 = \int_{0}^{\infty} l_x m_x \, dx \tag{3}$$

Alternatively, α and ω could be substituted for the 0 and ∞ limits. Once again, the equation on the left is for discrete age groups, while that on the right is for continuous ones. Table 4.1 illustrates the calculation of R_0 from a pair of discrete l_x and m_x schedules, and Figure 4.4 diagrams its calculation for continuous ones.

When R_0 is greater than 1 the population is increasing, when R_0 equals 1 it is stable, and when R_0 is less than 1 the population is decreasing. Because of this, the net reproductive rate has also been called the *replacement rate* of the population. A stable population, at equilibrium, with a steep l_x curve must have a correspondingly high m_x curve in order to replace itself (when death rate is high, birth rate must also be high). Conversely, when l_x is high, m_x must be low in order for R_0 to equal unity.

Another important concept, first elaborated by Fisher (1930), is reproductive value. To what extent, on the average, do members of a given age

TABLE 4.1 Illustration of the Calculation of E_x, T, R_0, and v_x in a Hypothetical Stable Population with Discrete Age Classes, Using Equations (1) to (4)

Age (x)	l_x	m_x	$l_x m_x$	$x l_x m_x$	E_x (see below)	v_x (see below)
0	1.0	0.0	0.00	0.00	3.40	1.00
1	0.8	0.2	0.16	0.16	3.00	1.25
2	0.6	0.3	0.18	0.36	2.67	1.40
3	0.4	1.0	0.40	1.20	2.50	1.65
4	0.4	0.6	0.24	0.96	1.50	0.65
5	0.2	0.1	0.02	0.10	1.00	0.10
6	0.0	0.0	0.00	0.00	0.00	0.00
Sums		2.2 (GRR)	1.00 (R_0)	2.78 (T)		

Expectation of life:

$E_0 = (l_0+l_1+l_2+l_3+l_4+l_5)/l_0 = (1.0+0.8+0.6+0.4+0.4+0.2)/1.0 = 3.4/1.0$

$E_1 = (l_1+l_2+l_3+l_4+l_5)/l_1 = (0.8+0.6+0.4+0.4+0.2)/0.8 = 2.4/0.8 = 3.0$

$E_2 = (l_2+l_3+l_4+l_5)/l_2 = (0.6+0.4+0.4+0.2)/0.6 = 1.6/0.6 = 2.67$

$E_3 = (l_3+l_4+l_5)/l_3 = (0.4+0.4+0.2)/0.4 = 1.0/0.4 = 2.5$

$E_4 = (l_4+l_5)/l_4 = (0.4+0.2)/0.4 = 0.6/0.4 = 1.5$

$E_5 = l_5/l_5 = 0.2/0.2 = 1.0$

Reproductive value:

$$v_0 = \frac{l_0}{l_0}m_0+\frac{l_1}{l_0}m_1+\frac{l_2}{l_0}m_2+\frac{l_3}{l_0}m_3+\frac{l_4}{l_0}m_4+\frac{l_5}{l_0}m_5 = 0.0+0.16+0.18+0.40+0.24+0.02 = 1.00$$

$$v_1 = \frac{l_1}{l_1}m_1+\frac{l_2}{l_1}m_2+\frac{l_3}{l_1}m_3+\frac{l_4}{l_1}m_4+\frac{l_5}{l_1}m_5 = 0.20+0.225+0.50+0.30+0.025 = 1.25$$

$$v_2 = \frac{l_2}{l_2}m_2+\frac{l_3}{l_2}m_3+\frac{l_4}{l_2}m_4+\frac{l_5}{l_2}m_5 = 0.30+0.67+0.40+0.03 = 1.40$$

$$v_3 = \frac{l_3}{l_3}m_3+\frac{l_4}{l_3}m_4+\frac{l_5}{l_3}m_5 = 1.0+0.6+0.05 = 1.65$$

$$v_4 = \frac{l_4}{l_4}m_4+\frac{l_5}{l_4}m_5 = 0.60+0.05 = 0.65$$

$$v_5 = \frac{l_5}{l_5}m_5 = 0.10$$

group contribute to the next generation between now and when they die? *Reproductive value* (v_x) in a stable population (one that is neither increasing or decreasing) is defined as the *age-specific expectation of **future** offspring*. Its mathematical definition in a stable population at equilibrium is:

$$v_x = \sum_{t=x}^{\infty} \frac{l_t}{l_x} m_t \quad \text{or} \quad v_x = \int_x^{\infty} \frac{l_t}{l_x} m_t \, dt \qquad \textbf{(4)}$$

As before, the equation on the left is for discrete age groups and that on the right for continuous age groups. The term l_t/l_x represents the probability of living from age x to age t, and m_t is the average reproductive success of an individual at age t. Clearly for newborn individuals of age 0 in a stable population, v_0 is exactly equal to the net reproductive rate, R_0. A postreproductive

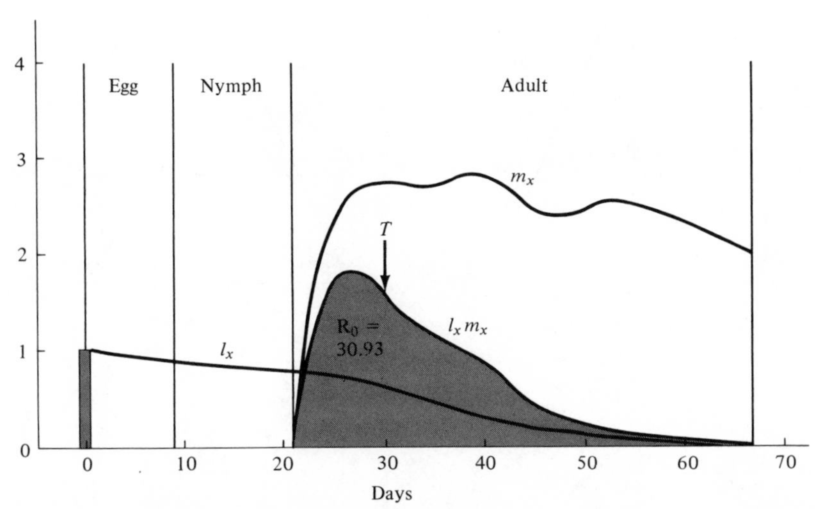

FIGURE 4.4 Diagram showing how continuous l_x and m_x schedules are multiplied together over all ages to obtain the area (shaded) under the $l_x m_x$ product curve, which is equal to the net reproductive rate, R_0. [From F. E. Smith (1954) after Evans and Smith (1952) from data for the human louse, *Pediculus humanus*.]

individual has a reproductive value of zero because it can no longer expect to produce any future offspring; moreover, since natural selection operates only by differential reproductive success (Chapter 1), such a postreproductive organism is no longer subject to the direct effects of natural selection (see also section on Evolution of Death Rates and Old Age). Under most l_x and m_x schedules, reproductive value is maximal around the onset of reproduction and falls off after that both because the organism has already produced some of its offspring and because fecundity often decreases with age (but fecundity is subject to natural selection, too—see section on Evolution of Reproductive Rates). Figure 4.5 shows how reproductive value changes with age in a variety of populations. In the expanding human population shown (Figure 4.5*a*), reproductive value of very young individuals is low for two reasons: (1) there is a finite probability of death before reproduction, and (2) because the future breeding population will be larger, offspring to be produced later will contribute less to the total gene pool than offspring currently being born (similarly, in a declining population, offspring expected at some future date are worth relatively more than current progeny because the total future population will be smaller). The latter component of reproductive value is tedious to calculate and applies only in populations changing in size. Table 4.1 illustrates

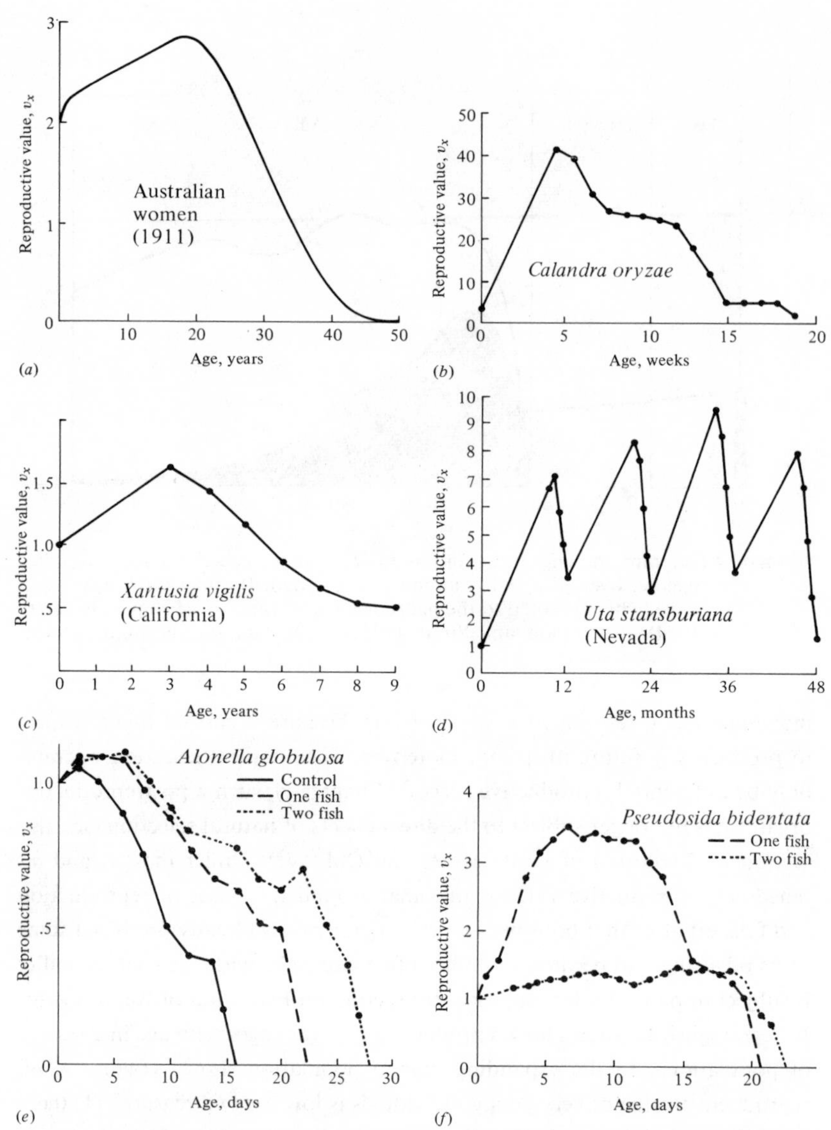

FIGURE 4.5 Reproductive value plotted against age for a variety of populations. (*a*) Australian women, about 1911. [From Fisher (1958a).] (*b*) *Calandra oryzae*, a beetle, in the laboratory. [From data of Birch (1948).] (*c*) *Xantusia vigilis*, a viviparous lizard with a single litter each year. [From data of Zweifel and Lowe (1966).] (*d*) *Uta stansburiana*, an egg-laying lizard that lays several clutches each reproductive season. [From data of Turner *et al.* (1970).] (*e*) *Alonella globulosa*, a microscopic aquatic crustacean, under three different competitive and predatory regimes in laboratory microcosms. (*f*) *Pseudosida bidentata*, a microscopic aquatic crustacean, under two different laboratory situations. [*e*, *f* after Neill (1972.)]

the calculation of reproductive value in a stable population. The general equations for reproductive value in any population, either stable or changing, are:

$$\frac{v_x}{v_0} = \frac{e^{rx}}{l_x} \sum_{t=x}^{\infty} e^{-rt} l_t m_t \quad \text{or} \quad \frac{v_x}{v_0} = \frac{e^{rx}}{l_x} \int_x^{\infty} e^{-rt} l_t m_t \, dt \tag{5}$$

where r is the instantaneous rate of increase per individual (see also below). The exponentials, e^{rx} and e^{-rt}, weight offspring according to the direction in which the population is changing. In a stable population, r is zero. Remembering that e^0 and e^{-0} equal unity, and that v_0 in a stable population is also 1, the reader can verify that (5) reduces to (4) when r is zero. Reproductive value does not directly take into account social phenomena, such as parental care or a grandmother caring for her grandchildren and thereby increasing their probability of survival.

In summary, any pair of age-specific mortality and fecundity schedules has its own implicit T, R_0, r (see below), and v_x curve.

Stable Age Distribution

Another important aspect of a population's structure is its age distribution (Figure 4.6), indicating the proportions of its members belonging to each age class. Two populations with identical l_x and m_x schedules, but with different age distributions, will behave differently and may even grow at different rates if one population has a higher proportion of reproductive members. Lotka (1922) proved that any pair of unchanging l_x and m_x schedules eventually gives rise to a population with a *stable age distribution*. When a population reaches this equilibrium age distribution, the percentage of organisms in each age group remains constant. Recruitment into every age class is exactly balanced by its losses due to mortality and aging. Provided l_x and m_x schedules are not changing, a kind of stable age distribution is quickly reached even in an expanding population, in which the per capita rate of growth of each age class is the same (equal to the intrinsic rate of increase per head, r, below), with the consequence that proportions of various age groups also stay constant. Equations (2) through (6) and (8) through (12) assume that a population is in its stable age distribution. [However, the intrinsic rate of increase (r), generation time (T), and reproductive value (v_x) are conceptually *independent* of these specific equations, since these concepts can be defined equally well in terms of the age distribution of a population, although they are somewhat more complex mathematically

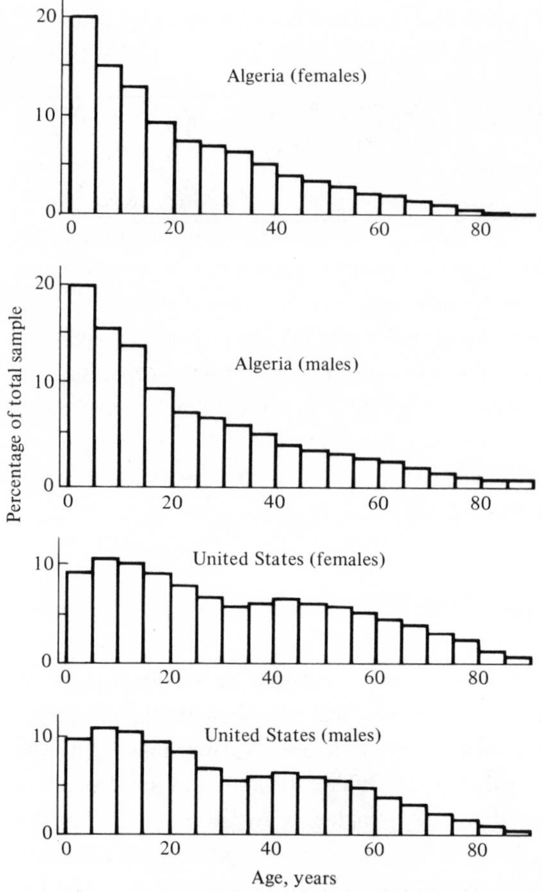

FIGURE 4.6 Age distributions, by sex, in two human populations. The Algerian population is increasing rapidly whereas the United States population is growing more slowly. [After Krebs (1972) after 1968 Demographic Yearbook of the United Nations. Copyright, United Nations (1969). Reproduced by permission.]

(Vandermeer, 1968).] The stable age distribution of a stable population with a net reproductive rate equal to 1 is called the "stationary age distribution." Computation of the stable age distribution is somewhat tedious, and so no more will be said about it here. The interested reader is referred to Lotka (1956), Mertz (1970), and/or Wilson and Bossert (1971).

Intrinsic Rate of Natural Increase

Still another parameter implicit in every pair of schedules of births and deaths is the *intrinsic rate of natural increase*, sometimes called the Malthusian parameter. Usually designated by r, it is a measure of the instantaneous rate of change of population size (per individual). The intrinsic rate of increase is defined as the instantaneous birth rate (per individual), b, minus the instantaneous death rate (per individual), d, or more precisely, as (births + immigration) − (deaths + emigration). When per capita births exceed per capita deaths ($b > d$) the population is increasing and r is positive; when deaths exceed births ($b < d$), r is negative and the population is decreasing. The intrinsic rate of increase is inversely related to generation time, T, by the following approximate formula (see also Figure 4.20, p. 102):

$$r = \frac{\log_e R_0}{T} \tag{6}$$

where R_0 is the net reproductive rate and e is the base of the natural logarithms. From (6) we see that r is positive when R_0 is greater than 1, and negative when R_0 is less than 1. Because $\log_e 1$ is zero, an R_0 of unity corresponds to an r of zero. Under theoretical optimal conditions, when R_0 is as high as possible, the maximal rate of natural increase is realized and is designated by r_{max}.

The maximal instantaneous rate of increase per head, r_{max}, varies among animals by several orders of magnitude (Table 4.2). Small short-lived organisms such as the common human intestinal bacterium *Escherichia coli* have a relatively high r_{max} value, whereas larger and longer-lived organisms such as man have, comparatively, very low r_{max} values. The components of r_{max} are the instantaneous birth rate per head, b, and the instantaneous death rate per head, d, under theoretical optimal environmental conditions. The evolution of rates of reproduction and death rates are taken up later in this chapter.

A population whose size increases linearly in time would have a constant populational growth rate given by

$$\begin{matrix}\text{growth rate}\\ \text{of population}\end{matrix} = \frac{N_t - N_0}{t - t_0} = \frac{\Delta N}{\Delta t} = \text{constant} \tag{7}$$

where N_t is the number at time t, N_0 the initial number, and t_0 the initial time. But at any positive value of r, the per capita rate of increase is constant, and a population grows exponentially as illustrated in Figure 4.7. Its growth rate is a function of population size, with the population growing faster as

TABLE 4.2 Estimated Maximal Instantaneous Rates of Increase (r_{max}, Per Capita Per Day) and Mean Generation Times (in Days) for a Variety of Organisms

Taxon	Species	r_{max}	Generation Time (T)
Bacterium	*Escherichia coli*	ca. 60.0	0.014
Protozoa	*Paramecium aurelia*	1.24	0.33–0.50
Protozoa	*Paramecium caudatum*	0.94	0.10–0.50
Insect	*Tribolium confusum*	0.120	ca. 80
Insect	*Calandra oryzae*	0.110 (.08–.11)	58
Insect	*Rhizopertha dominica*	0.085 (.07–.10)	ca. 100
Insect	*Ptinus tectus*	0.057	102
Insect	*Gibbium psylloides*	0.034	129
Insect	*Trigonogenius globulus*	0.032	119
Insect	*Stethomezium squamosum*	0.025	147
Insect	*Mezium affine*	0.022	183
Insect	*Ptinus fur*	0.014	179
Insect	*Eurostus hilleri*	0.010	110
Insect	*Ptinus sexpunctatus*	0.006	215
Insect	*Niptus hololeucus*	0.006	154
Mammal	*Rattus norwegicus*	0.015	150
Mammal	*Microtus aggrestis*	0.013	171
Mammal	*Canis domesticus*	0.009	ca. 1000
Insect	*Magicicada septendecim*	0.001	6050
Mammal	*Homo sapiens*	0.0003	ca. 7000

N gets larger. Suppose you wanted to estimate the rate of change of the population shown in Figure 4.7 at an instant in time, say at time t. As a first approximation, you might look at N immediately before and immediately after time t, say 1 hour before and 1 hour after, and apply the above $\Delta N/\Delta t$ equation. But examination of Figure 4.7 reveals that at time t_1 (say $t-1$ hr) the true rate is less than, and at time t_2 (say, $t+1$ hr), greater than, your straight-line estimate. Differential calculus was developed to handle just such cases, and allows us to calculate the rate of change at an *instant* in time. As ΔN and Δt are made smaller and smaller, $\Delta N/\Delta t$ gets closer and closer to the true rate of change at time t (Figure 4.7). In the limit, as the Δ's approach zero, $\Delta N/\Delta t$ is written as dN/dt which is calculus shorthand for the instantaneous rate of change of N at t. Exponential population growth is described by the following simple differential equation:

$$\frac{dN}{dt} = bN - dN = (b-d)N = rN \tag{8}$$

where, again, b is the instantaneous birth rate per individual and d the instantaneous death rate per individual (remember that $r = b-d$).

Using calculus and integrating (8), it can be shown that the number of organisms at some time t, N_t, under exponential growth is a function

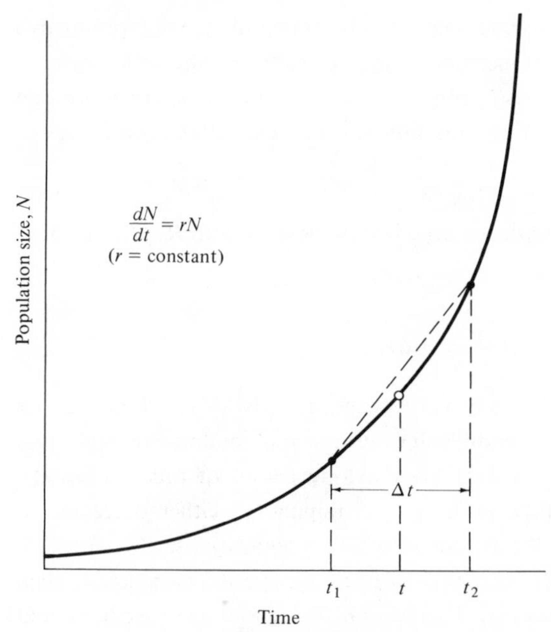

FIGURE 4.7 Exponential population growth under the assumption that the rate of increase per individual, r, remains constant with changes in population density. Note that a straight-line estimate of the rate of population growth at time t becomes more and more accurate as t_1 and t_2 converge; in the limit, as t_1 and t_2 approach t, or $\Delta t \to 0$, the rate of population growth equals the slope of a line tangent to the curve at time t (open circle).

of the initial number at time zero, N_0, r, and the time available for growth since time zero, t, as follows:

$$N_t = N_0 \, e^{rt} \tag{9}$$

here again, e is the base of the natural logarithms. By taking logarithms of the above equation, which is simply an integrated version of (8), we get

$$\log_e N_t = \log_e N_0 + \log_e e^{rt} = \log_e N_0 + rt \tag{10}$$

This equation indicates that $\log_e N$ changes linearly in time; that is, a semilog plot of $\log_e N$ against t gives a straight line with a slope of r and a y-intercept of $\log_e N_0$.

Setting N_0 equal to 1 (i.e., a population initiated with a single organism), after one generation, T, the number of organisms in the population is equal to the net reproductive rate of that individual, or R_0. Substituting these values in (10):

$$\log_e R_0 = \log_e 1 + rT \tag{11}$$

Since $\log_e 1$ is zero, (11) is identical with (6).

Another population parameter closely related to the net reproductive rate and the intrinsic rate of increase is the so-called *finite rate of increase*, λ, defined as the rate of increase per individual per unit time. The finite rate of increase is measured in the same time units as the instantaneous rate of increase, and

$$r = \log_e \lambda \quad \text{or} \quad \lambda = e^r. \tag{12}$$

In a population without age structure, λ is thus identical with R_0 [T equal to 1 in (6) and (11)].

Population Growth and Regulation

In a finite world, no population can grow exponentially for very long. Sooner or later it must encounter either difficult environmental conditions or shortages of its requisites for reproduction. Over a long period of time, unless the average actual rate of increase is zero, a population either decreases to extinction or increases to the extinction of other populations.

So far our populations have had fixed age-specific parameters, such as their l_x and m_x schedules. In this section, we ignore age specificity and instead allow R_0 and r to vary with population density. To do this, we define *carrying capacity*, K, as the density of organisms (i.e., the number per unit area) at which the net reproductive rate (R_0) equals unity and the intrinsic rate of increase (r) is zero. At "zero density" (only one organism, or a perfect ecological vacuum), R_0 is maximal and r becomes r_{max}. For any given density above zero density, both R_0 and r decrease, until, at K, the population ceases to grow. A population initiated at a density above K decreases until it reaches the steady state at K (Figure 4.8). Thus we define r_a as the *actual* instantaneous rate of increase; it is zero at K, negative above K, and positive when the population is below K.

The simplest assumption we can make is that r_a decreases linearly with N and becomes zero at an N equal to K (Figure 4.8); this assumption leads to the classical Verhulst-Pearl logistic equation, which may be written:

$$\frac{dN}{dt} = rN - rN\left(\frac{N}{K}\right) = rN - \frac{rN^2}{K} \tag{13}$$

Alternatively, by factoring out an rN, (13) may be written

$$\frac{dN}{dt} = rN\left(1 - \frac{N}{K}\right) = rN\left(\frac{K-N}{K}\right) \tag{14}$$

Or, simplifying by setting r/K in (13) equal to z,

$$\frac{dN}{dt} = rN - zN^2 \tag{15}$$

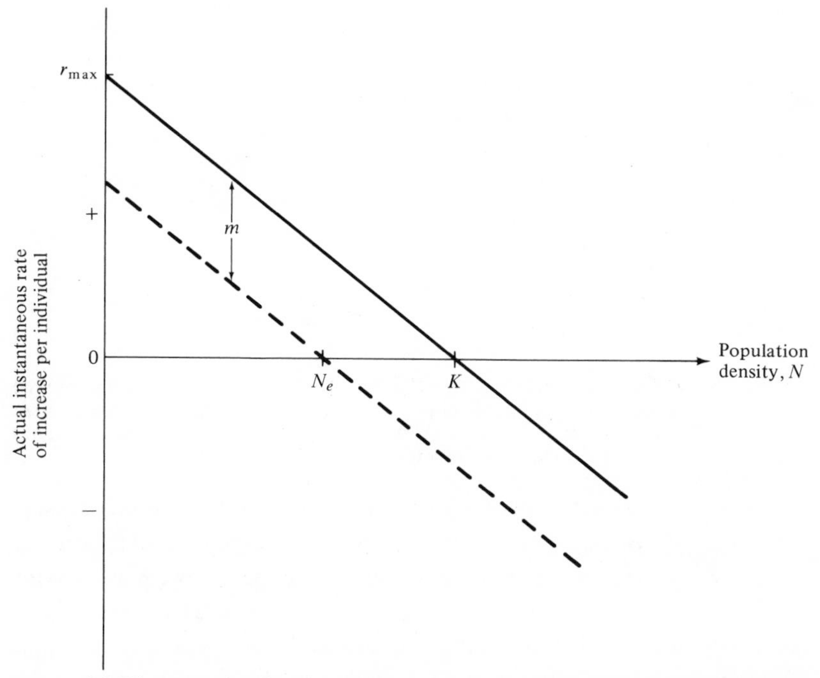

FIGURE 4.8 Diagram showing how the actual instantaneous rate of increase per individual, or r_a, decreases linearly with increasing population density under the assumptions of the Pearl-Verhulst logistic equation. The dashed line shows how the actual rate of increase decreases with N for a given amount of density independent rarefaction, m; equilibrium population size, N_e, is then less than carrying capacity, K.

The term $rN(N/K)$ in (13) and the term zN^2 in (15) represent the density-dependent reduction in the rate of population increase. Thus, at N equal to unity (an ecologic vacuum), dN/dt is nearly exponential, while at N equal to K, dN/dt is zero and the population is in a steady state at its carrying capacity. Logistic equations (there are many more besides the Verhulst-Pearl one above) generate so-called sigmoid (S-shaped) population growth curves (Figure 4.9). Implicit in the Verhulst-Pearl logistic equation are three assumptions: (1) that all individuals are equivalent—that is, that the addition of every new individual reduces the actual rate of increase by the same fraction, $1/K$, at every density (Figure 4.8); (2) that r_{max} and K are immutable constants; and (3) that there is no time lag in the response of the actual rate of increase per individual to changes in N.

All three assumptions are unrealistic, and so the logistic has been strongly criticized (Allee *et al.*, 1949; Smith, 1952, 1963a; Slobodkin, 1962). For instance, much more plausible curvilinear relationships between the rate

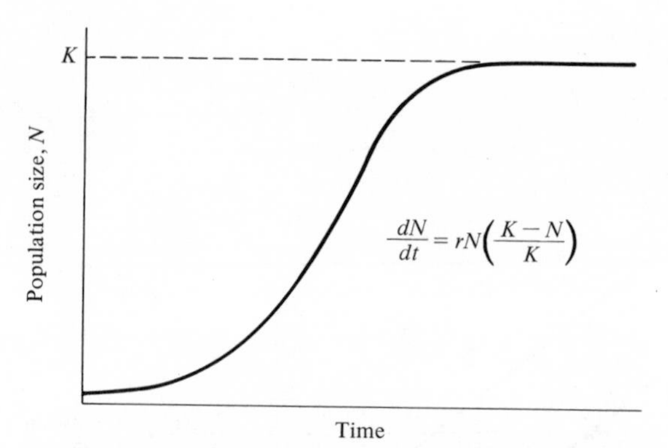

FIGURE 4.9 Population growth under the Pearl-Verhulst logistic equation is sigmoidal (S-shaped), reaching an upper limit termed the carrying capacity, K. Populations initiated at densities above K decline exponentially until they reach K, which represents the only stable equilibrium.

of increase and population density are shown in Figure 4.10. Note that density-dependent effects on birth rate and death rate are combined by the use of r (these effects are separated later in this section). Moreover, carrying capacity is also an extremely complicated and confounded quantity, for it necessarily includes both renewable and nonrenewable resources, as well as the limiting effects of predators and competitors, all of which are variables themselves. Carrying capacity almost certainly varies a great deal from place to place and from time to time for the majority of organisms. There is also, of course, some inevitable lag in feedback between population density and the actual instantaneous rate of increase. All these assumptions can be relaxed

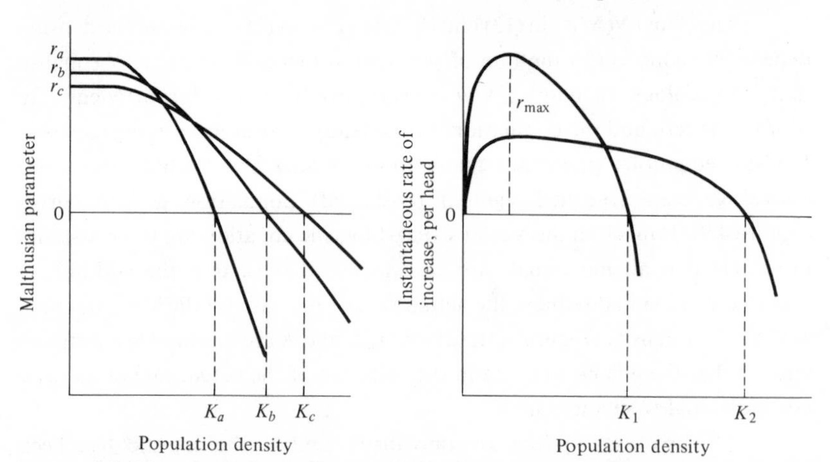

FIGURE 4.10 Hypothetical curvilinear relationships between instantaneous rates of increase and population density. [From Gadgil and Bossert (1970) and Pianka (1972).]

and more realistic equations developed, but the mathematics quickly become extremely complex and unmanageable. Nevertheless, a number of populational phenomena can be nicely illustrated using the simple Verhulst-Pearl logistic, and a thorough understanding of it is a necessary prelude to the equally simple Lotka-Volterra competition equations, which are taken up in Chapter 5. However, the numerous flaws of the logistic must be recognized and it should be taken only as a first approximation for small changes in population growth, most likely to be valid near equilibrium and over short time periods (i.e., situations in which linearity should be approximated).

Equations (13) to (15) are easily modified to incorporate density-independence as well as density-dependence by simply adding one new term, as follows:

$$\frac{dN}{dt} = rN\left(\frac{K-N}{K}\right) - mN \quad \text{or} \quad \frac{dN}{dt} = rN - zN^2 - mN \qquad \textbf{(16)}$$

where m is the instantaneous rate of density-independent removal (or rarefaction) per individual. In effect, the addition of the $-mN$ term merely lowers the maximal rate at which the population can increase, rN. The equilibrium population density with rarefaction is less than K by the ratio of m/r (Figure 4.8). To see this, factor N out of equation (16):

$$\frac{dN}{dt} = N\left\{r - \left(\frac{rN}{K}\right) - m\right\} \qquad \textbf{(17)}$$

When the term in brackets (r_a, see below) equals zero, $dN/dt = 0$ and the population is at equilibrium, or N_e. Thus

$$0 = r - \left(\frac{rN_e}{K}\right) - m \quad \text{or} \quad r - m = \frac{rN_e}{K} \qquad \textbf{(18)}$$

dividing both sides by r and multiplying by K,

$$N_e = K - \left(\frac{m}{r}\right)K \qquad \textbf{(19)}$$

For example, in the absence of rarefaction $N_e = K$, and with a rarefaction rate of one-third the intrinsic rate of increase, $N_e = 2/3K$.

The reader should have noticed that the r in the logistic equation is actually $r_{\max}$. The equation can be solved for the *actual* rate of increase, r_a, which is a variable and a function of r, N, K, and m, as follows:

$$r_a = \frac{dN}{Ndt} = r\left(\frac{K-N}{K}\right) - m = r - \left(\frac{N}{K}\right)r - m \qquad \textbf{(20)}$$

The actual instantaneous rate of increase per individual, r_a, is always less than or equal to r_{max} (r in the logistic); Equation (20) and Figure 4.8 show how r_a decreases linearly with increasing density under the assumptions of the Verhulst-Pearl logistic equation.

The two components of the actual instantaneous rate of increase per individual, r_a, are the actual instantaneous birth rate per individual, b, and the actual instantaneous death rate per individual, d. The difference between b and d (i.e., $b-d$) is r_a. Under theoretical ideal conditions when b is maximal and d is minimal, r_a is maximized at r_{max}. In the logistic, this is realized at a density of zero, or a perfect ecological vacuum. To be more precise, we subscript b and d, which are functions of density. Thus $b_N - d_N = r_N$ (which is r_a at density N), and $b_0 - d_0 = r_{max}$. When $b_N = d_N$, r_a and dN/dt are zero, and the population is at equilibrium. Figure 4.11 diagrams the way in which b and d vary linearly with N under the logistic. At any given density, b_N and d_N are given by the following linear equations:

$$b_N = b_0 - xN \quad \textbf{(21)}$$

$$d_N = d_0 + m + yN \quad \textbf{(22)}$$

where x and y represent, respectively, the slopes of the lines plotted in Figure 4.11 [see also Bartlett (1960) and Wilson and Bossert (1971)]. The instantaneous death rate, d_N, clearly has both density-dependent and density-independent components; in (22) and Figure 4.11, yN measures the density-dependent component of d_N, while m determines the density-independent

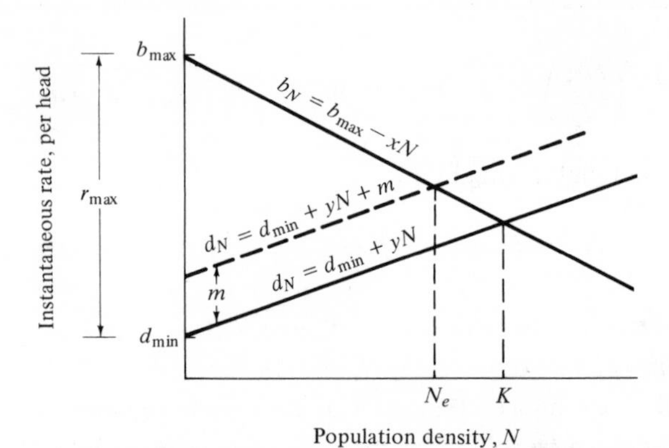

FIGURE 4.11 Diagram showing how the instantaneous birth rate per individual decreases linearly with population density under the logistic equation, whereas the instantaneous death rate per head rises linearly as population density increases. Two death rate lines are plotted, one with (dashed line) and one without (solid line) rarefaction. Note that equilibrium population density, N_e, is lower with rarefaction than without it (K).

component (d_0 can be interpreted as either density-independent or density-dependent).

At equilibrium, b_N must equal d_N, or:

$$b_0 - xN = d_0 + m + yN \tag{23}$$

substituting N_e for N at equilibrium, r for $(b_0 - d_0)$, and rearranging terms

$$r = (x+y)N_e + m \tag{24}$$

or

$$N_e = \frac{r-m}{x+y} \tag{25}$$

Note that the sum of the slopes of the birth and death rates $(x+y)$ is equal to z, or r/K. Substituting this for $(x+y)$ in (25) gives equation (19). Clearly z is the density-dependent constant which is analogous to the density-independent constants r (r_{max}) and m.

Density Dependence and Density Independence

Various factors can influence populations in two fundamentally different ways. If their effects on a population do not vary with population density, but the same *proportion* of organisms are affected at any density, factors are said to be *density-independent*. Climatic factors often, though by no means always, affect populations in this manner (Table 4.3). If, on the other hand, a factor's effects vary with population density, so that the proportion of organisms influenced actually changes with density, that factor is said to be *density-dependent*. Density-dependent factors and events can be either positive or negative. Death rate, which presumably often increases with increasing density, is an example of positive or direct density-dependence (Figure 4.11); whereas birth rate, which normally decreases with increasing density, is an example of negative or inverse density-dependence (Figure 4.12). Density-dependent influences on populations frequently result in an equilibrium density at which the population ceases to grow. Biotic factors, such as

TABLE 4.3 Fish Kills Following Sudden, Severe Cold Weather on the Texas Gulf Coast in the Winter of 1940, to Illustrate Density-Independent Mortality

	Commercial Catch		
Locality	before	after	Decline (%)
Matagorda	16,919	1,089	93.6
Aransas	55,224	2,552	95.4
Laguna Madre	2,016	149	92.6

Source: After Odum (1959) after Gunter.

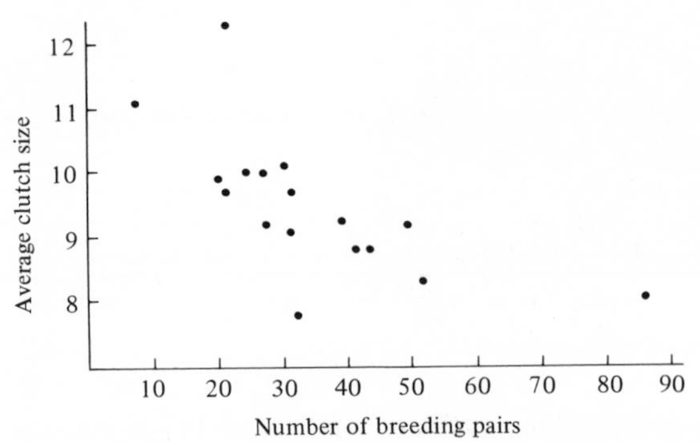

FIGURE 4.12 A plot of average clutch size against the density of breeding pairs of English great tits (birds) in a particular woods in a series of years over a 17-year period. [After Perrins (1965).]

competition, predation, and pathogens, often (though not always) act in this way.

Ecologists are divided in their opinions as to the relative importance of density-dependence and density-independence in natural populations (Andrewartha and Birch, 1954; Lack, 1954, 1966; Nicholson, 1957; Orians, 1962); a few even go so far as to deny categorically the existence of one or the other of these types of influences on populations.

As an example, in their studies on the population dynamics of *Thrips imaginis* (a small herbivorous insect), Davidson and Andrewartha (1948) found that they could predict population sizes of the insects fairly accurately using only past population sizes and recent climatic conditions. These workers could find no evidence of any density effects, and they therefore interpreted their data to mean that the populations of *Thrips* were controlled primarily by density-independent climatic factors. Interestingly enough, however, Smith (1961) reanalyzed their data and discovered pronounced density-dependent effects at high densities. He found a strong inverse correlation between population change and population size, which strongly suggests density-dependence. Also, Smith demonstrated a rapidly decreasing variance in population size during the later portion of the spring population increase. Further, Smith showed that these patterns persisted even after partial correlation analysis, which held constant the very climatic variables that Davidson and Andrewartha considered to be so important. This example illustrates the great difficulty ecologists frequently encounter in distinguishing cause from effect. There is now little real doubt that both density-dependent and density-independent events occur; their relative importance, however, may vary by many orders of magnitude from population to population, and even within the same population from time to time as the size of the population changes (Horn, 1968a).

r and K Selection

Density-independent and density-dependent factors and events differ significantly in their effects on natural selection and on populations. In highly variable and/or unpredictable environments, catastrophic mass mortality (such as that illustrated in Table 4.3) presumably often has relatively little to do with the genotypes and phenotypes of the organisms concerned or with the size of their populations. (Some degree of selective death and stabilizing selection has, however, been demonstrated in winter kills of certain bird flocks.) By way of contrast, under more stable and/or predictable environmental regimes, much mortality is more directed and favors individuals that are better able to cope with high densities and strong competition. Organisms in highly rarefied environments seldom deplete their resources to levels as low as do organisms living under less rarefied situations, and as a result, the former usually do not encounter such intense competition. In a "competitive vacuum" (or an extensively rarefied environment), the best reproductive strategy is to put maximal amounts of matter and energy into reproduction and to produce as many total progeny as possible as soon as possible. Because there is little competition, these offspring often can thrive even if they are quite small and therefore energetically inexpensive to produce. However, in a "saturated" environment, where density effects are pronounced and competition is keen, the best strategy is to put more energy into competition and maintenance, and to produce offspring with more substantial competitive abilities. This usually requires larger offspring; and, since they are energetically more expensive, it means that fewer can be produced.

MacArthur and Wilson (1967) designate these two opposing selective forces as r selection and K selection, after the two terms in the logistic equation. Of course, things are seldom so black and white, but there are usually all shades of gray. No organism is completely r selected or completely K selected, but rather all must reach some compromise between the two extremes. Indeed, one can think of a given organism as an "r-strategist" or a "K-strategist" only relative to some other organism; thus statements about r and K selection are invariably comparative. We think of an $r \rightarrow K$ selection continuum, and an organism's position along it in a particular environment at a given instant in time (Pianka, 1970, 1972). Table 4.4 lists a variety of correlates of r and K selection.

TABLE 4.4 Some of the Correlates of r and K Selection

	r Selection	K Selection
Climate	Variable and/or unpredictable; uncertain	Fairly constant and/or predictable; more certain
Mortality	Often catastrophic, nondirected, density independent	More directed, density dependent
Survivorship	Often Type III	Usually Types I and II
Population size	Variable in time, nonequilibrium; usually well below carrying capacity of environment; unsaturated communities or portions thereof; ecologic vacuums; recolonization each year	Fairly constant in time, equilibrium; at or near carrying capacity of the environment; saturated communities; no recolonization necessary
Intra- and interspecific competition	Variable, often lax	Usually keen
Selection favors	1 Rapid development 2 High maximal rate of increase, r_{max} 3 Early reproduction 4 Small body size 5 Single reproduction	1 Slower development 2 Greater competitive ability 3 Delayed reproduction 4 Larger body size 5 Repeated reproductions
Length of life	Short, usually less than 1 year	Longer, usually more than 1 year
Leads to	Productivity	Efficiency

Source: After Pianka (1970).

Population "Cycles": Cause and Effect

Ecologists have long been intrigued by the regularity of certain population fluctuations, such as those of the snowshoe hare, the Canadian lynx, the ruffed grouse, and many microtine rodents (voles and lemmings) as well as their predators, including the arctic fox and the snowy owl (Keith, 1963; Elton, 1942). These population fluctuations (sometimes called "cycles") are of two types: voles, lemmings, and their predators display roughly a 4-year periodicity, whereas the hare, lynx, and grouse have approximately a 10-year cycling time. Lemming population eruptions and the fabled, but very rare suicidal marches of these rodents into the sea have frequently been popularized and are well known to the layman.

Many different hypotheses for the explanation of population cycles have been offered and the literature on them is extensive (references at end of chapter). Here, as elsewhere in ecology, it is often extremely difficult or even impossible to devise tests that separate cause from effect, and many of the putative causes of population cycles may in fact merely be side effects of the cyclical changes in the populations concerned. Several of the currently more popular hypotheses, which are not necessarily mutually exclusive, are briefly outlined below. Descriptions and discussion of others, which have now fallen out of vogue, such as the "sunspot" and the "random peaks" hypotheses, can be found in the references. The reader should bear in mind that two or more of these hypothetical mechanisms could act together in any given situation.

Stress Phenomena Hypothesis. At the extremely high densities occurring during population peaks of voles, a great deal of fighting occurs among these rodents. The so-called stress syndrome is manifested by the animals, their adrenal weights increase and they become extremely aggressive, so much so that successful reproduction is almost completely curtailed. Eventually, "shock disease" may set in and large numbers of animals may die off, apparently because of the physiological stresses on them. Christian and Davis (1964) review evidence pertaining to this hypothesis.

Predator–Prey Oscillation Hypothesis. It is well known both theoretically and empirically that, in simple ecological systems, predator and prey populations can oscillate because of the interaction between them (see also Chapter 5). When the predator population is low, the prey increase, which then allows the predators to increase, although this increase lags behind that of the prey. Eventually, the predators overeat their prey and the prey population begins to decline; but, because of time lag effects, the predator population continues to increase for a period, driving the prey to an even lower density. Finally at low enough prey densities many of the predators starve and the cycle repeats itself. However, prey populations oscillating for reasons other than predation pressures obviously constitute cyclical food supplies for their predators, which should in turn lag behind and oscillate with prey availability; it is thus extremely difficult to determine whether or not changes in predator populations are causally related to changes in prey population density without manipulating predator populations experimentally.

Nutrient Recovery Hypothesis. According to this hypothesis (Pitelka, 1964; Schultz, 1964, 1969), one reason for the periodic decline of rodent populations (especially lemmings) is that the quality of their plant food changes in a cyclical way. During a lemming "high," the ground is blanketed with lemming fecal pellets, and many important chemical elements such as nitrogen and phosphorus are tied up and unavailable to growing plants. In the cold arctic tundra, decomposition of fecal materials takes a long time. During this period, the lemmings decline due to inadequate nourishment. Eventually, after a lapse of a few years, the feces are decomposed and their nutrients recycled once again and taken up by the plants. Now, because their plant food is especially nutritious, the lemmings increase in numbers and the cycle repeats. Schultz (1969) has evidence of such cyclical changes, but whether they are causing, or merely effects of, the lemming population fluctuations has not been definitely established. In any case, the nutrient recovery mechanism does not apply to microtine cycles in general, for Krebs and DeLong (1965) provided supplemental food to a declining population of *Microtus* which failed to reverse the decline. More experiments like this one are badly needed to assess the importance of various mechanisms of population control.

Genetic Control Hypothesis. This hypothesis, credited to Chitty (1960, 1967a), explains population fluctuations in terms of changing genetic composition of the population concerned. During troughs, the animals experience little competition and are relatively r selected, whereas at peaks competition is intense and they are more K selected. Thus directional selection, related to population density, is always occurring; modal phenotypes are never the most fit individuals in the population, and each generation the gene pool changes. The population always lags somewhat behind the changing selective pressures and so no stable equilibrium exists. Some evidence exists for such genetic changes in populations of *Microtus* (Tamarin and Krebs, 1969), but here again cause and effect are extremely difficult to disentangle.

Needless to say, one must always be wary of oversimplification and "single-factor thinking"; most or even all of the above hypothetical mechanisms could work in concert to produce observed population "cycles." The extreme difficulty of separating cause from effect, illustrated above, plagues much of ecology. Simple tests, such as that of Krebs and DeLong (1965), which actually refute a hypothesis are badly needed. Indeed, for scientific understanding to progress rapidly and efficiently, a logical framework of *refutable* hypotheses, complete with alternatives, is absolutely essential (Platt, 1964).

Evolution of Reproductive Rates

Birth and death processes are intimately interrelated and interdependent; in a stable population, or almost any population averaged over a long enough period of time, they must be equal and opposite, exactly balancing each other. Since births and deaths must balance, patterns of reproduction and mortality have evolved together. However, for convenience and clarity, we first consider these processes separately and later integrate them.

Because natural selection operates by differential reproductive success, reproductive rates are of considerable interest in population ecology. An organism can increase its reproductive output in at least three ways, all of which require either a larger total energy budget or a reapportionment of the energy budget: (1) by breeding more often, (2) by increasing clutch (or litter) size, and/or (3) by breeding over a longer period of time. [In an expanding population, breeding earlier results in greater reproductive success, other things being equal; see Mertz (1971a).]

Organisms vary widely in the patterns of reproduction they employ. Some breed only once, but produce many offspring (as in the seed crops of many annual plants); still others produce rather few offspring at a time but breed repeatedly (as in some perennial plants). The onset of reproduction also varies greatly among organisms, with some reproducing very soon after hatching or birth and others delaying reproduction for several to many years. This great range in reproductive patterns is doubtlessly adaptive, and population ecologists have devoted much effort to studying them. Cole (1954b) first developed the theory of the evolution of reproductive patterns. Gadgil and Bossert (1970) extended and improved upon some of Cole's preliminary conclusions, which were based on exponentially expanding populations with rectangular survivorship (Type I, p. 72 and Figure 4.3). These theoretical explorations have produced the following important principles: (1) the absolute gain in the Malthusian parameter which an annual species can achieve by becoming perennial in reproduction and immortal is approximately equivalent to doubling its average litter size; (2) repeated reproduction is more advantageous for long-lived organisms and/or those with delayed reproduction than it is for rapidly maturing and/or short-lived species; (3) increased litter size is more advantageous for short-lived and/or rapidly-maturing organisms than for long-lived organisms with delayed reproduction; and (4) when adult survivorship is low, a large first litter can offset the advantage of repeated reproduction.

So far in this chapter we have treated all individuals within a population as having equivalent fitnesses. To be more realistic, we must allow for differences in fitness between them. In general, it is reasonable to assume that the more energy a parent expends upon an individual offspring, the more fit that offspring will be. Given a fixed amount of reproductive energy, there is an inverse relationship between the total number of offspring produced and their average fitness. The best reproductive strategy must therefore be a compromise between conflicting demands: (1) for production of the largest possible total number of progeny (r selection) and (2) for production of offspring of the highest possible individual fitness (K selection). Exactly where this compromise lies is a function of many variables, including body size, length of life, survivorship of adults and juveniles, population density, and spatial and temporal patterns of resource availability.

If fitness were exactly proportional to energy expenditure in a linear fashion (Figure 4.13, curve *A*), the fitness of individual offspring declines exponentially with increased litter size as depicted in Figure 4.14, curve *A*. However, because initial energy outlays on a particular offspring should contribute more to its fitness than subsequent ones, fitness may often not be linearly related to expenditure but rather might usually be bowed somewhat as in Figure 4.13, curve *B*. Hence total fitness, the sum of the fitnesses of individual offspring, is also greatest at some intermediate clutch size (Figure 4.14, curve *B*). A massive body of data has now been accumulated demonstrating just such optimality in clutch sizes, particularly in birds (Lack, 1954, 1966).

The elegant studies of Lack and his students and colleagues have repeatedly shown that, compared with very small and very large clutches, intermediate sized clutches leave proportionately more offspring which survive to breed in the next generation. This is an excellent example of stabilizing selection. Young birds from large clutches leave the nest at a lighter

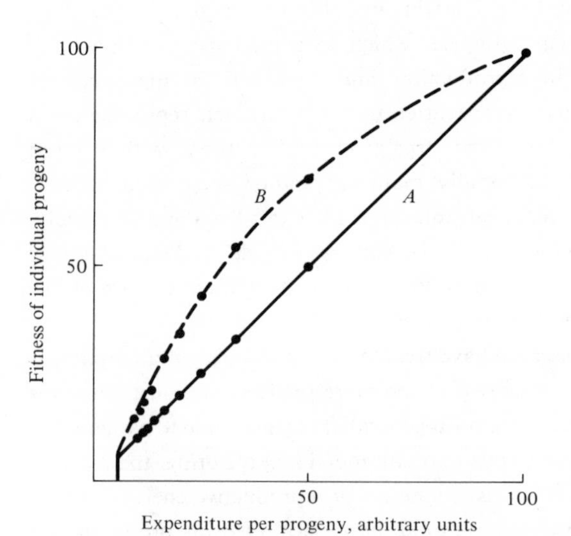

FIGURE 4.13 Diagram showing how the fitness of an individual offspring might vary with expenditure per progeny, all else being equal (genetic background, etc.). Because initial outlays upon an offspring should contribute more to its fitness than subsequent ones, curve *B* is biologically more realistic than the line *A*. Dots represent different clutch or litter sizes as depicted in Figure 4.14.

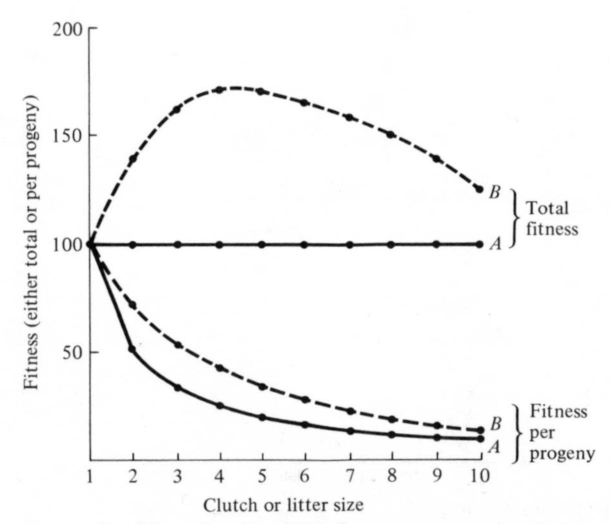

FIGURE 4.14 Plots showing how fitness per progeny (lower two curves) and total fitness, the sum of the fitnesses of all offspring produced (upper curve and line), vary with clutch or litter size under the assumptions of Figure 4.13. Total energy devoted to reproduction, or reproductive effort, is assumed to be constant. Note that total fitness peaks at an intermediate clutch size under assumption *B*. Optimal clutch size in this example is 4.

weight (Figure 4.15) and have a substantially reduced postfledging survivorship. Moreover, Lack contends that the optimal clutch represents the number of young for which the parents can, on the average, provide just enough food. Perrins (1965) has good evidence for this in a population of Great Tits, *Parus major*, which varied their average clutch size from 8 to 12 over a 17-year period, apparently in response to the density of their major food, caterpillars (Figure 4.16). Wynne-Edwards (1962) interprets the optimal clutch as that which produces a net number of young just replacing the parents during their lifetime of reproduction. As such, his explanation involves group selection (see Chapter 1), because individual birds do not necessarily raise as many young as possible, but rather produce only as many as are required to replace themselves. Clearly a "cheater" which produced more offspring would soon swamp the gene pool. We return to this point at the end of this chapter.

Even within the same widely ranging species, many birds and some mammals produce larger clutches (or litters) at higher latitudes than they do at lower latitudes (Figure 4.17). Such latitudinal increases in clutch size are widespread and have intrigued many population ecologists because of their general occurrence. The following hypotheses, which are not mutually

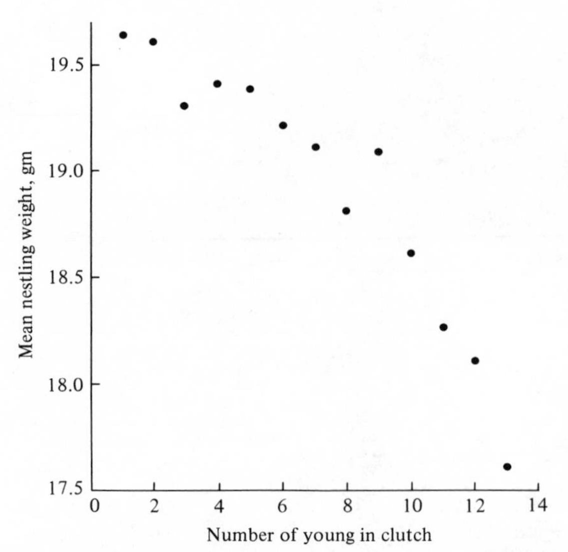

FIGURE 4.15 Average weight of nestling great tits plotted against clutch size, showing that individual young in larger clutches weigh considerably less than those in smaller clutches. [From data of Perrins (1965).]

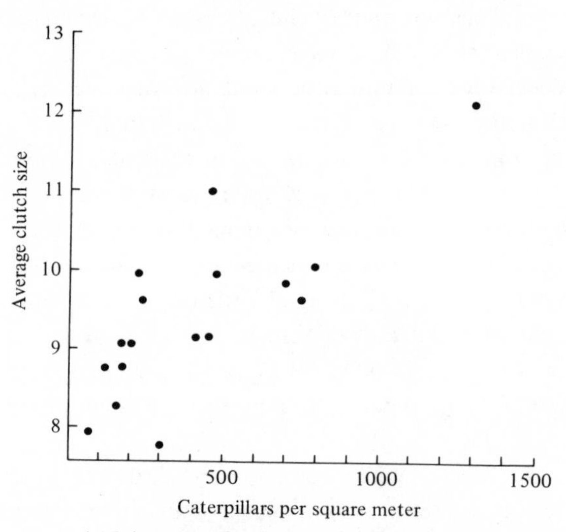

FIGURE 4.16 Average clutch size of great tit populations in 17 years plotted against caterpillar density. Clutches tend to be larger when the caterpillars are abundant than they are when this insect food is sparse. [After Perrins (1965).]

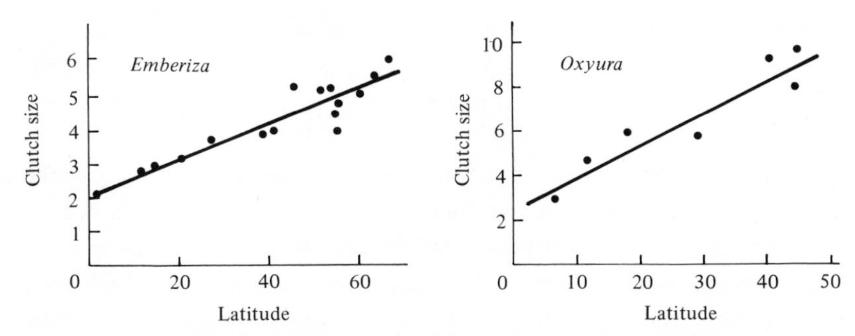

FIGURE 4.17 Graphs of clutch size against latitude for the avian genera *Emberiza* and *Oxyura*. [After Cody (1966).]

exclusive, have been proposed to explain latitudinal gradients in avian clutch sizes.

The Daylength Hypothesis. As indicated in Chapter 2, during the late spring and summer, days are longer at higher latitudes than they are at lower latitudes. Diurnal birds therefore have more daylight hours in which to gather food; thus they can gather more food, and are able to feed larger numbers of young.

The Spring Bloom or Competition Hypothesis. Many temperate zone birds are migratory, while few tropical birds migrate. During the spring months at temperate latitudes, there is a great surge of primary production and the insects dependent upon these sources of matter and energy rapidly increase in numbers. Winter losses of both resident and migratory birds are often heavy so that spring populations may be relatively small. Hence returning individuals find themselves in, relatively speaking, a competitive vacuum with abundant food and little competition for it. In the tropics wintering migrants ensure that competition is keen all year long, whereas in the temperate zones competition is distinctly reduced during the spring months. Thus, because birds at higher latitudes are able to gather more food per unit time they are able to raise larger numbers of offspring to an age at which the young can fend for themselves.

The Predator Hypothesis. There seem to be proportionately more predators, both individuals and species, in tropical than in temperate habitats (see also Chapter 7). Nest failure due to nest predation is extremely frequent in the tropics. Skutch (1949) proposed that many nest predators locate bird nests by watching and following the parents. Since the parents must make more trips to the nest if they have a large clutch, larger clutches should suffer heavier losses than smaller ones (this effect, however, has not yet been demonstrated empirically). A fact in support of the hypothesis is that hole-nesting birds, which are relatively free of nest predators, do not show as great an increase in clutch size with latitude as birds that do not nest in holes (Cody, 1966). Moreover, on tropical islands, known to support fewer predators than adjacent mainland areas, birds tend to have larger clutches than mainland populations.

An observation that is somewhat difficult to reconcile with any of the three hypotheses is that clutch size often increases with altitude [as it does in the song sparrow on the Pacific coast (Johnston, 1954)], because neither daylength, competition, nor predation need necessarily vary altitudinally. Cody (1966) suggested that climatic uncertainty, both instability and/or unpredictability, may well result in reduced competition at higher elevations.

Using the principle of allocation, Cody (1966) merged the above hypotheses and developed a more general theory of clutch size involving a compromise between the conflicting demands of predator avoidance, competitive ability, and clutch size. His model fits the observed facts reasonably well, as one would expect of a more complex model with more parameters.

Reproductive strategies are considered further in Chapter 6.

Evolution of Death Rates and Old Age

Why do organisms become senile as they grow old? One might predict quite the opposite, since older organisms have had more experience and should therefore have learned how to avoid predators, have more antibodies, etc., and in general should be wiser and better adapted, both behaviorally and immunologically. The physiological processes of aging have long been of interest and have received considerable attention, but only fairly recently has the evolutionary process been examined (Medawar, 1957; Williams, 1957; Hamilton, 1966; Emlen, 1970). Here again, Fisher (1930) foreshadowed thought on the subject, and paved the way for its development with his concept of reproductive value.

Medawar nicely illustrated the evolution of senescence by setting up the following inanimate model. A chemist's laboratory has a stock of 1000 test tubes and a monthly breakage rate of 10 percent. Every month 100 test tubes are broken completely at random and 100 new ones are regularly added to replace them. (Although the rate of breakage is thus deterministic, the model could easily be rephrased in probabilistic or stochastic terms.) All new test tubes are marked with the date of acquisition so that their age (in months) can be determined at a later date. Every test tube has exactly the same probability of survival from any one month to the next: 900/1000 or 0.9. Thus initially the older test tubes have the same mortality as younger ones and there is no senility. All test tubes are potentially immortal. The probability of surviving two months is the product of the probability of surviving each month separately, or 0.9 times 0.9 ($0.9^2 = 0.81$), while that of surviving three months is 0.9^3, and that of surviving x months is 0.9^x. After some years, the population of test tubes reaches a stable age distribution with 100 age 0 months, 90 age 1, 81 age 2, 73 age 3, . . . , 28 age 12, . . . , 8 age 24, . . . , 2.25 age 36, . . . , and less than one test tube in age groups over 48 months, . . . etc., totaling 1000 test tubes. (Actually, of course, these numbers are merely the *expected* numbers of tubes of a given age; random sampling and stochastic variations will result in some of the numbers in the various age groups being slightly above, and others slightly below, the expected values.) Figure 4.18 shows part of the expected stable age distribution, with the younger test tubes greatly outnumbering older ones. Virtually no test tubes are over five years old, even though individual test tubes are potentially immortal.

Next, Medawar assigned to each of the 900 test tubes surviving every month an equal share of that month's "reproduction" (i.e., the 100 tubes added that month). Hence each surviving test tube reproduces 1/9 of one tube a month. Fecundity does not change with age but the proportions of tubes reproducing does. From Figure 4.18 it is apparent that the younger age groups contribute much more to each month's reproduction than the older

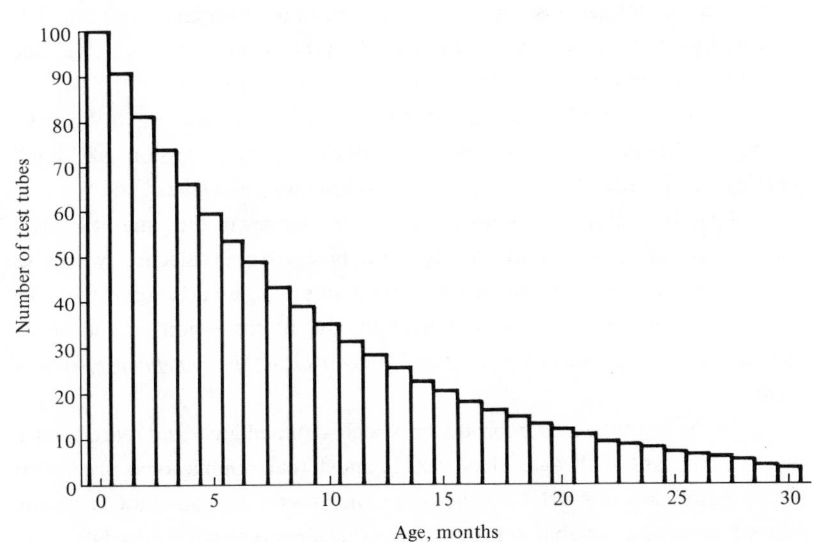

FIGURE 4.18 Stable age distribution of test tubes with a 10 percent breakage rate per month. Because very few test tubes survive longer than 30 months, the distribution is arbitrarily truncated at this age.

ones, simply because there are more of them; moreover, as a group their age-specific expectation of future offspring, or total reproductive value, is also higher (Figure 4.18), because their total expectation of future life is greater. In such an equally fecund and potentially immortal population, reproductive value of individuals, however, remains constant with age.

Now, pretending test tubes have "genes," consider the fate of a mutant whose phenotypic effect is to make its bearer more brittle than an average test tube. This gene is clearly detrimental; it reduces the probability of survival and therefore the fitness of its carrier. This mutant is at a selective disadvantage and will eventually be eliminated from the population. Next consider the fate of another set of mutant alleles at a different locus which control the time of expression of the first gene for brittleness. Various alleles at this second locus alter the time of expression of the brittle gene differently, with some causing it to be expressed early, and others late. Obviously, a test tube with the brittle gene and a "late" modifier gene is at an advantage over a tube with the brittle gene and an "early" modifier, because it will live longer on the average and therefore produce more offspring. Thus, even while the brittle gene is being eliminated by selection, "late" modifiers accumulate at the expense of "early" modifiers. The later the time of expression of brittleness, the more nearly normal a test tube is in its contribution to future generations of test tubes. In the extreme, after reproductive value decreases to zero, natural selection, which operates only by differential reproductive success, can no longer postpone the expression of a detrimental trait and it is expressed as senescence. Thus traits that have been postponed into old age by selection of modifier genes have effectively been removed from the population gene pool. For this reason, old age has been referred to as a "genetic dustbin." The process of selection postponing the expression of "bad" genetic traits is termed *recession of the overt effects of an allele*.

Exactly analogous arguments apply to changes in the time of expression of "good" genetic traits, except that here natural selection works to move the time of expression of such characters to the early ages, with the result that their bearers benefit maximally from possession of the allele. Such a sequence of selection is termed *precession of the beneficial effects of an allele*.

In the test tube case, reproduction begins immediately and reproductive value is constant with age. However, in most real populations, organisms are not potentially immortal; furthermore, the onset of reproduction is usually delayed somewhat so that reproductive value first rises and then falls with increasing age (see Figure 4.5). Thus individuals at an intermediate age have the highest expectation of future offspring. Under this situation, detrimental traits first expressed *after* the period of peak reproductive value can readily be postponed by selection of appropriate modifiers, but those first expressed *before* the period of maximal reproductive value can be quite another case, particularly if they prevent their bearers from reproducing at all. It is difficult to see how selection acting upon the bearer of such a trait could postpone the time of its expression.

The Joint Evolution of Rates of Reproduction and Mortality

A species with high mortality obviously must also possess a correspondingly high fecundity in order to persist in the face of its inevitable mortality. Similarly, a very fecund organism must on the average suffer equivalently heavy mortality or else its population increases until some balance is reached. Likewise, organisms with low fecundities enjoy low rates of mortality, and those with good survivorship have low fecundities. Figure 4.19 shows the inverse relationship between survivorship and fecundity in a variety of lizard populations. Rates of reproduction and death rates evolve together and must stay in some kind of balance. Changes in either of necessity affect the other.

When expectation of future offspring (reproductive value) is very low, natural selection favors an immediate all-out reproductive effort, even at the expense of an organism's own survival. In contrast, under circumstances where expectation of future offspring is high, optimal phenotypes are likely to

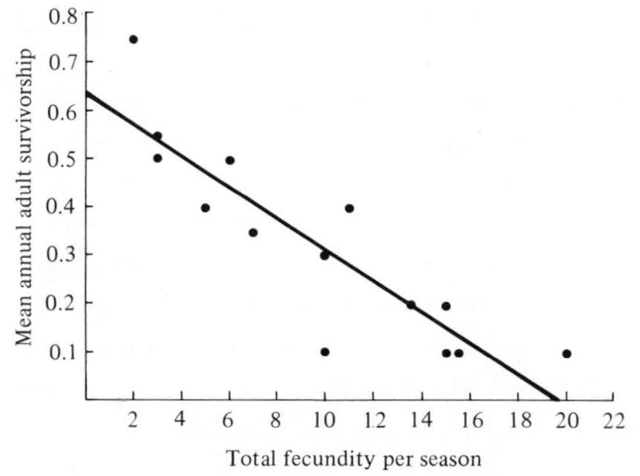

FIGURE 4.19 Total fecundity per reproductive season plotted against the probability of surviving to a subsequent reproductive year for 14 lizard populations. [After Tinkle (1969).]

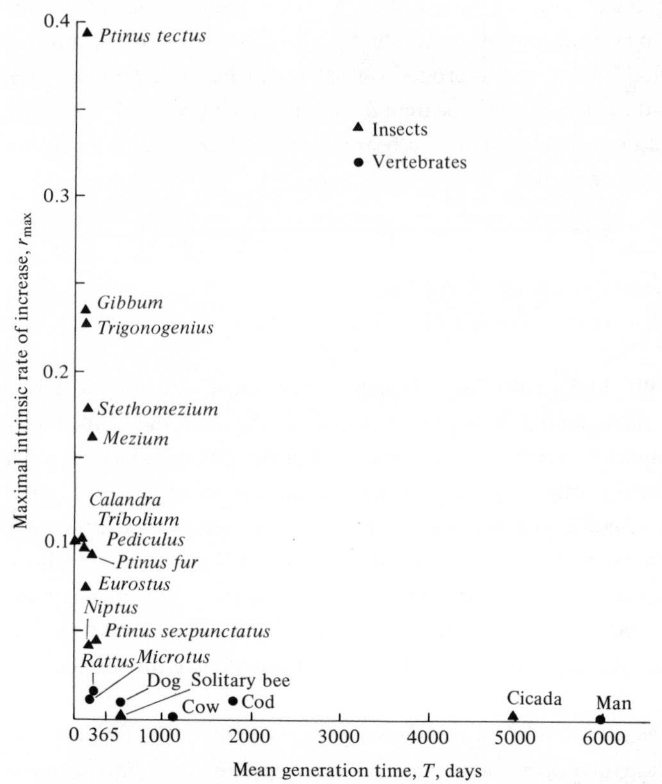

FIGURE 4.20 Maximal intrinsic rate of increase, r_{max}, is plotted against mean generation time, T, in days, to show the inverse hyperbolic relationship between these two population parameters. [From Pianka (1970).]

reserve a certain portion of their matter and energy for maintenance, requiring a lower immediate reproductive effort. Such "strategies" of reproduction and the concept of reproductive effort are discussed by Williams (1966a, 1966b), who points out that natural selection should favor organisms that behave as though they were weighing their short-term against their long-term expectation of successful reproduction. If there is little chance of surviving long enough to reproduce again, the optimal reproductive effort is high; whereas, if survivorship is higher, immediate reproductive effort may be reduced in order to enhance survivorship further, thereby maximizing an organism's lifelong overall contribution to the next generation. Thus there should generally be an inverse relationship between reproductive value and reproductive effort (see also Chapter 6, pp. 212–214).

A convenient indicator of an organism's potential for increase in numbers is the maximal instantaneous rate of increase, r_{max}, which takes into account simultaneously processes of both births and deaths. Table 4.2 shows the great range of values r_{max} takes. Since the actual instantaneous rate of increase averages zero over a sufficiently long time period, organisms with high r_{max} values such as *Escherichia coli* are farther from realizing their maximal rate of increase than organisms with lower r_{max} values such as man. In fact, as Smith (1954) points out, r_{max} is a measure of the "harshness" of an organism's average natural environment. As such, it is one of the better indicators of an organism's position along the $r-K$ selection continuum. Furthermore, organisms with high r_{max} values usually have much more variable actual rates of increase (and decrease) than organisms with low r_{max} values.

The hyperbolic inverse relationship between r_{max} and generation time T was mentioned earlier and is illustrated in Figure 4.20. Small organisms tend to have much higher r_{max} values than larger ones, primarily because of their shorter generation times. Figure 4.21 shows the positive correlation between body size and generation time. The causal basis for this correlation is obvious; it takes time for an organism to attain a large body size, and delays in reproduction invariably reduce r_{max}.

Nevertheless, the gains of increased body size must sometimes outweigh the losses of reduced r_{max}, or large organisms would never have evolved. The frequent tendency of phyletic lines to increase in body size during geological time as evidenced by the fossil record (Newell, 1949) has given rise to the notion of "phyletic size increase." Many of the advantages of large size are patently obvious, but so are some disadvantages. Certainly a larger organism is less likely to fall victim to predators and would therefore have fewer potential predators and better survivorship. Small organisms are at the mercy of their physical environment and very slight changes in it can be devastating; larger organisms, on the other hand, are comparatively better buffered and therefore better protected. Some disadvantages of large size are (1) larger organisms require more matter and energy per individual per unit time than smaller ones and (2) there are fewer refuges, hiding places, and safe sites for large animals than there are for small ones. Reproductive strategies and body size are treated further in Chapter 6.

Use of Space: Home Range, Territoriality, and Foraging Strategy

Most habitats consist of a spatial–temporal mosaic of many different, often intergrading, elements, each with its own complement of organisms and

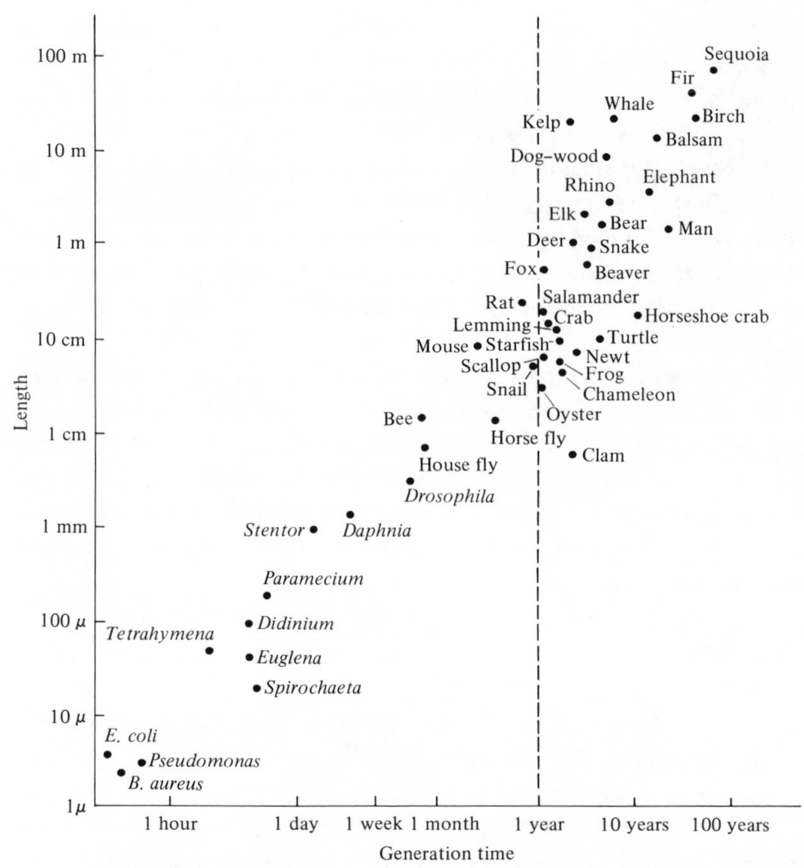

FIGURE 4.21 Log-log plot of organism length against generation time for a wide variety of organisms. [From John Tyler Bonner, *Size and Cycle: An Essay on the Structure of Biology* (Copyright © 1965 by Princeton University Press), Fig. 1, p. 17. Reprinted by permission of Princeton University Press.]

other resources. Because of this extensive environmental heterogeneity, an individual's exact location can be a major determinant of its immediate fitness. Thus, members of a prey species that are well protected from their predator(s) in one environmental patch type may be extremely vulnerable to the same predator(s) in another patch. Natural selection, by favoring those individuals that select better microhabitats, should produce a correlation between preference for a given patch type and fitness within it. The density of other individuals in a particular patch, of course, strongly influences the suitability of a given patch. Moreover, because most animals use several to many microhabitats for different purposes and/or at different times of day, the fitness relations of the use of space are usually quite complex. Some consequences of patchy environments are discussed here; others are examined in Chapter 5 (pp. 147–149, 165–169) and Chapter 6 (pp. 206–210).

There are two extreme ways in which organisms can be spaced about the landscape: they may occur in groups (*clumped* or *contagious spatial distributions*) or individuals may be evenly spread out (*dispersed spatial distributions*). Intermediate between these extremes are *random spatial distributions* in which the organisms are spread randomly over the landscape. Statistical techniques have been developed to quantify the spatial relationships of individuals in a population; one such technique uses the ratio of mean to variance in the numbers of individuals per quadrat. When this ratio is unity, the distribution of organisms in the quadrats fits the so-called Poisson distribution and the organisms are randomly distributed with respect to the quadrats. Ratios greater than unity indicate dispersion, while those less than unity reflect clumped distributions. Dispersed spatial distributions are generally indicative of competition and K selection; however, random and clumped distributions, in themselves, do not indicate very much about the factors influencing the distribution of the organisms concerned.

Organisms vary widely in their degree of mobility (frequent ecological synonyms are *motility* and *vagility*). Some, such as terrestrial plants and sessile marine invertebrates like barnacles, spend their entire adult life at one spot, with their gametes (and/or larvae) being the dispersal stages. Others, such as earthworms and snails, although vagile, seldom move very far. Still others, such as the monarch butterfly and migratory birds, regularly move distances of many kilometers during their lifetimes.

The area of volume over which an individual animal roams during the course of its usual daily wanderings and in which it spends most of its time is the animal's *home range*. Often home ranges of several individuals overlap; home ranges are not defended and are not used to the exclusion of other animals. *Territories*, on the other hand, are defended and used exclusively by an individual, a pair, a family, or a small inbred group of individuals. Nonoverlapping territories normally give rise to dispersed spacing systems and are almost invariably indicative of competition for some resource in short supply.

There are several different kinds of territories, classified by the functions they serve. Many sea birds, such as gulls, defend only their nest and the area immediately adjacent to it, that is, they have a *nesting territory*. Some male birds and mammals, such as grouse and sea lions, defend territories which are used solely for breeding termed *mating territories*. By far the most widespread type of territory, however, is the *feeding territory*, which occurs in a few insects, some fish, numerous lizards, many mammals, and most birds.

In order for territoriality to evolve, some resource must be in short supply and that resource must also be defendable (Figure 4.22). Food items are generally not defendable, since most animals eat their prey as soon as they encounter it. However, the space in which prey occurs can often be defended with a reasonable amount of effort. Sometimes even space is not easily defended, especially when food items are very sparse or extremely mobile; under such circumstances feeding territories cannot be evolved (Brown, 1964). Often birds that nest in colonies (i.e., seabirds and swallows) defend nesting territories, but feed in flocks exploiting very mobile foods and therefore have no feeding territories. Territoriality is particularly prominent in insectivorous and carnivorous birds, probably largely as a result of flight and their great mobility, which make territorial defense economically feasible. Typically, males of these birds set up territories during the early spring, often even before the wintering females return. During this period, many disputes over territories occur and there is much fighting among the males. Once breeding has begun, however, disputes over boundaries are usually greatly reduced, with the males advertising that they are still "on territory" only briefly during the morning and evening hours.

It has been demonstrated that male ovenbirds recognize individual neighbors by their territorial songs (Weedon and Falls, 1959). These investigators played tape-recorded songs back in different places and at different rates and times, and watched the responses of various males to the playbacks. When a tape-recording of a nonneighbor's call is played from a neighboring

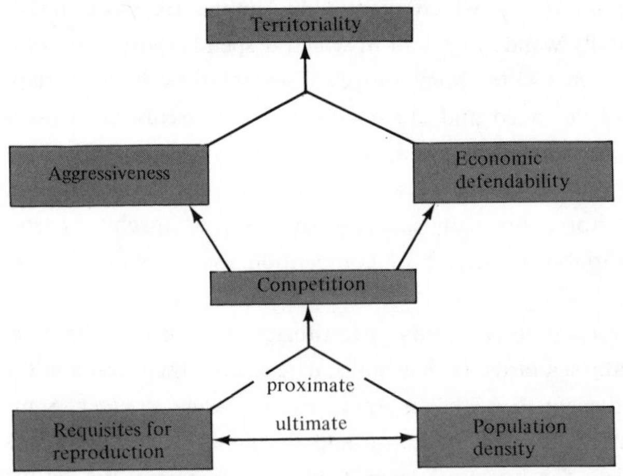

FIGURE 4.22 Diagrammatic representation of the various factors influencing the evolution of territoriality. [From Brown (1964).]

territory (after the original neighbor has been removed), a male ovenbird reacts by responding vigorously and singing frequently and loudly. This shows that the male recognizes the substitution of calls, because the same male does not react strongly to the call of his original neighbor. Falls (1969) used this technique to study the function of the territorial song of the white-throated sparrow; a greater frequency of playback elicits a stronger response which suggests that highly motivated birds sing more often than birds that are less likely to win a territorial encounter.

There are definite advantages to recognizing the territorial calls of established neighbors and to not reacting strongly to them; it would be a waste of time and energy to respond vigorously to such calls once territorial boundaries have been well established. Both individuals stand only to benefit from a "gentleman's agreement" over where such boundaries lie.

Possession of home ranges and territories also serves other important functions; for example, by becoming familiar with a small local area, an animal can learn: (1) when and where food is likely to be found; (2) the locations of safe retreats from predators; and (3) in some cases, when and where those predators are likely to be encountered. Thus resident individuals will normally have distinct advantages over nonresidents (so-called *vagrants*).

An enormous literature on territoriality exists; the interested reader can find an entry to it through the references at the end of this chapter.

There are important differences between organisms in exactly how they use space. The distinction between two-dimensional versus three-dimensional patterns of utilization is fundamental (one-dimensional use of space may also be approximated, as along the shore of a stream, lake, or ocean). Thus we find certain ecologic similarities between organisms as diverse as plankton, pelagic fish, flying insects, many birds, and bats, all of which live in a three-dimensional world.

Returns on many species of banded birds have shown that, even after migrating thousands of kilometers, individuals often return to the same general area where they were raised. Similarly, fruit flies (*Drosophila*) labeled with radioactive tracers do not usually move very far. Such restricted movements presumably allow individuals and populations to become genetically adapted to local conditions. We distinguish two extreme types of populations, although, of course, there are all degrees of intermediates: *viscous populations* in which individuals do not usually move very far, and *fluid populations* in which individuals cover great distances. In viscous populations there is little gene flow and great genetic variability can occur from place to place, whereas the opposite is true of fluid populations. At a local level, however, inbreeding in a viscous population may lead to reduced genetic variability.

Another essential aspect of the use of space is what might be termed foraging "strategy," which involves the ways in which an animal gathers matter and energy. This matter and energy constitute the profits gained from foraging, in that they are used in growth, maintenance, and reproduction. But foraging has its costs as well; thus, a foraging animal may often expose itself to potential predators and much of the time spent in foraging is rendered unavailable for other activities, including reproduction. An optimal foraging strategy maximizes the difference between foraging profits and their costs. Presumably natural selection, acting as an efficiency expert, has often favored such optimal foraging behavior. Consider, for example, prey of different sizes and what might be termed "catchability." How great an effort should a foraging animal make to obtain a prey item with a given catchability and of a particular size (and, therefore, matter and energy content)? Clearly, an optimal consumer should be willing to expend more energy to find and capture food items that return the most energy per unit of expenditure upon them. Moreover, an optimal forager should take advantage of natural feeding routes and should not waste time and energy looking for prey either in inappropriate places or at inappropriate times. What is optimal in one environment is seldom optimal in another, and an animal's particular anatomy strongly influences its optimal foraging strategy. There is considerable evidence that animals actually do maximize their foraging efficiencies, and a substantial body of theory on optimal foraging strategies exists (see also Chapter 6, pp. 202–212).

Many aspects of optimal foraging theory are concisely summarized in an excellent chapter entitled "The Economics of Consumer Choice" by MacArthur (1972). He makes several preliminary assumptions to reduce chaos: (a) Environmental structure is assumed to be repeatable, with some statistical expectation of finding a particular resource (such as a habitat, microhabitat, and/or prey item). (b) There is a continuous and unimodal spectrum of food items, such as is known for size distributions of insects (Schoener and Janzen, 1968; Hespenhide, 1971). [This assumption is clearly violated by the foods of some animals, such as monophagous insects or herbivores generally, because plant chemical defenses are typically discrete (Chapter 5, pp. 174–176).] (c) Similar animal phenotypes are usually closely equivalent in their harvesting abilities; an intermediate phenotype is thus best able to exploit foods intermediate between those which are optimal for two neighboring phenotypes (see also Chapter 6, pp. 199–201). Conversely, similar foods are gathered with similar efficiencies; a lizard with a jaw length that adapts it to exploit best 5-mm long insects is only slightly less efficient at eating 4-mm and 6-mm insects. (d) The principle of allocation applies, and no one phenotype can be maximally efficient on all prey types; improving harvesting efficiency on one food type necessitates reducing the efficiency of exploiting other kinds of items. (e) Finally, the economic "goal" of a species is to maximize its total intake of food resources.

MacArthur then breaks foraging down into four phases: (1) deciding where to search, (2) searching for palatable food items, (3) upon locating a potential food item, deciding whether or not to pursue it, and (4) the pursuit, with possible capture and eating. Search and pursuit efficiencies for each food type in each habitat are entirely determined by the above assumptions about morphology (assumption c) and environmental repeatability (assumption a); moreover, these efficiencies dictate the probabilities associated with the searching and pursuing phases of foraging (2 and 4, respectively). Thus, MacArthur considers only the two decisions—where to forage and what prey items to pursue (phases 1 and 3 of foraging). Clearly an optimal consumer should forage where its expectation of yield is greatest, an easy decision to make, given knowledge of the above efficiencies and the structure of its environment. The decision as to which prey items to pursue is also easy. Upon finding a potential prey item, a consumer has only two options: either pursue it or go on searching for a better item and pursue that one instead. Both decisions end in the forager beginning a new search, so the best choice is clearly the one which returns the greatest yield per unit time. Thus an optimal consumer should pursue an item only when it cannot expect both to locate and catch a better item during the time required to capture and ingest the first item.

Many animals, such as foliage-gleaning insectivorous birds, spend much of their foraging time searching for prey, but expend relatively little time and energy pursuing, capturing, and eating small sedentary insects which are usually easy to catch. In such "searchers," mean search time per item eaten is large compared to the average pursuit time per item; hence the optimal strategy is to eat essentially all palatable insects encountered. Other animals ("pursuers") which expend little energy in finding their prey but a great deal of effort in capturing it (such as, perhaps, a falcon or a lion) should select prey with small average pursuit times (and energetic costs). Hence, pursuers should generally be more specialized than searchers. Moreover, because a food-dense environment offers a lower mean search time per item than does a food-sparse area, an optimal consumer should restrict its diet to only the better types of food items in the former habitat. Optimal foraging strategies are considered further in Chapter 6 (pp. 202–212).

Sex, Sex Ratio, Sexual Selection, Sexual Dimorphism, and Mating Systems

Sexual reproduction is probably present in many or even most organisms, although a number of plants and invertebrate animals employ it only at very infrequent intervals. The evolutionary origin and selective advantage(s) of sexual reproduction are still major unresolved problems in biology (Williams, 1971). Sexual processes allow the genes in a gene pool to be mixed up each generation and recombined in various new combinations; as such, sex is an important way in which genetic variability is maintained. Moreover, the potential rate of evolution of a sexual population is greater than that of a group of asexual organisms, because a variety of beneficial mutations are readily combined into the same individual in a sexual species. Certainly sexual reproduction is very basic in diploid organisms and is doubtlessly quite an ancient and primitive trait. Considered from an individual's standpoint, however, sex is expensive because an individual's genes are thereby mixed with those of another organism and hence each of its offspring carries only one-half of its genes (i.e., heritability is halved). A female reproducing asexually (including parthenogenesis), in contrast, duplicates only her own genome in each of her offspring. Even Fisher (1930) believed that sex could have evolved for the benefit of the group by way of group selection. Strangely enough, although many temporary losses of sexuality have been secondarily evolved, relatively few known organisms seem to have completely lost the capacity to exchange genes with other organisms for any geologically long period of time. Presumably, evolutionary benefits of genetic recombination and increased variability more than offset the disadvantage of one organism perpetuating the genes of another. Still another advantage to an individual might be that, by reproducing sexually, an organism is able to mix its genes with other very fit genes, thereby increasing the fitness of its progeny. Of course, this can work both ways, for by mating with a less fit partner, an organism could decrease its own fitness (to the extent that heterozygosity in itself confers increased fitness, however, sexual reproduction is clearly advantageous to individuals).

Numerous varieties of sexual reproduction exist. Organisms may be *monoecious*, or hermaphroditic, ranging from *simultaneous hermaphrodites* (in which one individual has both male and female gonads at the same time—as in many invertebrates and many plants) to *sequential hermaphrodites* (an example is *protandry*, in which a given individual is a male when young and subsequently develops into a female). Perhaps most familiar are so-called *dioecious* organisms, in which the sexes are separate (as in most vertebrates and some plants).

In populations of many (perhaps most) diecious organisms there are approximately equal numbers of males and females. The *sex ratio* is defined as the proportion of males in the population. To be more precise, we distinguish the sex ratio at conception, or the *primary sex ratio*, from that at the end of the period of parental care which is the *secondary sex ratio*. The sex ratio of newly independent nonbreeding animals (as recently fledged birds) is the *tertiary sex ratio*, while that of the older breeding adult population is the *quaternary sex ratio*. Why these various sex ratios are often near equality (i.e., 50:50, or 0.5) is an intriguing problem.

Darwin (1871) speculated that sex ratios of 1:1 might benefit groups in that they would minimize intrasexual fighting over mates. Other workers have reasoned that since one male can easily serve a number of females, it might "be better for the *species*" if the population sex ratio were biased in favor of females, since this would increase the total number of offspring produced. Both the above interpretations invoke group selection and it is preferable to look for an explanation of sex ratio in terms of selection at the level of the individual. Here again, Fisher (1930) first solved the problem, by noting that, in sexually reproducing diploid species, exactly half the genes (or more precisely, half those on the autosomal chromosomes) come from males and half from females each and every generation. This statement merely asserts that every individual organism has a mother and a father, but its implications regarding the sex ratio are extensive. Since it would be very difficult to improve upon it, Fisher's concise early statement on the sex ratio is quoted below (Fisher, 1930; pp. 142–143);

> In organisms of all kinds the young are launched upon their careers endowed with a certain amount of biological capital derived from their parents. This varies enormously in amount in different species, but, in all, there has been, before the offspring is able to lead an independent existence, a certain expenditure of nutriment in addition, almost universally, to some expenditure of time or activity, which the parents are induced by their instincts to make for the advantage of their young. Let us consider the reproductive value of these offspring at the moment when this parental expenditure on their behalf has just ceased. If we consider the aggregate of an entire generation of such offspring it is clear that the total reproductive value of the males in this group is exactly equal to the total value of all the females, because each sex must supply half the ancestry of all future generations of the species. From this it follows that the sex ratio will so adjust itself, under the influence of Natural Selection, that the total parental

expenditure incurred in respect of children of each sex, shall be equal; for if this were not so and the total expenditure incurred in producing males, for instance, were less than the total expenditure incurred in producing females, then since the total reproductive value of the males is equal to that of the females, it would follow that those parents, the innate tendencies of which caused them to produce males in excess, would, for the same expenditure, produce a greater amount of reproductive value; and in consequence would be the progenitors of a larger fraction of future generations than would parents having a congenital bias towards the production of females. Selection would thus raise the sex-ratio until the expenditure upon males became equal to that upon females. If, for example, as in man, the males suffered a heavier mortality during the period of parental expenditure, this would cause them to be more expensive to produce, for, for every hundred males successfully produced expenditure has been incurred, not only for these during their whole period of dependence but for a certain number of others who have perished prematurely before incurring the full complement of expenditure. The average expenditure is therefore greater for each boy reared, but less for each boy born, than it is for girls at the corresponding stages, and we may therefore infer that the condition toward which Natural Selection will tend will be one in which boys are the more numerous at birth, but become less numerous, owing to their higher death-rate, before the end of the period of parental expenditure. The actual sex-ratio in man seems to fulfil these conditions somewhat closely, especially if we make allowance for the large recent diminution in the deaths of infants and children; and since this adjustment is brought about by a somewhat large inequality in the sex ratio at conception, for which no *a priori* reason can be given, it is difficult to avoid the conclusion that the sex-ratio has really been adjusted by these means.

The sex-ratio at the end of the period of expenditure thus depends upon differential mortality during that period, and if there are any such differences, upon the differential demands which the young of such species make during their period of dependency; it will not be influenced by differential mortality during a self-supporting period; the relative numbers of the sexes attaining maturity may thus be influenced without compensation, by differential mortality during the period intervening between the period of dependence and the attainment of maturity. Any great differential mortality in this period will, however, tend to be checked by Natural Selection, owing to the fact that the total reproductive value of either sex, being, during this period, equal to that of the other, whichever is the scarcer, will be the more valuable, and consequently a more intense selection will be exerted in favour of all modifications tending towards its preservation. The numbers attaining sexual maturity may thus become unequal if sexual differentiation in form or habits is for other reasons advantageous, but any great and persistent inequality between the sexes at maturity should be found to be accompanied by sexual differentiations, having a very decided bionomic value.

In short, Fisher concludes that, at equilibrium, an optimal organism should allocate exactly half its reproductive effort to progeny of each sex; thus if each male offspring costs approximately as much to produce as each female offspring, the optimal family sex ratio is near 50:50, provided that the population is in or near equilibrium. It is extremely important to note that Fisher's argument does *not* depend upon competition for mates in any way, as it assumes that every male has the same probability of mating as every other male (likewise all females are assumed to be of equal fitness).

Figure 4.23 and Table 4.5 illustrate Fisher's principle in two hypothetical populations. In the first case, there is no sexual dimorphism in energy demands of the progeny and the energetic argument can be translated directly into numbers; thus, this case leads to an optimal family sex ratio at equilibrium of 0.5. This is not true in the second case, where there is an energy differential such that individual offspring of one sex require twice as much parental expenditure as individuals of the other sex; this case leads to an optimal family sex ratio at equilibrium of either 0.33 or 0.67, depending upon which sex is more expensive to produce. In both cases, the optimal

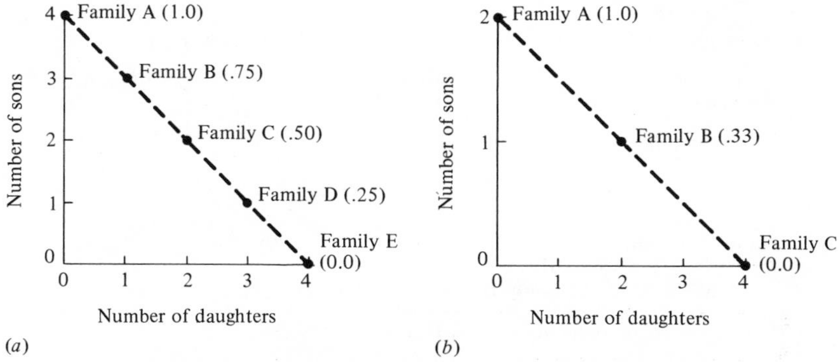

FIGURE 4.23 Two hypothetical situations illustrating the possible family structures without a sexual dimorphism in energy cost to the parent (*a*) and with such a sexual differential (*b*). Parents are assumed to have a given amount of energy available to devote to reproduction. In case (*a*), males cost parents the same amount of energy that females do, and the optimal family sex ratio (provided the population is at equilibrium) is 0.50 at which sex ratio, energy expenditure is equalized on offspring of the two sexes. In case (*b*), males cost their parents twice as much as female progeny, and energy expenditure on offspring of the two sexes is equalized at a family sex ratio of 0.33 (again, provided the population at large is near equilibrium). Table 4.5 and the text develop the reasons why parental energy should be divided equally between offspring of each sex when the population is at equilibrium.

TABLE 4.5 Comparison of the Contributions to Future Generations of Various Families in Case *a* and Case *b* of Figure 4.23 in Populations with Different Sex Ratios

CASE *a*:

	Number of Males	Number of Females
Initial Population	100	100
Family A	4	0
Family C	2	2
Subsequent Population (sum)	106	102

$C_A = 4/106 = 0.03773$
$C_C = 2/106+2/102 = 0.03846$ (family C has a higher reproductive success)

CASE *b*:

	Number of Males	Number of Females
Initial Population	100	100
Family A	2	0
Family B	1	2
Subsequent Population	103	102

$C_A = 2/103 = 0.01942$
$C_B = 1/103+2/102 = 0.02932$ (family B is more successful)

	Number of Males	Number of Females
Initial Population	100	100
Family B	1	2
Family C	0	4
Subsequent Population	101	106

$C_B = 1/101+2/106 = 0.02877$
$C_C = 4/106 = 0.03773$ (family C is more successful than family B)

Natural selection will favor families with an excess of females until the population reaches its equilibrium sex ratio (below)

	Number of Males	Number of Females
Initial Population	100	200
Family B	1	2
Family C	0	4
Subsequent Population	101	206

$C_B = 1/101+2/206 = 0.01971$
$C_C = 4/206 = 0.01942$ (family B now has the advantage)

Note: The contribution of family x is abbreviated C_x.

family sex ratio differs when the population sex ratio deviates from the optimal family sex ratio at equilibrium. In such a circumstance, families producing the sex in deficit (compared to the equilibrium sex ratio) have a selective advantage; this results in an excess production of the underrepresented sex which forces the population sex ratio to the equilibrium sex ratio. Sex ratio is a special case of so-called *frequency-dependent selection*, which arises whenever the selective value of a trait is a function of its frequency of occurrence. Figure 4.24 shows the way in which the optimal family sex ratio varies with population sex ratio under the circumstance in which the

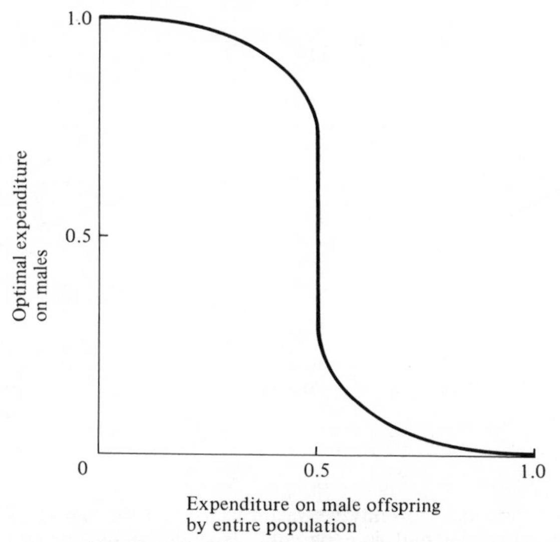

FIGURE 4.24 Plot showing how the optimal family expenditure on males varies with the expenditure on male offspring in the population at large.

two sexes are equally expensive to produce. This plot is more general in that it shows the way in which the optimal energy expenditure of a family varies as the average energy expenditure of the population at large changes. Note that when the average energy expenditure on offspring of the two sexes by the entire population is near the 1:1 equilibrium, a relatively broad range of family strategies are near optimal; as the overall population expenditure deviates more from this equilibrium, the optimal family rapidly converges on producing families containing only the underrepresented sex.

In summary, the only factor which can influence the primary and secondary sex ratios is a sexual difference in energy costs of progeny to the parents. A special case is differential mortality of the sexes *during the period of parental care* (Figure 4.25). Differential energetic requirements and differential mortality after this period can not directly alter the primary and secondary sex ratios, unless their effects are manifest during the period of parental care. Sexual dimorphisms, or physiological, morphológical, and/or behavioral differences between the sexes, are of paramount importance in any discussion of sex ratio. We next discuss the great variety of ecological factors that can influence sexual dimorphism, especially sexual selection and mating systems, which are complexly intertwined.

Fisher (1958a) quotes an unnamed "modern" biologist as having asked the following question: "Of what advantage could it be *to any species* for the males to struggle for the females and for the females to struggle for the males?" (italics mine). As Fisher points out, this is really a pseudo-question

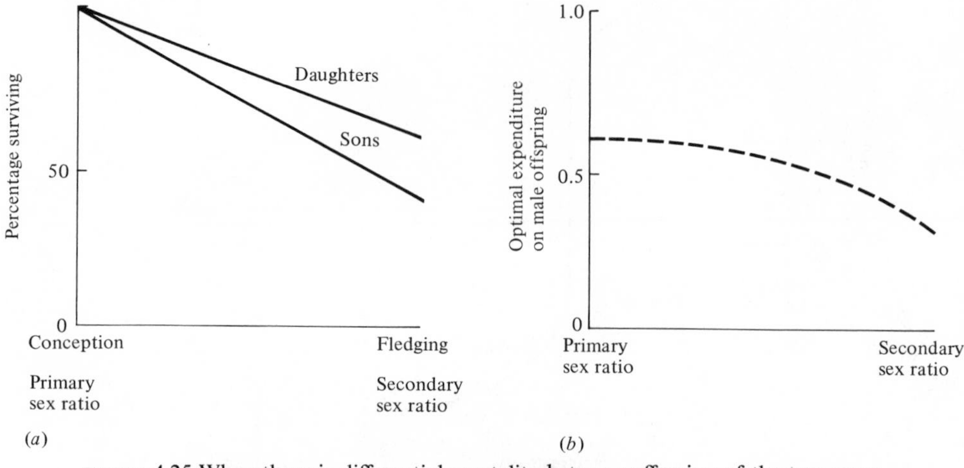

FIGURE 4.25 When there is differential mortality between offspring of the two sexes during the period of parental care, parental expenditure on the two sexes can only be equalized through skewed primary and secondary sex ratios. For example, if sons die off more rapidly than daughters (*a*), the optimal primary sex ratio will be biased toward sons while the optimal secondary sex ratio is biased in favor of daughters (*b*). The line curves because the proportion of males to females decreases more and more rapidly as surviving daughters accumulate.

because the fundamental units of natural selection are individuals, rather than species. The question thus reflects the attitude of a group selectionist. In the next few pages, we see exactly why this is not a biologically meaningful question.

Given that an organism is to mix its genes with those of another individual (i.e., that it is to reproduce sexually), just which other individual those genes are mixed with can make a substantial difference. An organism mating with a very fit partner by virtue of associating his genes with "good" genes passes his own genes on to future generations more effectively than another genetically identical (twin) individual mating with a less fit partner. Thus, those members of any population that make the best matings leave a statistically greater contribution to future generations. As a result, *within* each sex, there is competition for the best mates of the opposite sex; this leads to *intrasexual sexual selection.* Intrasexual selection usually generates antagonistic and aggressive interactions between the members of a sex, with those individuals best able to dominate other individuals of their own sex being at a relative advantage. Often direct physical battle is unnecessary and mere gestures (and/or various other signals of "strength") are enough to determine which individual "wins" an encounter. This makes some selective sense, for if the outcome of a fight is relatively certain there is little if anything to be gained from actually fighting, and, in fact, some disadvantage because there is a finite risk of injury to both contestants. Similar considerations apply in the defense of territories.

Maynard Smith (1956) convincingly demonstrated mating preferences in laboratory populations of the fruitfly *Drosophila subobscura.* Females of these flies usually mate only once during their lifetime, and store sperm in a seminal receptacle. Males breed repeatedly. Maynard Smith mated genetically similar females to two different strains of males, one inbred (homozygous) and one outbred (heterozygous), and collected all the eggs laid by these females during their entire lifetimes. He found that similar total numbers of eggs were laid by both groups of females, but the percentage of eggs which hatched differed rather markedly. Females mated to inbred males laid an average of only 264 fertile eggs each, whereas those bred to outbred males produced an average of 1134 fertile eggs per female (and hence produced over 4 times as many viable offspring). Maynard Smith reasoned that there should therefore be strong selection for females to mate with outbred rather than inbred males. When virgin females were placed in a bottle with outbred males, mating occurred within an hour in 90 percent of the cases; however, when similar virgins were offered inbred males, only 50 percent of them mated during the first hour. Both kinds of males courted females vigorously and repeatedly attempted to mount females, but outbred males were much more successful than inbred ones. By carefully watching the elaborate courtship behavior of these little flies, Maynard Smith discovered that inbred males responded more slowly than outbred males to the rapid side-step dance of the females. Presumably as a result of this lagging, females often rejected the advances of inbred males and flew away before being inseminated. These observations clearly show that females exert a preference as to which males they will accept. Similar mating preferences presumably exist in most natural populations, although they are usually very difficult to demonstrate. Over evolutionary time, natural selection operates to produce a correlation between male fitness and female preference because those females preferring the fittest males associate their own genes with the best male genes and therefore produce the fittest male offspring.

As a result of such mating preferences, populations have *breeding structures*. At one extreme is inbreeding in which genetically similar organisms mate with one another (so-called *homogamy*); at the other extreme is outbreeding in which unlikes mate with each other (*heterogamy*). Outbreeding leads to association of unlike genes and thus generates genetic variation. Inbreeding produces genetic uniformity at a local level, although variability may persist over a broader geographic region. Both these extremes represent nonrandom breeding structures; the randomly mating *panmictic* populations described by the Hardy-Weinberg equation of population genetics lie midway between them. It is doubtful, however, that any natural population is truly panmictic.

Animal populations also have *mating systems.* Most insectivorous birds and carnivorous birds and mammals are *monogamous*, with a pair bond between one male and one female. In such a case, both parents typically care for the young. *Polygamy* refers to mating systems in which one individual maintains simultaneous pair bonds with more than one member of the opposite sex. There are two kinds of polygamy depending upon which sex maintains the multiple pair bonds. In some birds, such as marsh wrens and yellowheaded blackbirds, for instance, one male may have pair bonds with two or more females at the same time (*polygyny*). Much less common is *polyandry*, in which one female has simultaneous pair bonds with more than one male; polyandry is, however, thought to occur in a few bird species, such as some jacanas, rails, and tinamous. A fourth type of mating system is *polybrachygyny* (Selander, 1972), in which one male has several short pair bonds with different females in sequence, typically each such pair bond lasts only long enough for completion of copulation and insemination. Polybrachygyny occurs in a variety of birds (including some grackles, hummingbirds, and grouse) and mammals (many pinnepeds and some ungulates). Finally, an idealized mating system (or perhaps more appropriately, a lack of a mating system) is *promiscuity*, in which each organism has an equal probability of mating with every other organism. True promiscuity is extremely unlikely and probably nonexistent; it would result in a panmictic population. It may be approached in organisms like barnacles, which shed their gametes into the sea, or terrestrial plants that release pollen to the wind, where they are mixed up by currents of water and air. However, various forms of chemical discrimination of gametes and therefore mating preferences could well occur even in such sessile organisms.

The intersexual component of sexual selection, that is, that occurring between the sexes, is termed *epigamic selection.* It is often defined as "the reproductive advantage accruing to those genotypes which provide the stronger heterosexual stimuli," but is also aptly described as the "battle of the sexes." Epigamic selection operates by mating preferences. Of prime importance is the fact that what maximizes the fitness of an individual male is not necessarily coincident with what is best for an individual female and vice versa. As an example, in most vertebrates, individual males can usually leave more genes under a polygynous or polybrachygynous mating system, whereas an individual female is more likely to maximize her reproductive success under a monogamous or polyandrous system. Sperm are small, energetically inexpensive to produce, and are produced in large numbers. As a result, vertebrate males have relatively little invested in each act of reproduction and can and do mate frequently and rather indiscriminately (i.e., males tend toward promiscuity). Vertebrate females, on the other hand, often or usually have much more invested in each act of reproduction because eggs and/or offspring are usually energetically expensive. Because these females have so much more at stake in each act of reproduction, they tend to exert much stronger mating preferences than males and to be more selective as to acceptable mates. By refusing to breed with promiscuous and polygynous males, vertebrate females can sometimes "force" males to become monogamous and to contribute their share toward raising the offspring. In effect, polygyny is the outcome of the battle of the sexes when the males win out (patriarchy), whereas polyandry is the outcome when females win out (matriarchy). Monogamy is a compromise between these two extremes. Under a monogamous mating system, a male must be certain that the offspring are his own; otherwise he might expend energy raising offspring of another male (note that females do not have this problem). It is no wonder that monogamous males jealously guard their females against stolen copulations!

Let us now examine the ecological determinants of mating systems. It has sometimes been asserted that sex ratios "drive" mating systems; under such an interpretation polygyny arises when males are in short supply and polyandry occurs when there are not enough females to go around. According to this explanation, many species are monogamous simply *because* sex ratios are often near equality. It is now generally accepted that quite the reverse is true, with sexual selection and mating systems indirectly and directly determining sexual dimorphisms and hence various sex ratios. In many birds and some mammals, there are so-called floating populations of nonbreeding males. These can be demonstrated by simply removing breeding individuals; typically they are quickly replaced with younger and less experienced animals (Stewart and Aldrich, 1951; Hensley and Cope, 1951; Orians, 1969b).

Only 14 of the 291 species (5 percent) of North American passerine birds are regularly polygynous (Verner and Willson, 1966). Some 11 of these 14 (nearly 80 percent) breed in prairies, marshes, and savannah habitats. Verner and Willson suggested that, in these extremely productive habitats, insects are continually emerging and thus food supply is rapidly renewed; as a result several females can successfully exploit the same feeding territory. However, a similar review of the nesting habitats used by polygynous passerines in Europe (which also constitute about 5 percent of the total number of species) showed no such prevalence toward grassland or marshy habitats (Haartman, 1969). Indeed, for elusive reasons, Haartman suggested that closed-in, safe nests were a more important determinant of polygynous mating systems than were breeding habitats. Crook (1962, 1963, 1964, 1965) has suggested that among African weaverbirds monogamy is evolved when food is scarce and both parents are necessary to raise young, whereas polygyny evolves in productive habitats with abundant food where male assistance is less essential. This argument, of course, ignores entirely the "battle of the sexes" (epigamic selection).

One of the best field studies of polygyny to date is that of Verner (1964), who studied long-billed marsh wrens in Washington state. He discovered that some males possessed two females (one male had three), whereas other males on adjacent territories had only one female or none at all. Verner was able to demonstrate that the territories of bigamous and trigamous males were not only larger than those of bachelors and monogamous males, but that they also contained more emergent vegetation (where the female wrens forage). He reasoned that females must be able to raise more young by pairing with a mated male on a superior territory than with a bachelor on an inferior territory (even though she obtains less help from her mate). Moreover, Verner noted that the evolution of polygyny is dependent upon males being able to defend territories containing enough food to support more than one female and her offspring; this condition for the evolution of polygyny requires fairly productive habitats. It is noteworthy that female wrens are antagonistic toward each other and as a result males cannot make a second mating until their first female is incubating; this produces a temporal staggering of the females (Verner, 1965). Building on Verner's work and studies on blackbirds, Verner and Willson (1966) defined the *polygyny threshold* as the minimum difference in habitat quality of territories held by males in the same general region which is sufficient to favor bigamous matings by females (Figure 4.26).

Polygyny is much more prevalent in mammals than in birds, presumably because in most mammals, females nurse their young, and, at least among herbivorous species, the males can do relatively little to assist the females in raising young (such species typically have a pronounced sexual dimorphism). A notable exception is carnivorous mammals that are often monogamous during the breeding season with the males participating in feeding the young (typically sexual dimorphisms are slight in such species). Similarly, most carnivorous and insectivorous birds are monogamous, and males can and do gather food for the nestlings. Often there is relatively little

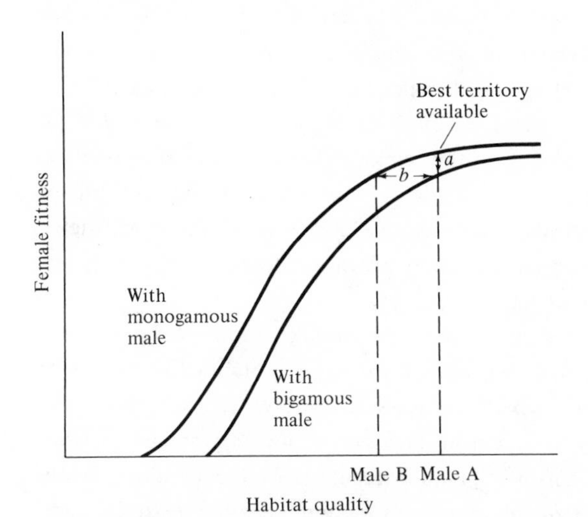

FIGURE 4.26 Graphic model of the conditions necessary for the evolution of polygyny. Reproductive success of females is correlated with environmental quality and females select mates which give them the highest individual fitness. Distance *a* is the difference in fitness between a female mated monogamously and a female mated bigamously in the same environment. Distance *b* is the *polygyny threshold*, or the minimum difference in territory quality held by males in the same region which is sufficient to favor bigamous matings by females. [After Orians (1969b).]

sexual dimorphism in such species, and those that are dimorphic are usually migratory [Hamilton (1961) suggested that sexual dimorphism promotes rapid pairing, as well as species recognition]. Birds whose young are well developed at hatching (so-called *precocial* as opposed to *altricial* birds), typically have little male parental care and are frequently polygynous or polybrachygynous with pronounced sexual dimorphisms.

Given a population sex ratio near equality and a monogamous mating system, every individual (both male and female) has a substantial opportunity to breed and to pass on its genes; however, with an equal population sex ratio and polygyny, a select group of the fittest males make a disproportionate number of matings. Dominant battle-scarred males of the northern sea lion, *Eumetopias jubata*, which have "won" the rocky islets where most copulations occur, often have harems of 10 to 20 females. Under such circumstances, these males father most progeny and their genes constitute half the gene pool of the subsequent generation. Because those heritable characteristics making them good fighters and dominant animals are passed on to their sons, contests over the breeding grounds may be intensified in the next generation. It is essentially an "all or none" proposition, in that only the winning males are able to perpetuate their genes. As a result of this intense competition between males for the breeding grounds, intrasexual selection has favored a striking sexual dimorphism in size. Whereas adult females usually weigh less than 500 kg, adult males may weigh as much as 1000 kg. The sexual dimorphism in size is even more pronounced in the California sea lion, *Zalophus californianus*, where females attain weights of only about 100 kg, while males reach nearly 500 kg. Presumably, the upper limit on such a differential size escalation is set by various other ecological determinants of body size, such as predation pressures, foraging efficiency, and food availability (see also Chapter 6).

Somewhat analogous situations occur among various polygynous and polybrachygynous birds. Many species of grouse are polybrachygynous with the males displaying their sexual prowess in groups termed "leks." The dominant males occupy the central portion of the communal breeding grounds and make a disproportionate percentage of the matings. Strong sexual dimorphisms in size, plumage, color, and behavior exist in many grouse. In addition to intrasexual selection, epigamic selection operating through female choice can also produce and maintain sexual dimorphisms; usually both types of sexual selection occur simultaneously and it is often difficult to separate their effects. Indeed, by choosing to mate with gaudy and conspicuous males, females have presumably forced the evolution of some bizarre male sexual adaptations, such as the long tails of some male birds of paradise. Bower birds have avoided becoming overly gaudy (and hence dangerously conspicuous) by evolving a unique behavioral adaptation; males build highly ornamented bowers that are used to attract and to court females and that signal the male's intersexual attractiveness. Frequently, if not usually, the same sexual characters (such as size, color, plumage, song, behavior) advertise both intrasexual prowess and intersexual attractiveness. This makes evolutionary sense because an individual's overall fitness is determined by its success at coping with both types of sexual selection, which should usually be positively correlated; moreover, economy of energy expenditure is also obtained by such a consolidation of sexual signals.

Sexual dimorphisms sometimes serve still another ecologic function: namely, to reduce niche overlap and competition between the sexes. In certain island lizards (Schoener, 1967, 1968a) and some birds (Selander, 1966), there are strong sexual dimorphisms in the feeding apparatus (jaws and beaks, respectively), which are correlated with differential utilization of food resources (see also Chapter 6).

Fitness and the Individual's Status in the Population

An organism's fitness is determined by the interaction between its phenotype and the totality of its environment. In *K*-selected organisms, fitness is determined largely by the biotic environment, especially the individual's status within his own population; fitness of *r*-selected organisms, on the other hand, is often less dependent upon the biotic environment and more strongly influenced by the physical environment.

Compared with breeding individuals, members of a nonbreeding floating population have very low immediate fitnesses (although their fitness rises if they are able to breed later). Even within a breeding population, various individuals may often differ substantially in fitness; for instance, among long-billed marsh wrens, there is great variation between males in their individual fitness, with bigamous and trigamous males being more fit than monogamous ones. Predators tend to crop mainly the "excess" members of bobwhite quail and muskrat populations (Errington, 1956, 1963); presumably resident prey individuals know their own territories and home ranges making them more difficult to capture than vagrant individuals with a less stable status in the population (see also earlier section on Use of Space).

Maintenance of Variability

The fundamental source of variation between individuals is sexual reproduction; reassortment and recombination of genes each generation assure that new genotypes will arise regularly in any population with genetic variability. In fact, in most higher organisms no two individuals are genetically identical (except identical twins and progeny produced asexually). Population biologists are very interested in understanding the factors that create and maintain genetic variability in natural populations (Selander and Johnson, 1972). Numerous genetic mechanisms operate to produce genetic variability, both within and between populations, including linkage, chromosomal inversions, translocations, and heterosis (see also Chapter 1).

At the outset, we must distinguish phenotypic from genotypic variation. The phenotypic component of variability is the total observable variability, whereas the genotypic component is that with a genetic basis. It is usually difficult to distinguish genetically induced variation from that which is environmentally induced. However, by growing clones of genetically identical individuals (i.e., with the same genotype) under differing environmental conditions, biologists have been able to determine how much inter-individual variation is due to the developmental plasticity of a particular genotype in different environments. It is now thought that approximately half the phenotypic variation observed in natural populations has a genetic basis and the remaining variation is environmentally induced. Since natural selection can act only on heritable traits, many phenotypic variants may have little direct selective value (but see also below). The degree of developmental flexibility of a given phenotypic trait strongly influences an organism's fitness; such a trait is said to be *canalized* when the same phenotypic character is produced in a wide range of genetic and environmental backgrounds. Presumably some genes are rather strongly canalized, such as those that produce "wild-type" individuals, whereas others are less determinant, allowing individuals to adapt and regulate via developmental plasticity. Such environmentally induced phenotypic varieties are common in plants; they are less common among animals, probably because these mobile organisms can easily select an appropriate environment. It may be selectively advantageous for certain genetically induced traits to be under tight control, while others operate to increase the fitness of the individual by allowing some flexibility of response to differing environmental influences.

Genotypic and phenotypic variation between individuals is probably seldom selected for directly, in itself, although it may often arise and be maintained in a number of more or less indirect ways. Especially important are changing environments; in a temporally varying environment, selective pressures vary from time to time and the phenotype of highest fitness is always changing. There is inevitably some lag in response to selection and organisms adapted to tolerate a wide range of conditions are frequently at an advantage. (Heterozygotes may often be better able to perform under a wider range of conditions than homozygotes.) Indeed, in unpredictably changing environments, reproductive success may usually be maximized by production of offspring with a broad spectrum of phenotypes.

Similar considerations apply to spatially varying environments, because the phenotypes best able to exploit various "patches" usually differ (see also Chapter 6). On a broader geographic level, differences from one habitat to the next presumably often result in different selective milieus, and therefore, in different gene pools that are adapted to local conditions. Gene flow between and among such divergent populations can result in substantial amounts of genetic variability, even at a single spot (this has been termed the "gene flow-variation hypothesis"—Chapter 8, pp. 266–267).

Competition among the members of a population for preferred resources (Chapter 5) may often confer a relative advantage on variant individuals that are better able to exploit marginal resources; thus competition within a population can directly favor an increase in its variability. By virtue of such variation between individuals, the population exploits a broader spectrum of resources more effectively and has a larger populational "niche breadth"; the "between phenotype" component of niche breadth is great (Roughgarden, 1972; see also Chapter 6, pp. 199–201 and Chapter 8, pp. 264–266). Because such increased phenotypic variability between individuals promotes a broader populational niche, this has been termed the "niche-variation hypothesis" (Soulé and Stewart, 1970). Similarly, environments with a low availability of resources usually require that individuals exploiting them make use of a wide variety of available resources; in this case, however, because each individual must possess a broad niche, variation between individuals is not great (i.e., the "between-phenotype" component of niche breadth is slight while the "within-phenotype" component is great).

One further way in which variability can be advantageous involves the interactions between individuals belonging to different species (Chapter 5), especially interspecific competition and predation. Fisher (1958b) likened such interspecific interactions and coevolution to a giant evolutionary game in which moves alternate with countermoves, and he suggested that it may well be more difficult to evolve against an unpredictable and variable polymorphic species than against a better standardized and more predictable monomorphic species. As a possible example, foraging birds tend to develop a "search image" for prey items commonly encountered, often bypassing other kinds of suitable prey.

Social Behavior and Kin Selection

A wide variety of ecological phenomena have been interpreted as having been evolved for the benefit of the population, rather than for the benefit of the individuals comprising a particular population. Clutch size, sex ratio, and sexual selection are examples which have already been mentioned in this context; predator alarm calls and so-called "prudent predation" are discussed in Chapter 5, and "selection at the level of the ecosystem" is considered in Chapter 7. Another broad area latent with opportunities for interpretation of events and phenomena as having arisen for the advantage of a group is social behavior. For example, why does a worker honey bee sacrifice her own reproduction for the good of the colony? As is well known, a worker bee will even give up her own life in defense of the hive. The probable answer to the above question is presented later in this section, but first we must develop some basic considerations and definitions.

True altruism occurs only when an individual behaves in such a manner that he suffers a net loss while a neighbor (or neighbors) somehow gain from his loss (Figure 4.27). Obviously selfish behavior will always be of selective advantage; the problem is to explain the occurrence of apparently altruistic behavior such as the worker honey bee's.

The most significant and sophisticated treatment of the evolution of sociality is that of Hamilton (1964) who points out that the very existence of social behavior implies that *individuals* living in cooperating groups are in fact leaving more genes in the population gene pool than hermits. Thus, social behavior may be expected to evolve when distinct advantages are inherent in group participation. To develop his thesis, Hamilton defines *kin selection* as selection operating between closely related individuals to produce cooperation. As an extreme example, an individual should theoretically sacrifice his own life if he can thereby save the lives of more than two siblings, each of which shares half his genes. Such behavior furthers his own genotype even more effectively than living to reproduce himself. This is not true altruistic behavior, because the individual making the "sacrifice" actually gains more than he loses. Kin selection operates at a much more subtle level than that in the above example, the major factor being that closely related relatives are much more likely to benefit from such pseudo-altruistic behavior than distantly related ones; in order for the latter to occur, the loss to the benefactor should either be very small and/or the number of distant relatives benefited must be large. Thus the condition necessary for the evolution of such pseudo-altruistic behavior is that the total gain(s) to the relative(s) must be greater

	Neighbor(s) gain	Neighbor(s) lose
Individual gains	Pseudo-altruistic behavior (kin-selection)	Selfish behavior (selected)
Individual loses	True altruistic behavior (counter-selected)	Mutually disadvantageous behavior (counter-selected)

FIGURE 4.27 Diagram showing the four possible situations involving an individual's behavior and its influence upon a neighbor. Kin selection occurs when an individual actually gains more than he loses (due to the advantage he obtains in perpetuating his genes among his neighbors which are relatives); thus, it falls in the box labeled "pseudo-altruistic behavior." True altruistic behavior, in which an individual actually loses while his neighbor gains, is virtually unknown (except, perhaps, in man); moreover, it would require group selection to evolve.

than the loss to the pseudo-altruist. Parental care is, of course, a special case of kin selection.

Here again, ideas were long ago foreshadowed by Fisher (1930), in his consideration of the evolution of distastefulness and warning coloration in insects. Many distasteful or even actually poisonous insects, especially the larvae of some moths and butterflies, are brightly colored; vertebrate predators, especially birds, quickly learn not to eat such warningly colored insects. Fisher noted the difficulty of explaining the origin of this gaudy coloration since the new ultraconspicuous mutants would be expected to suffer the first attacks of inexperienced predators and therefore be at a selective disadvantage to less gaudily colored genotypes. He suggested that benefits to siblings on the same branch accruing under a gregarious family system could favor the evolution of warning coloration; hence, such ultraconspicuous mutants would be "pseudo-altruists." Moreover, Fisher even made his argument quantitative as he noted that although each sibling shares only half of the individual's genes, their numbers ensure that the total number of shared genes benefited greatly exceeds the number destroyed in the individual's own genome.

Members of the insect order Hymenoptera, which includes ants, bees, wasps, and hornets, often form colonies and exhibit apparent altruistic social behavior. Hymenoptera have a unique and peculiar genetic system; males are produced asexually and are haploid. Thus all sperm produced by any given male are genetically exactly identical (barring somatic mutation), and carry that male's entire genome. (Males have no father but do have a grandfather.) Females are normal diploids with each ovum carrying only one-half their genomes. In many hymenopteran species, a queen mates just once during her entire lifetime and stores one male's sperm in a spermatotheca; thus all her progeny have the same father. A result of this strange genetic system is that sisters are more closely related to one another than mothers are to their own daughters; the former share $\frac{3}{4}$ of their genes, the latter only $\frac{1}{2}$. Hence, one would predict that worker bees should help raise their sisters to reproductive maturity in preference to mating themselves, which in fact they often do. Interestingly, these same workers can and sometimes do lay haploid male-producing eggs. Males share only $\frac{1}{4}$ of their genes with siblings (both brothers and sisters) and $\frac{1}{2}$ with their daughters; as would be expected, males never work for the benefit of the colony.

Termites (order Isoptera) also form highly organized colonies, complete with queens and "kings" and both male and female workers with distinct castes. But because both sexes of termites have a normal diploid genetic system, the above kin selection argument is inadequate to explain the evolution of sociality in termites. To date, the best "explanation" for this apparent evolutionary enigma involves the fact that termites are highly dependent upon one another for continual replenishment of their intestinal protozoa. (These symbionts, which produce the cellulases that enable termites to digest wood, are lost with each molt and must be replenished continually during the lifetime of an individual.) Presumably, a pair of termites (the king and queen) maximize their own reproductive success by producing many non-reproductive progeny (workers) which in turn allow the production of many successful reproductive offspring (new kings and queens). Natural selection among workers, however, should operate to release them from such parental "control." Doubtlessly, termites also enjoy other advantages from group cooperation, such as protection from predators and the elements.

Another form of pseudo-altruism, termed "reciprocal altruism" (Trivers, 1971), does not require genetic affinity or kin selection to operate. In reciprocal altruism, some behavioral act incurs a relatively minor loss to a donor but provides a recipient with a large gain; thus two entirely unrelated animals can both benefit from mutual assistance. An example best explained by reciprocal altruism is the posting of sentinels. A sentinel crow spends a brief time period sitting in a tree watching for predators while the rest of the flock forages, and in turn he receives continual sentinel protection from the remainder of the flock during the much longer period of time he forages. Reciprocity is, of course, absolutely essential for the evolution and maintenance of this sort of altruistic behavior unless kin selection is also operating to promote it.

Selected References

Introduction

Cole (1954b, 1958); Gadgil and Bossert (1970); Harper (1967); MacArthur and Connell (1966); Mettler and Gregg (1969); Slobodkin (1962); Wilson and Bossert (1971).

Life Tables and Tables of Reproduction

Bogue (1969); Caughley (1966); Cole (1965); Deevey (1947); Fisher (1930); Lotka (1925, 1956); Mertz (1970); Pearl (1928); Slobodkin (1962); Spinage (1972); Zweifel and Lowe (1966).

Net Reproductive Rate and Reproductive Value

Emlen (1970); Fisher (1930, 1958a); Hamilton (1966); Mertz (1970, 1971a, 1971b); Slobodkin (1962); Turner *et al.* (1970); Vandermeer (1968); Wilson and Bossert (1971).

Stable Age Distribution

Krebs (1972); Lotka (1922, 1925, 1956); Mertz (1970); Vandermeer (1968); Wilson and Bossert (1971).

Intrinsic Rate of Natural Increase

Andrewartha and Birch (1954); Birch (1948, 1953); Cole (1954b, 1958); Evans and Smith (1952); Fisher (1930); Gill (1972); Goodman (1971); Leslie and Park (1949); Mertz (1970); F. E. Smith (1954, 1963a).

Population Growth and Regulation

Allee *et al.* (1949); Andrewartha and Birch (1954); Ayala (1968); Bartlett (1960); Beverton and Holt (1957); Chitty (1960, 1967a, 1967b); Christian and Davis (1964); Clark *et al.* (1967); Cole (1965); Ehrlich and Birch (1967); Errington (1946, 1956); Fretwell (1972); Gadgil and Bossert (1970); Gibb (1960); Green (1969); Grice and Hart (1962); Hairston, Smith and Slobodkin (1960); Horn (1968a); Krebs (1972); Lack (1954, 1966); McLaren (1971); Murdoch (1966a, 1966b, 1970); Nicholson (1933, 1954, 1957); Pearl (1927, 1930); Pimentel (1968); Slobodkin (1962); F. E. Smith (1952, 1954, 1963a); Solomon (1949, 1972); Southwood (1966); Williamson (1971).

Density Dependence and Density Independence

Andrewartha (1961, 1963); Andrewartha and Birch (1954); Davidson and Andrewartha (1948); Gunter (1941); Horn (1968a); Lack (1954, 1966); McLaren (1971); Nicholson (1957); Orians (1962); Pianka (1972); F. E. Smith (1961, 1963b); Solomon (1972).

r and K Selection

Anderson (1971); Charlesworth (1971); Clarke (1972); Dobzhansky (1950); Force (1972); Gadgil and Bossert (1970); Gadgil and Solbrig (1972); King and Anderson (1971); Lewontin (1965); MacArthur (1962); MacArthur and Wilson (1967); Pianka (1970, 1972); Roughgarden (1971); Wilson and Bossert (1971).

Population "Cycles": Cause and Effect

Chitty (1960, 1967a); Christian and Davis (1964); Cole (1951, 1954a); Elton (1942); Keith (1963); Krebs (1964, 1966, 1970); Krebs and DeLong (1965); Krebs, Keller, and Myers (1971); Krebs, Keller and Tamarin (1969); Pitelka (1964); Platt (1964); Schaeffer and Tamarin (1973); Schultz (1964, 1969); Tamarin and Krebs (1969); Wellington (1960).

Evolution of Reproductive Rates

Ashmole (1963); Baker (1938); Chitty (1967b); Cody (1966, 1971); Cole (1954b); Gadgil and Bossert (1970); Harper and Ogden (1970); Istock (1967); Johnson and Cook (1968); Johnston (1954); Klomp (1970); Lack (1954, 1966, 1968, 1971); Mertz (1970); Millar (1973); Murphy (1968); Perrins (1964, 1965); Royama (1969); Salisbury (1942); Skutch (1949); Tinkle (1969); Tinkle, Wilbur, and Tilley (1970); Williams (1966a, 1966b); Wynne-Edwards (1955, 1962).

Evolution of Death Rates and Old Age

Emlen (1970); Fisher (1930); Hamilton (1966); Medawar (1957); Pearl (1922, 1928); Sokal (1970); Williams (1957); Willson (1971).

The Joint Evolution of Rates of Reproduction and Mortality

Bonner (1965); Cole (1954b); Frank (1968); Gadgil and Bossert (1970); Lack (1954, 1966); Newell (1949); F. E. Smith (1954); Tinkle (1969); Williams (1966a, 1966b).

Use of Space: Home Range, Territoriality, and Foraging Strategy

Ardrey (1966); Brown (1964, 1969); Brown and Orians (1970); Carpenter (1958); Emlen (1966, 1968a); Falls (1969); Howard (1920); Hutchinson (1953); Kohn (1968); MacArthur (1959, 1972); MacArthur and Pianka (1966); McNab (1963); Menge (1972); Morse (1971); Orians and Horn (1969); Orians and Willson (1964); Pielou (1969); Rapport (1971); Royama (1970); Schoener (1969a, 1969b, 1971); C. C. Smith (1968); Tinbergen (1957); Weedon and Falls (1959).

Sex, Sex Ratio, Sexual Selection, Sexual Dimorphism, and Mating Systems

Crook (1962, 1963, 1964, 1965, 1972); Darwin (1871); Emlen (1968b); Fisher (1930, 1958a); Haartman (1969); Hamilton (1961); Hensley and Cope (1951);

Kolman (1960); Lack (1968); Maynard Smith (1956, 1958, 1971); Orians (1969b); Schoener (1967, 1968a); Selander (1965, 1966, 1972); Smouse (1971); Stewart and Aldrich (1951); Trivers (1972); Trivers and Willard (1973); Verner (1964, 1965); Verner and Engelsen (1970); Verner and Willson (1966); Williams (1971); Willson and Pianka (1963).

Fitness and the Individual's Status in the Population

Errington (1946, 1956, 1963); Fretwell (1972); C. C. Smith (1968); Verner (1964, 1965); Verner and Engelsen (1970); Wellington (1957, 1960).

Maintenance of Variability

Ehrlich and Raven (1969); Fisher (1958b); Mettler and Gregg (1969); Selander and Johnson (1972); Somero (1969); Soulé (1971); Soulé and Stewart (1970); Van Valen (1965); Wilson and Bossert (1971).

Social Behavior and Kin Selection

Brown (1966); Crook (1965); Fisher (1930, 1958a); Hamilton (1964, 1967, 1970, 1972); Horn (1968b); Maynard Smith (1964); N. Smith (1968); Trivers (1971); Wallace (1973); Wiens (1966); E. O. Wilson (1971); Wynne-Edwards (1962).

5 *Interactions Between Populations*

Introduction

Two populations may or may not affect each other; if they do, the influence may be beneficial or adverse. By designating a detrimental effect with a minus, no effect with a zero, and a beneficial effect with a plus, all possible population interactions can be conveniently classified. Thus, when neither of two populations affects the other, the interaction is designated as (0, 0). Similarly, a mutually beneficial relationship is (+, +), and a mutually detrimental one is (−, −). Other possible interactions are (+, −), (−, 0), and (+, 0), making a total of six fundamentally different ways in which populations can interact (Table 5.1).

Competition (−, −) takes place when each of two populations affects the other adversely. Typically both require the same resource(s) which is (are) in short supply; the presence of each population inhibits the other. *Neutralism* (0, 0) occurs when the two populations do not interact and neither affects the other in any way whatsoever; it is thus of little ecological interest. Moreover, true neutralism is likely to be very rare or even nonexistent in nature, because there are probably indirect interactions between all the populations in any given ecosystem. Interactions that benefit both populations (+, +) are classified as *mutualism* if the association is obligatory, in which neither population can exist without the other, and as *protocooperation* when the interaction is not an essential condition for survival of either population. *Predation* (+, −) occurs when one population affects another adversely but

TABLE 5.1 Summary of the Various Sorts of Interactions that May Occur Between Two Populations

Type of Interaction	Species A	Species B	Nature of the Interaction
Competition	−	−	Each population inhibits the other
Neutralism	0	0	Neither population affects the other
Mutualism	+	+	Interaction is favorable to both and is obligatory
Protocooperation	+	+	Interaction is favorable to both but is not obligatory
Predation	+	−	Population A, the predator, kills and consumes members of population B, the prey
Parasitism	+	−	Population A, the parasite, exploits members of population B, the host, which is affected adversely
Commensalism	+	0	Population A, the commensal, benefits whereas B, the host, is not affected
Amensalism	−	0	Population A is inhibited, but B is unaffected

Source: Adapted from Odum (1959).

benefits itself from the interaction. Usually a predator kills its prey and consumes part or all of the prey organism. (Exceptions include lizards losing their tails to predators, and plants losing their leaves to herbivores.) *Parasitism* (+, −) is essentially identical to predation, except that the host (a member of the population being adversely affected) is usually not killed outright, but is exploited over some period of time. Thus parasitism might be best considered as a "weak" form of predation; herbivory and some cases of mimicry could be placed here. When one population benefits while the other is unaffected, the relationship is termed *commensalism* (+, 0). *Amensalism* (−, 0) is said to occur when one population is affected adversely by another but the second is unaffected. Two of the six population interactions, competition and predation, are of overwhelming importance and most of this chapter is devoted to them.

Competition

Competition occurs when two or more organismic units are using the same resources and when those resources are in short supply. In addition, the interaction between them reduces the fitness and/or equilibrium population size of each organismic unit. This can occur in either or both of two ways. First, by requiring that the organismic unit expend some of its time and/or matter and/or energy on competition or the avoidance of competition, a competitor may effectively reduce the amounts left for maintenance and reproduction. Second, by using up or occupying some of a scarce resource, competitors directly reduce the amount available to other organismic units. Competition by way of direct interaction, such as aggressive encounters between competitors, is termed *interference competition*; more indirect inhibitory effects, such as those arising from reduced availability of a common resource, are known as *exploitation competition. Intra*specific competition, which was considered in Chapter 4, is competition between individuals belonging to the same species, usually to the same population. *Inter*specific competition is that occurring between individuals that belong to different species and is of most interest in the present chapter. It is always advantageous, *when possible*, for either party in a competitive relation to avoid the interaction. Because of this fact, competition has been an important evolutionary force that has led to niche separation, specialization, and diversification. If, however, avoidance of a competitive interaction is impossible, selection may sometimes favor convergence.

Competition is not an on–off process, but rather its level presumably varies continuously as the ratio of demand over supply changes; thus, there is little, if any, competition in an ecological vacuum while competition is keen in a fully saturated environment.

Lotka-Volterra Equations and Competition Theory

Competition was placed on a fairly firm, although somewhat oversimplified, theoretical basis nearly 50 years ago by Lotka (1925) and Volterra (1926, 1931). Their equations describing competition have strongly influenced the development of modern ecological theory and provide a nice illustration of a mathematical model of an important ecological phenomenon. They also help to develop a number of important concepts, such as competition coefficients, that are independent of the equations.

The Lotka-Volterra competition equations are a modification of the Verhulst-Pearl logistic equation (Chapter 4) and share its assumptions. Consider two competing species N_1 and N_2, with carrying capacities K_1 and K_2 in the absence of one another. Similarly, each species has its own maximal instantaneous rate of increase per head, r_1 and r_2. The simultaneous growth of the two competing species occurring together can be described by the following pair of differential logistic equations:

$$\frac{dN_1}{dt} = r_1 N_1 \left(\frac{K_1 - N_1 - \alpha N_2}{K_1} \right) \tag{1}$$

$$\frac{dN_2}{dt} = r_2N_2\left(\frac{K_2 - N_2 - \beta N_1}{K_2}\right) \quad (2)$$

where α and β are competition coefficients; α is a characteristic of species 2 which measures its competitive inhibition (per individual) on the species 1 population; and β is a similar characteristic of species 1 that measures its inhibitory effects on species 2. The reader should verify that, in the absence of any interspecific competition [α or N_2 equal zero in Equation (1); β or N_1 equal zero in Equation (2)], both populations grow sigmoidally according to the Verhulst-Pearl logistic equation and reach an equilibrium population size at their carrying capacity.

By definition, the inhibitory effect of each individual in the N_1 population on its own population's growth is $1/K_1$ (see also Chapter 4); similarly the inhibition of each N_2 individual on the N_2 population is $1/K_2$. Likewise, from inspection of (1) and (2), the inhibitory effect of each N_2 individual upon the N_1 population is α/K_1 and the inhibitory effect of each N_1 individual on the N_2 population is β/K_2. Competition coefficients are normally, though not always (see below), numbers less than 1. The outcome of competition depends upon the relative values of K_1, K_2, α, and β. There are four possible cases of competitive interaction corresponding to different combinations of values for these constants (below).

To see this, we ask at what density of N_1 individuals is the N_2 population held exactly at zero, and vice versa? In other words, what density of each species will always prevent the other from increasing? The equations can easily be solved for these values, by simply noting that at an N_2 of K_1/α, N_1 can never increase, and that when N_1 reaches K_2/β, N_2 can never increase.

Summing up the preceding two paragraphs, in the absence of the other species, populations of both species increase at any density below their own carrying capacity and decrease at any value above it. As developed above, in the presence of K_1/α individuals in the N_2 population, N_1 decreases at all densities; and, in the presence of K_2/β individuals in the N_1 population, N_2 decreases at every density.

Recall that, under the Verhulst-Pearl logistic equation, r_a decreases linearly with increasing N, reaching a value of zero at density K (see Figure 4.8). Exactly the same relationships hold for the Lotka-Volterra competition equations, except here a family of straight lines relate r_1 to K_1 and r_2 to K_2; each of these lines corresponds to a different population density of the competing species (Figure 5.1*a* and *b*).

The r axis is omitted in Figure 5.2 and N_1 is simply plotted against N_2. Points in this N_1–N_2 plane thus correspond to different proportions of

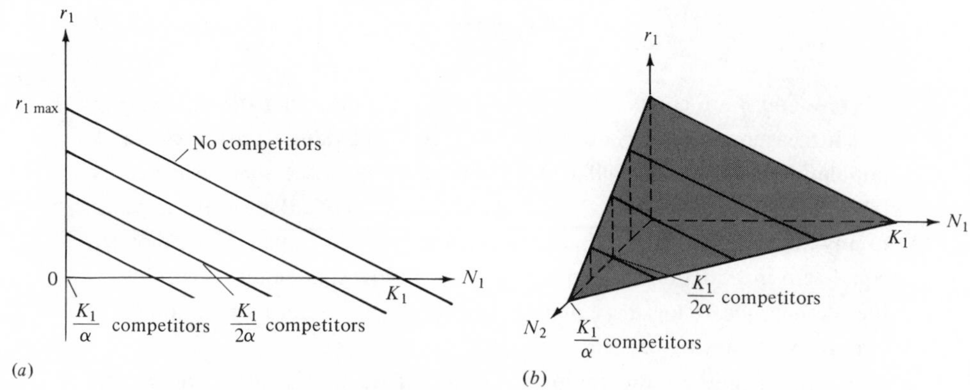

FIGURE 5.1 Two plots showing how the actual instantaneous rate of increase per individual (r_{a_1}) varies with the densities of a population (N_1) and its competitor (N_2) under the Lotka-Volterra competition equations. (*a*) Two dimensional graph with four lines, each of which represents a given density of competitors (compare with Figure 4.8). (*b*) Three-dimensional graph with an N_2 axis showing the plane on which each of the four lines in (*a*) lie. At densities of N_1 and N_2 above K_1 and K_1/α, respectively, the plane continues with r_{a_1} becoming negative [compare with (*a*)].

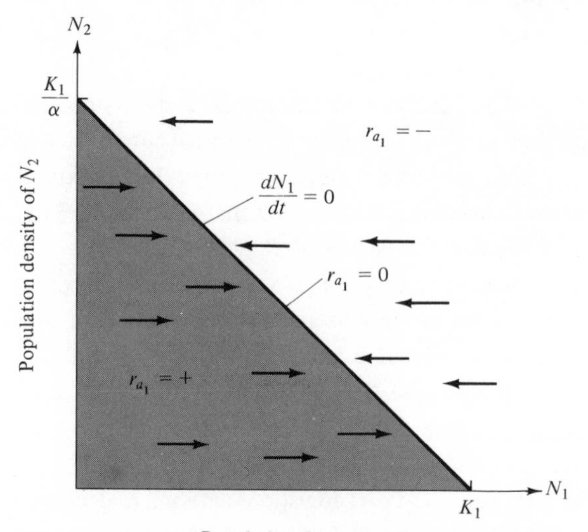

FIGURE 5.2 A plot identical to Figure 5.1(*b*), but only the N_1-N_2 plane is shown at r_{a_1} equal to zero. The line represents all equilibrium conditions ($dN_1/dt = 0$) at which the N_1 population just maintains itself ($r_{a_1} = 0$). In the shaded area below the line, r_{a_1} is positive and the N_1 population increases (arrows), whereas above the isocline r_{a_1} is always negative and the N_1 population decreases, as depicted by the arrows. An exactly equivalent plot could be drawn on the same axes for the competitor, N_2, except that the intercepts of the N_2 isocline ($dN_2/dt = 0$) are K_2 and K_2/β and arrows parallel the N_2 axis rather than the N_1 axis.

the two species and to different population densities of each. Setting dN_1/dt and dN_2/dt equal to zero in Equations (1) and (2), and solving, gives us equations for the boundary conditions between increase and decrease for each population:

$$\frac{K_1 - N_1 - \alpha N_2}{K_1} = 0 \quad or \quad N_1 = K_1 - \alpha N_2 \tag{3}$$

$$\frac{K_2 - N_2 - \beta N_1}{K_2} = 0 \quad or \quad N_2 = K_2 - \beta N_1 \tag{4}$$

These two equations are plotted in Figure 5.3, giving dN/dt isoclines for each species; below these isoclines populations of each species increase, above them they decrease. Thus the lines represent saturation values and neither species can increase when the combined densities of the two lie above its isocline.

The four cases of competitive interaction are summarized in Table 5.2 and are shown graphically in Figure 5.3. Only one combination of values (Case 4) leads to a stable equilibrium of the two species; this case arises when neither species is able to reach densities high enough to eliminate the other, that is, when both $K_1 < K_2/\beta$ and $K_2 < K_1/\alpha$. Implicit in these inequalities are the conditions necessary for coexistence: each population must inhibit its *own* growth more than that of the other species. For this to occur, if K_1 is equal to K_2, α and β must be numbers less than 1. When carrying capacities are not equal, α or β can take on a value greater than 1 and coexistence may still be possible. Note that population sizes at equilibrium

TABLE 5.2 Summary of the Four Possible Cases of Competition Under the Lotka-Volterra Competition Equations

	Species one can contain species two ($K_2/\alpha_{21} < K_1$)	Species one cannot contain species two ($K_2/\alpha_{21} > K_1$)
Species two can contain species one ($K_1/\alpha_{12} < K_2$)	Either species can win (Case 3)	Species two always wins (Case 2)
Species two cannot contain species one ($K_1/\alpha_{12} > K_2$)	Species one always wins (Case 1)	Stable coexistence (Case 4)

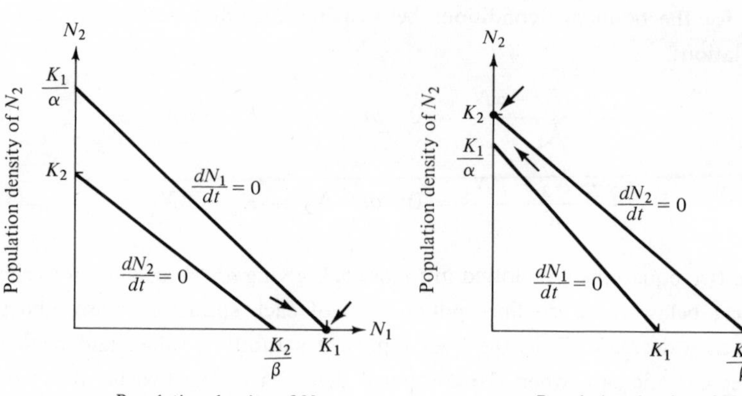

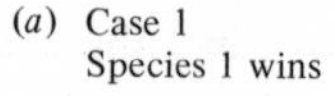
(*a*) Case 1
Species 1 wins

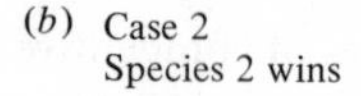
(*b*) Case 2
Species 2 wins

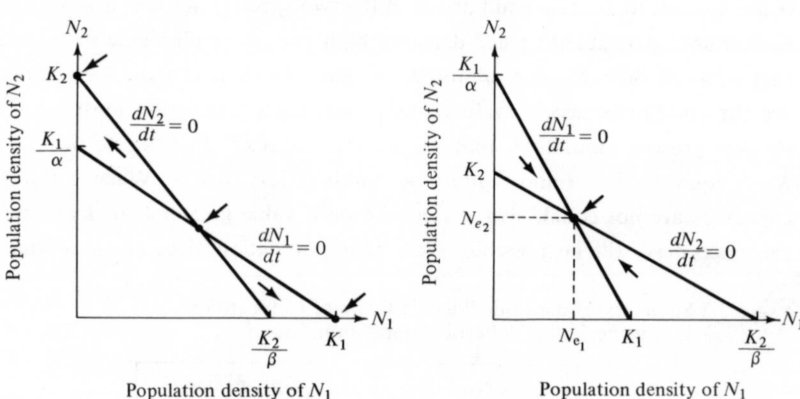

(*c*) Case 3
Unstable equilibrium

(*d*) Case 4
Stable equilibrium

FIGURE 5.3 Plots like Figure 5.2 but with the isoclines of both species superimposed upon one another. (*a*) Case 1: The N_1 isocline lies above the N_2 isocline and species 1 always wins in competition. The only stable equilibrium (dot) is at $N_1 = K_1$ and $N_2 = 0$. (*b*) Case 2: The reverse case in which species 2 is the superior competitor and excludes species 1. Here the only stable equilibrium (dot) is at $N_2 = K_2$ and $N_1 = 0$. (*c*) Case 3: Each species is able to contain the other (Table 5.2); that is, each inhibits the other population's growth more than its own. Three possible equilibria exist (dots), but the joint equilibrium of both species (where the two isoclines cross) is unstable. Alternate stable equilibria are $N_2 = K_2$ and $N_1 = 0$ or $N_1 = K_1$ and $N_2 = 0$. Depending upon initial proportions of the two species, either can win. (*d*) Case 4: Neither species can contain the other, but both inhibit their own population growth more than that of the other species. Only one equilibrium exists at N_{e_1} and N_{e_2}; both species thus *coexist* at densities *below* their respective carrying capacities.

(N_{e_1} and N_{e_2}) of the two species in Figure 5.3*d* are below their respective carrying capacities K_1 and K_2; thus, when in competition neither population reaches densities as high as it does without competition. None of the other three cases leads to stable coexistence of both populations and hence they are of less interest. However, in Case 3 an unstable equilibrium exists with each species inhibiting the *other's* growth rate more than its own; in this case the outcome of competition depends entirely upon the initial proportions of the two species.

It is often much more convenient to subscript competition coefficients to show at a glance which population is affected and which is having the effect; thus α is written as α_{12} and β as α_{21} (for α_{12} read "the inhibitory effect of one N_2 individual upon the growth of the N_1 population," and vice versa for α_{21}). Using this convention, the Lotka-Volterra equations can be written in a more general form than those given in (1) and (2), for a community composed of *n* different species:

$$\frac{dN_i}{dt} = r_i N_i \left\{ \frac{K_i - N_i - \left(\sum_{j \neq i}^{n} \alpha_{ij} N_j \right)}{K_i} \right\} - m_i N_i \tag{5}$$

where the *i*'s and *j's* subscript species, and m_i is the species-specific instantaneous rate of density-independent rarefaction. Notice that the more competitors a given species has, the larger the term $\left(\sum_{j \neq i}^{n} \alpha_{ij} N_j \right)$ becomes, and the farther that species' equilibrium population size is from its K value; this accords well with biological intuition.

Implicit in the Lotka-Volterra competition equations are a number of assumptions; some of them can be relaxed, although the mathematics rapidly become unmanageable. The maximal rates of increase, competition coefficients, and carrying capacities are all assumed to be constant and immutable; they do not vary with population densities. As a result, all inhibitory relationships within and between populations are strictly linear, and every N_1 individual is identical as is every N_2 individual. [Similar results can, however, be reached without assuming linearity by a purely graphical argument (Figure 5.4).] There is no lag in response to changes in density. In addition, the two species are assumed to be using the same "resource" because there is no way in which they can diverge; thus the environment is assumed to be completely homogeneous.

In real populations, rates of increase, competitive abilities, and carrying capacities *do* vary from individual to individual, with population density, and in space and time. Indeed, temporal variation in the environment may often allow coexistence by continually altering the competitive

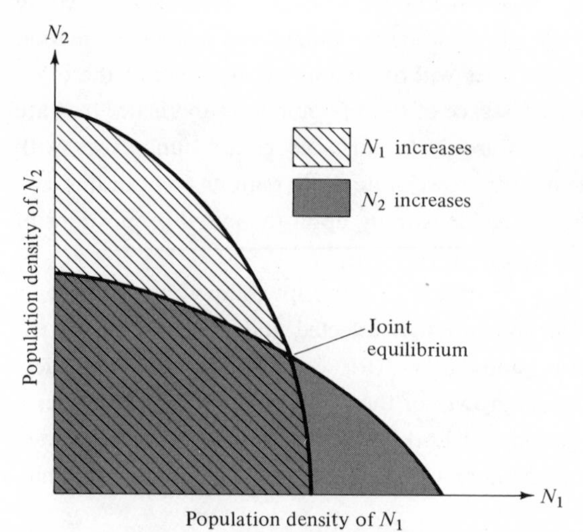

FIGURE 5.4 A pair of nonlinear isoclines which represent graphically the conditions for the stable coexistence of two competitors.

abilities of populations inhabiting it. Moreover, time lags are doubtlessly of some importance in real populations. Finally, a heterogeneous environment may allow real competitors to evolve divergent resource utilization patterns and to reduce interspecific competitive inhibition.

Still another shortcoming of the Lotka-Volterra competition equations is that, at equilibrium, they assume saturation. Real ecological systems (and portions thereof) may often be partially unsaturated while still at equilibrium, which in itself could allow coexistence of otherwise competitively intolerant populations. For instance, either predation or density-independent rarefaction might hold down population levels and thereby decrease competition. However, increasing rarefaction in the Lotka-Volterra equations does not reduce the intensity of competitive inhibition, except by reduction of population densities. Thus, a more realistic model might incorporate the ratio of instantaneous demand to supply; one such possibility might be to make competition coefficients variables and functions of the combined densities of both populations. At saturation values (K_1, K_2 or $N_{e_1}+N_{e_2}$ in Figure 5.3*d*), demand/supply is unity and the α's would take on maximal values, but as a perfect ecological vacuum is approached these same α's would become vanishingly small.

The Lotka-Volterra competition equations play an extremely prominent role in modern ecological theory (Levins, 1968; MacArthur, 1968, 1972; Vandermeer, 1970, 1972); however, their numerous biologically unrealistic assumptions are often stressed as indicating the inadequacy of competition theory. Although these equations have perhaps been overworked and overextended by enthusiastic theorists, the equations have contributed substantially to the development of many important ecological concepts, such as the actual rate of increase, r and K selection, intraspecific and interspecific competitive abilities, and the community matrix (Chapter 7), which in fact are *independent* of the equations. Thus, these equations help to provide an important conceptual framework.

Competitive Exclusion

How does one population drive another to extinction? In Cases 1, 2, and 3 of Figure 5.3, one species ultimately eliminates the other entirely when the two come into competition and the system is allowed to go to saturation; we say that competitive exclusion has occurred. Consider an ecologic vacuum inoculated with small numbers of each species. At first, both populations grow nearly exponentially at rates determined by their respective instantaneous maximal rates of increase. As the ecological vacuum is filled, the actual rates of increase become progressively smaller and smaller. Both populations are infinitely unlikely to have exactly the same rates of increase, competitive abilities, and carrying capacities. Hence, as the ecological vacuum is filled, a time comes when one population's actual rate of increase drops to zero while the other's rate of increase is still positive. This situation represents a turning point in competition, for the second population now increases still further and its competitive inhibition of the first is intensified, reducing the actual rate of increase of the first population to a negative value. The first population is now declining while the second is still increasing; barring changes in competitive parameters, competitive exclusion, or extinction of the first population, is only a matter of time. This process has been demonstrated experimentally (Figure 5.5).

Some rather strong statements concerning competitive exclusion have been made. Among them is a hypothesis, the *competitive exclusion principle* often stated as follows: two species with identical ecologies cannot live together in the same place at the same time. Ultimately, one must edge out the other; complete ecological overlap is impossible. The corollary is that, if two species coexist, there must be ecological differences between them. Since any two organismic units are infinitely unlikely to be *exactly* identical, mere observation of ecological differences between species does not constitute "verification" of the hypothesis. Untestable hypotheses like this one are of little scientific utility, and are generally gradually forgotten by the scientific community.

The competitive exclusion "principle" has, however, served a useful

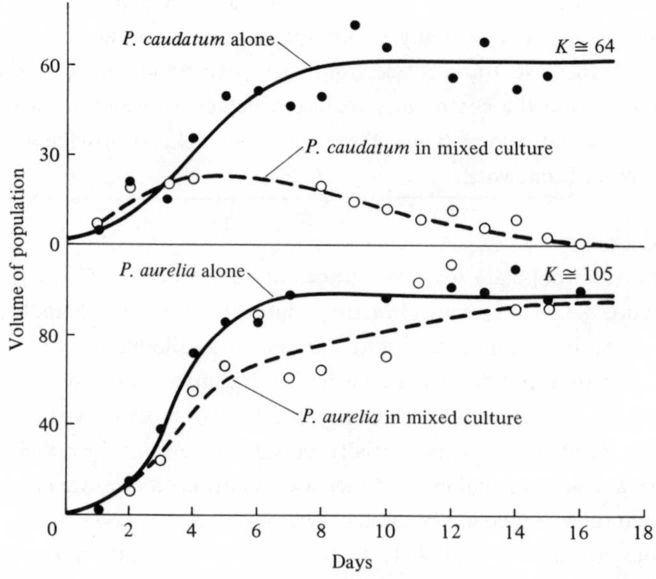

FIGURE 5.5 Competitive exclusion in a laboratory experiment with two protozoans, *Paramecium caudatum* and *P. aurelia*. [From Gause (1934). *The Struggle for Existence.* Hafner, New York.]

purpose in that it emphasizes that some ecological difference may be necessary for coexistence in *saturated* environments. Ecologists are now asking more sophisticated questions, such as "how much ecological overlap can two species tolerate and still coexist?" and "how does this maximal tolerable overlap vary as the ratio of demand to supply changes?"

The Balance Between Intraspecific and Interspecific Competition

Intraspecific and interspecific competition probably often have opposite effects on a population's tolerance, as depicted in Figure 1.5, as well as on its use of resources and its phenotypic variability. To see this, first consider an idealized case with no interspecific competition, and look at the effects of intrapopulational competition on the population's use of a resource or habitat. We will consider the former a resource continuum and the latter a habitat gradient, although an exactly parallel argument can easily be developed for discrete resources and habitats. Individuals should spread themselves out more or less evenly along such a continuum or gradient for the following reason. If all resources along the continuum are not being utilized to approximately the same degree, those individuals using relatively unused portions should encounter less intense intraspecific competition and, as a result, presumably will often have higher individual fitnesses. Hence, we would predict that, in using such a continuous resource in its entirety, individuals should behave so as to equalize the ratio of demand to supply along the continuum, which in turn equalizes the level of intraspecific competition. [MacArthur (1972) calls this the "principle of equal opportunity."]

Using the same argument, consider next the manner in which an *expanding* population makes use of a resource continuum or a habitat gradient (Figure 5.6). Again, the argument applies equally well to resource categories that are not continuous, but discrete. The first individuals will no doubt select those resources and/or habitats that are optimal in the absence of competition. However, as the density of individuals increases, competition among them reduces the benefits to be gained from these optimal resources and/or habitats, and favors deviant individuals that use less "optimal," but also less hotly contested, resources and/or habitats. By these means, intraspecific competition can often act to *increase* the variety of resources and habitats utilized by a population.

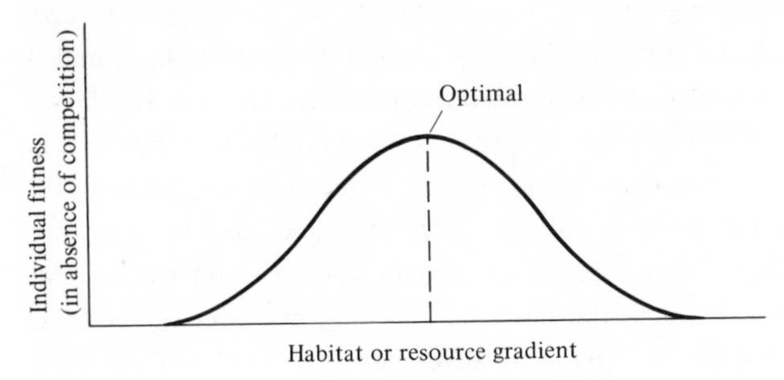

(*a*)

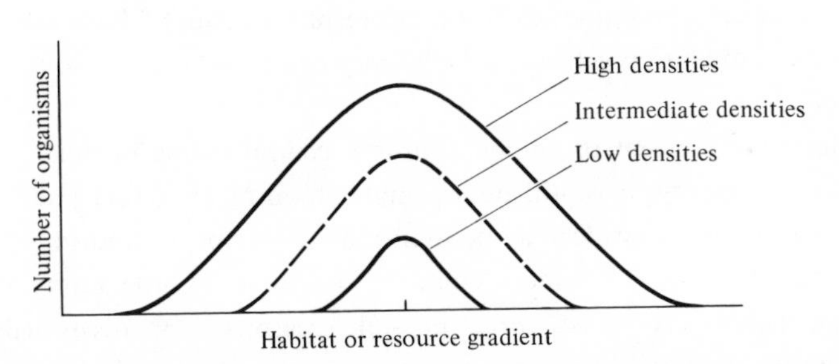

(*b*)

FIGURE 5.6 Diagrammatic representation of how an expanding population might use a gradient of habitats or resources. (*a*) Individual fitness as a function of the habitat or resource gradient in the absence of any intraspecific competition. (*b*) At low densities, most individuals select near optimal environmental conditions, but as density increases, intraspecific competition for more optimal habitats (or resources) increases, favoring individuals that exploit less optimal, but also less hotly contested, habitats or resources. The variety of habitats or resources exploited thus *increases* with increasing population density.

Interspecific competition, on the other hand, generally tends to *restrict* the range of habitats and resources a population uses, because different species normally have differing abilities at exploiting habitat types and harvesting resources. Individuals using marginal habitats presumably cannot compete as effectively against members of another population as can individuals that are exploiting more "optimal" habitats. Thus, in most communities, any given population is "boxed in" by other populations that are superior at exploiting adjacent habitats (see also Chapter 6). Indeed, since these two forces oppose each other, one could even hypothesize that, at equilibrium, the sum total of intraspecific competition should be exactly balanced by the sum total of all interspecific competition. (Actually, this is not quite true because inherent genetic and physiological limitations must also restrict the range of habitats and resources used by an organism.)

Evolutionary Consequences of Competition

Many of the long-term consequences of competition have been mentioned in Chapters 1 and 4; others are treated in Chapters 6, 7, and 8. For instance, natural selection in saturated environments (*K* selection) favors competitive ability. A raft of presumed populational products of intraspecific competition over evolutionary time include: rectangular survivorship, delayed reproduction, decreased clutch size, increased size of offspring, parental care, mating systems, dispersed spacing systems, and territoriality (see also Chapter 4). Perhaps the most far reaching evolutionary effect of interspecific competition is ecological diversification, also termed niche separation. This in turn has made possible, and has led to, the development of complex biological communities. Another presumed result of both intra- and interspecific competition is increased efficiency of utilization of resources in short supply.

Although the concept of competition is a central theme in much of modern ecological theory, it is still poorly understood as an actual phenomenon. Some ecologists consider competition among the most important of ecological generalizations, yet others maintain that it is of little utility in understanding nature. At least three possible reasons, not necessarily mutually exclusive, for this difference in viewpoint have been suggested: (1) competition in nature is often elusive and very difficult to study and to quantify, (2) ecologists with an evolutionary approach might often consider competition more important than workers concerned with explaining more immediate events (Orians, 1962),* and (3) there may be a natural dichotomy

* See Chapter 1, pp. 17–18, for a discussion of the difference between the "proximate" and the "ultimate" approaches to biological phenomena.

of relatively r selected and relatively K selected organisms, especially in terrestrial communities (Pianka, 1970).

Laboratory Experiments

Competition is often fairly easily studied by direct experiment and many such studies have been made. Gause (1934) was one of the first to investigate competition in the laboratory; his classic early experiments on protozoa verified competitive exclusion. He grew cultures of two species of *Paramecium* in isolation and in mixed cultures, under carefully controlled conditions and nearly constant food supply. "Carrying capacities" and respective rates of population growth of each species when grown separately and in competition were then calculated (Figure 5.5). Interestingly enough, the protozoan with the highest maximal instantaneous rate of increase per individual (*P. caudatum*) was the inferior competitor, as expected from considerations of r and K selection (Chapter 4).

An effect of environment on the outcome of competition was demonstrated by Park (1948, 1954, 1962) and his many colleagues. Working with two species of flour beetles (*Tribolium*), these investigators showed that, depending upon conditions of temperature and humidity, either species could eliminate the other (Table 5.3). In early experiments, they could not always predict the outcome of competition under particular environmental conditions; this set of environments was termed the "indeterminate zone." More recently, however, this zone has been substantially reduced by taking into account the genotypes of the beetles (Lerner and Ho, 1961; Park, Leslie, and Mertz, 1964). Under some environmental conditions, cultures begun with a numerical preponderance of a given species always resulted in the

TABLE 5.3 The Outcome of Competition Between Two Species of Flour Beetles, *Tribolium confusum* and *T. castaneum*, in Many Replicates of Laboratory Experiments at Different Temperatures and Humidities

Temperature (°C)	Relative Humidity (%)	Climate	Single Species Numbers	Mixed Species (% wins)	
				confusum	*castaneum*
34	70	Hot–moist	*confusum* = *castaneum*	0	100
34	30	Hot–dry	*confusum* > *castaneum*	90	10
29	70	Temperate–moist	*confusum* < *castaneum*	14	86
29	30	Temperate–dry	*confusum* > *castaneum*	87	13
24	70	Cold–moist	*confusum* < *castaneum*	71	29
24	30	Cold–dry	*confusum* > *castaneum*	100	0

Source: From Krebs (1972) after Park.

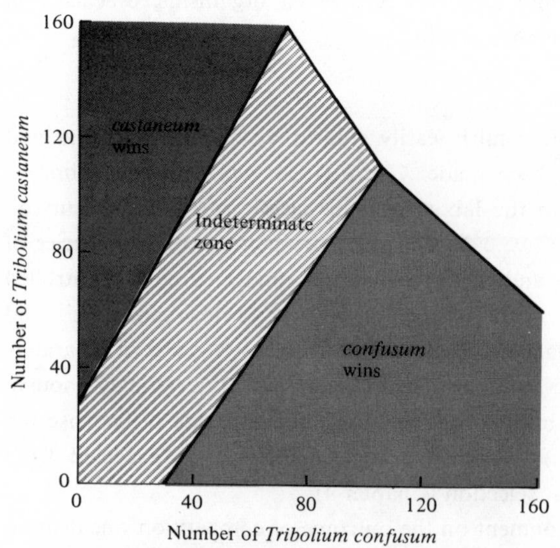

FIGURE 5.7 The outcome of competition between laboratory strains of two beetles, *Tribolium confusum* and *T. castaneum*, is a function of the initial densities of the species; an initial preponderance of each species increases the likelihood of its winning. The "indeterminate zone" represents initial conditions under which either species can win. Note the similarity with Case 3 of Figure 5.3*c*. [From Krebs (1972) after Neyman, Park, and Scott. Originally published by the University of California Press; reprinted by permission of The Regents of the University of California.]

extermination of the other species (Figure 5.7), thus constituting an empirical verification of Case 3 of the Lotka-Volterra equations (Figure 5.3*c*).

Competition and competitive exclusion have now been demonstrated in laboratory experiments on a wide variety of plants and animals. One potential flaw in these investigations is that, for practical reasons, they are invariably carried out on small, often relatively *r*-selected, organisms, which may not encounter high levels of competition regularly under natural circumstances.

Evidence from Nature

Competition is notoriously difficult to demonstrate in natural communities, but a variety of observations and studies suggest that it does indeed occur regularly in nature and that it has been an important force in molding the ecologies of many species of plants and animals. Even if competition did not occur on a day to day basis, it could nevertheless still be a significant force; active avoidance of interspecific competition in itself implies that competition has occurred sometime in the past and that the species concerned have adapted to one another's presence. Also, it might be difficult to find competition actually occurring in nature because inefficient competitors should be eliminated by competitive exclusion and therefore might not normally be observable. We might not expect to find abundant evidence of competition in small, short-lived organisms, such as insects and annual plants, but would look for it in larger, longer-lived organisms such as vertebrates and perennial plants.

Ecologists have several different sorts of evidence, much of which is circumstantial, suggesting that competition either has occurred or is occurring in natural populations. Among these are: (1) studies on the ecologies of closely related species living in the same area; (2) character displacement; (3) studies on "incomplete" floras and faunas and associated changes in niches, or "niche shifts"; and (4) taxonomic composition of communities. Each of these is taken up in greater detail below, with examples.

Closely related species, especially those in the same genus, or "congeneric" species, are often quite similar morphologically, physiologically, behaviorally, and ecologically. As a result, competition is intense between pairs of such species that live in the same area, known as *sympatric* congeners; and selection may be strong to render their ecologies more different, or to lead to ecological separation. Many groups of closely related sympatric species have been studied; almost without exception, detailed investigations on relatively *K*-selected organisms have revealed subtle, but important, ecological differences between such species. Usually, the differences are of one or more of the following types: the species exploit different habitats or microhabitats (differential spatial utilization of the environment), or they are active at different times (differential patterns of temporal activity), or they eat different foods. These parameters are known as "niche dimensions" because they are important in defining a species' role in its community and its interactions with other species (see Chapter 6).

Many examples of clear-cut habitat and microhabitat differences could be cited; two are briefly described here. MacArthur (1958) studied spatial utilization patterns in five species of warblers (genus *Dendroica*), by noting the time spent in precise locations by foraging individuals of each species. Each species has its own unique pattern of exploiting the forest (Figure 5.8).

Competition for space between two species of barnacles, *Balanus balanoides* and *Chthamalus stellatus*, was studied by Connell (1961a, 1961b) along a rocky Scottish seacoast. These two crustaceans occupy sharply defined horizontal zones, as do most sessile organisms in the marine intertidal, with *Chthamalus* occupying an upper, and *Balanus*, a lower zone (Figure 5.9). Larvae of both species settle and attach over a wider vertical range than the zone occupied by adults. By manipulation of wire cages to exclude a snail

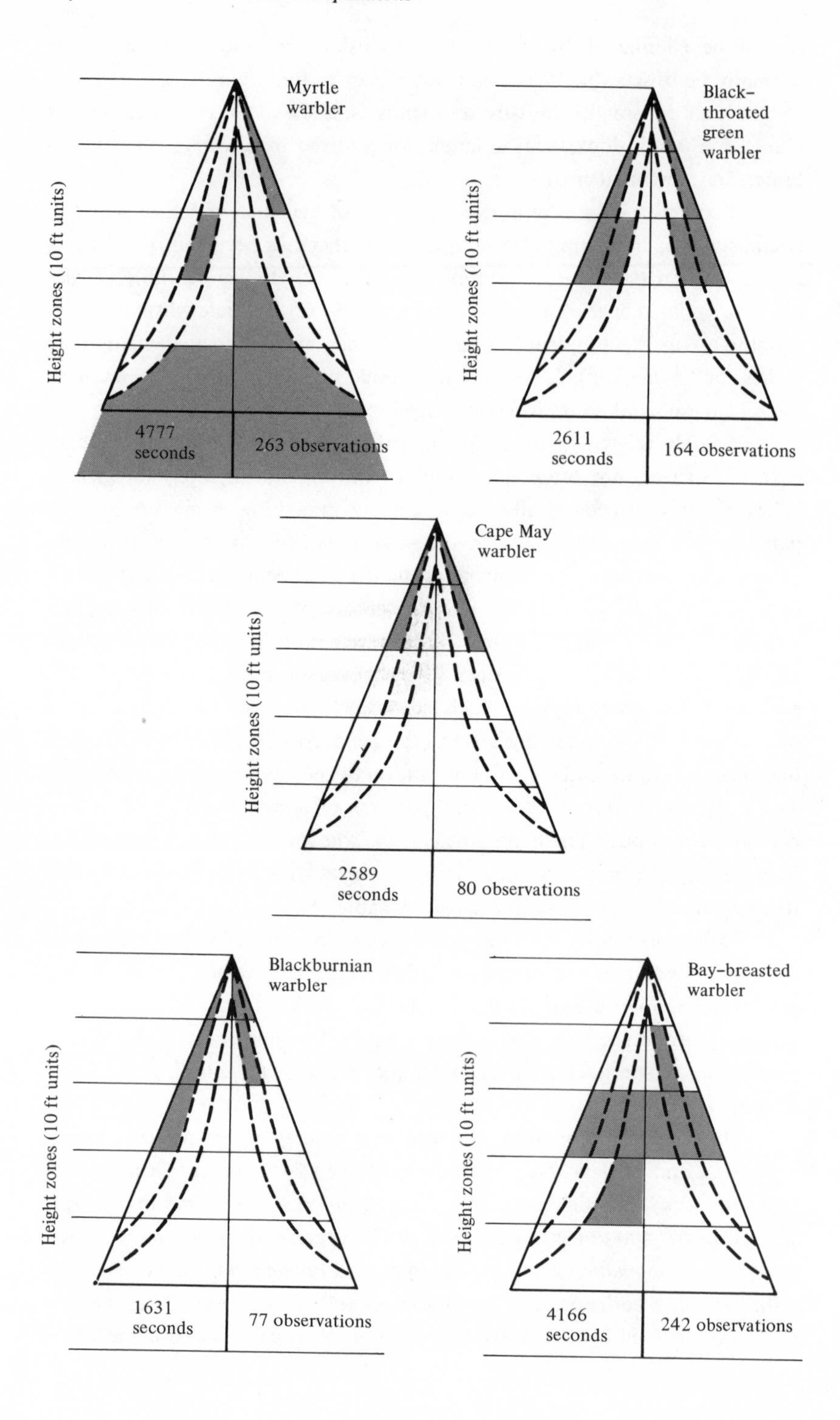
Myrtle warbler
Height zones (10 ft units)
4777 seconds
263 observations
Black-throated green warbler
Height zones (10 ft units)
2611 seconds
164 observations
Cape May warbler
Height zones (10 ft units)
2589 seconds
80 observations
Blackburnian warbler
Height zones (10 ft units)
1631 seconds
77 observations
Bay-breasted warbler
Height zones (10 ft units)
4166 seconds
242 observations

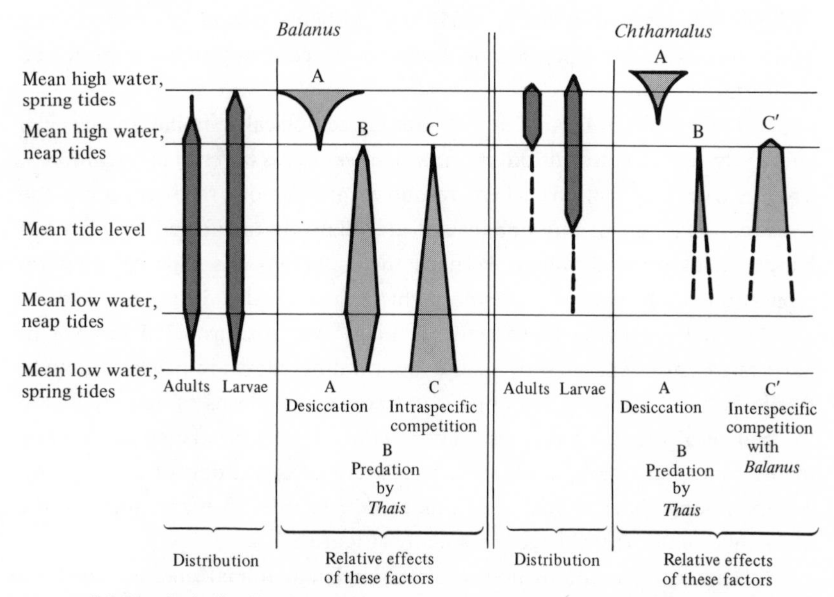

FIGURE 5.9 Vertical distributions of the larvae and adults of two barnacle species, *Balanus balanoides* and *Chthamalus stellatus*, in the rocky intertidal of Scotland. The relative intensities of various limiting factors are represented diagrammatically by the width of the stippled bars. [After Connell (1961b). By permission of Duke University Press.]

predator (*Thais*) and by periodic removal of barnacles, Connell elucidated the various forces causing the zonation. He demonstrated that, although *Chthamalus* loses in competition with *Balanus*, adults persist in a narrow band because *Chthamalus* are more tolerant of physical desiccation than *Balanus*; Connell suggests that the lower limit of distribution of intertidal organisms is usually determined mainly by biotic factors, such as competition with other species and predation, whereas the upper limit is more often set by physical factors, such as dry conditions prevailing during low tides.

Overdispersion in general, and territoriality in particular (Chapter 4), are indicative of competition in that they reduce its intensity. Overdispersion is widespread in plants, as is territoriality in vertebrate populations (Chapter 4). Indeed, *inter*specific territoriality has been documented in many species

---

FIGURE 5.8 Differential use of parts of trees in a coniferous forest by five sympatric species of congeneric warblers (genus *Dendroica*). Stippling denotes the parts of trees in which foraging activities of each species are most concentrated. The right side of each schematic tree represents use based on the total number of birds observed (sample size given below each "tree"), whereas the left side is based on the total number of seconds of observation (again, sample size in seconds is given at the bottom of each "tree"). [After MacArthur (1958). By permission of Duke University Press.]

of birds (Orians and Willson, 1964) and probably occurs in other taxa as well; thus both intraspecific and interspecific competition have produced territorial behavior.

Differences in time of activity among ecologically similar animals can effectively reduce competition, *provided that resources differ at different times.* This is true in situations where resources are rapidly renewed, since the resources available at any one instant are relatively unaffected by what has happened at previous times. Perhaps the most obvious type of temporal separation is that between day and night; animals active during the daytime are "diurnal," whereas those active at night are "nocturnal." Examples of pairs apparently separated by such temporal differences are hawks and owls, swallows and bats, or grasshoppers and crickets. Patterns of activity within the course of the day alone also differ, with some species being active early in the morning, others at midday, etc. Seasonal separation of activity also occurs among some animals, such as certain lizards. In many animals, the daily time of activity changes seasonally (Figure 5.10).

Dietary separation among closely related animal species has been shown repeatedly. For example, Table 5.4 shows that several sympatric congeneric species of the marine snail genus *Conus* (commonly called "cone shells"), eat distinctly different foods (Kohn, 1959). Among desert lizards, the diets of several sympatric species are composed predominantly of ants, termites, other lizards, and plants, respectively (Pianka, 1966b). Similar cases of food differences among related sympatric species are known in many birds and mammals.

Simultaneous differences in the use of space, time, and food have also been documented for some sympatric congeneric species. In lizards of the genus *Ctenotus* (Pianka, 1969), for instance, seven sympatric species forage at different times, in different microhabitats, and/or on different foods.

TABLE 5.4 Major Foods (Percentages) of Eight Species of Cone Shells, *Conus*, on Subtidal Reefs in Hawaii

Species	Gastropods	Enteropneusts	Nereids	Eunicea	Terebellids	Other polychaetes
flavidus		4			64	32
lividus		61		12	14	13
pennaceus	100					
abbreviatus				100		
ebraeus			15	82		3
sponsalis			46	50		4
rattus			23	77		
imperialis				27		73

Source: From data of Kohn (1959). By permission of Duke University Press.

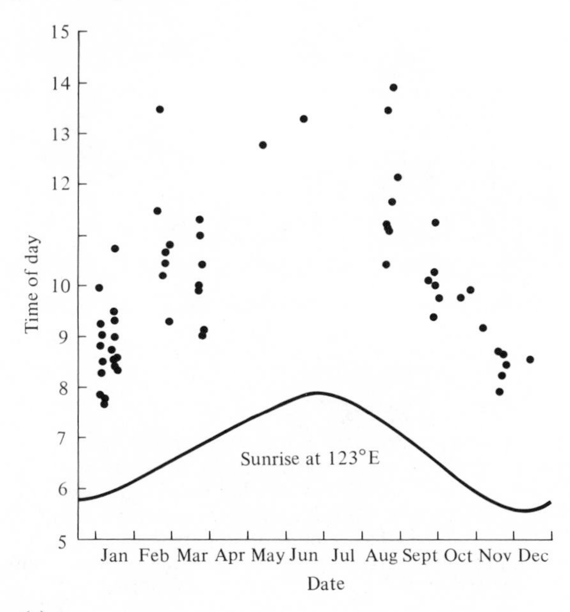

(*a*)

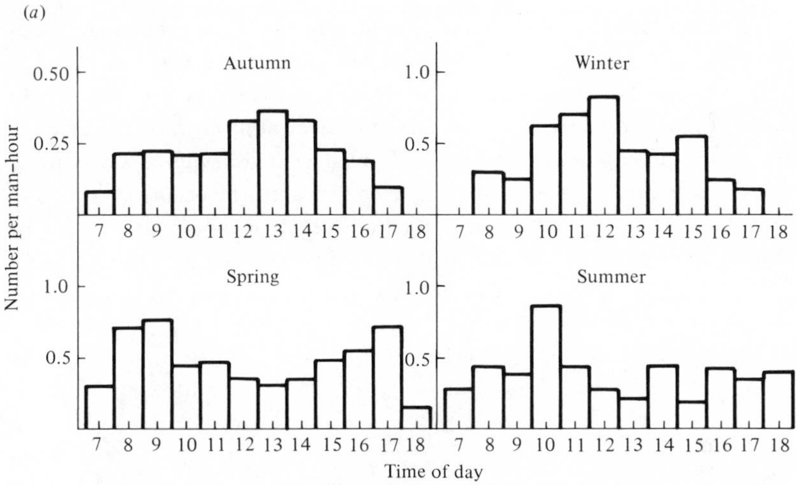

(*b*) FIGURE 5.10 Seasonal changes in daily activity patterns in two species of Australian desert lizards. (*a*) A small blue-tailed skink, *Ctenotus calurus*. Each dot represents one lizard. [From Pianka (1969). By permission of Duke University Press.] (*b*) The "military dragon," an agamid, *Amphibolurus isolepis*. Numbers observed per man-hour of observation at different times of day in each season. [After Pianka (1971b).]

Frequently, in such cases, pairs of species with high overlap in one niche dimension have low overlap along another (Pianka, 1973), presumably reducing competition between them.

The phenomenon of "character displacement," which refers to increased differences between species where they occur together, is also evidence

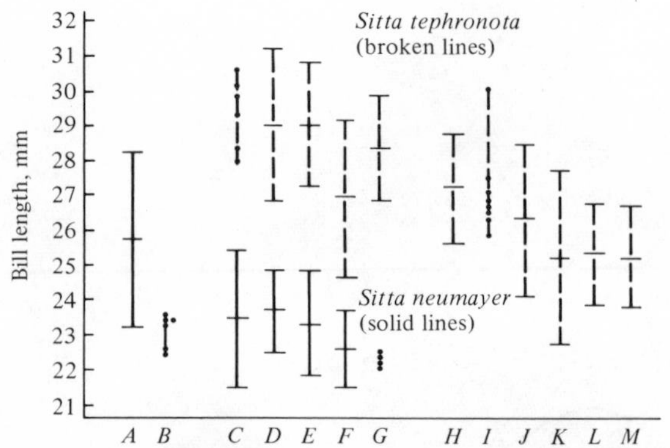

FIGURE 5.11 Bill lengths of 2 congeneric species of nuthatches, *Sitta tephronota* and *S. neumayer*, at 13 localities in Asia (arranged from east to west). In far allopatry (area *A* versus areas *L* and *M*), bill lengths are nearly the same size, whereas in areas where the 2 species are sympatric (areas *C*, *D*, *E*, *F*, and *G*), their bill lengths are nonoverlapping. Such "character displacement" in phenotypes presumably reduces interspecific competition. [After MacArthur and Connell (1966) after Vaurie.]

that competition occurs in nature. Frequently two widely ranging species are ecologically more similar in the parts of their ranges where each occurs alone without its competitor (i.e., in *allopatry*) than they are where both occur together (in *sympatry*). This sort of ecological divergence can take the form of morphological, behavioral, and/or physiological differences. One of the most common ways in which character displacement occurs is in the size of the food-gathering or "trophic" apparatus, such as mouthparts, beaks, or jaws (Figure 5.11). Prey size is usually strongly correlated with the size of an animal's beak or jaw as well as with its structure (Figure 5.12). Character

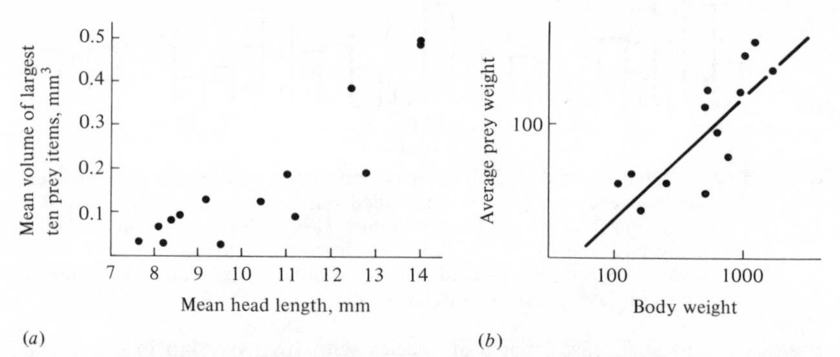

FIGURE 5.12 Two plots of prey size versus predator size. (*a*) Plot of the average volume of the 10 largest prey items (in cubic millimeters) against mean head length in 14 species of lizards (genus *Ctenotus*). [After Pianka (1969). By Permission of Duke University Press.] (*b*) Average prey weight plotted against mean body weight of 13 species of hawks (log-log plot). [After Schoener (1968). By Permission of Duke University Press.]

displacement in the size of the trophic apparatus has been demonstrated in many birds and mammals, and in some insects and lizards; it serves to separate food niches. Such niche shifts in the presence of a potential competitor strongly suggest that each population has adapted to the other by evolving a means to reduce interspecific competition.

Morphological character displacement in the size of mouthparts need not evolve if the populations concerned have diverged in other ways, and hence it is expected only in situations where both competitors occur side by side, exploiting identical microhabitats (i.e., true *syntopy*). Animals that forage in different microhabitats, such as the warblers of Figure 5.8, have adapted to one another primarily by means of behavioral, rather than morphological, character displacement.

It has been suggested that there is a definite limit on how similar two competitors can be and still avoid competitive exclusion; character displacement in average mouthpart sizes is typically about 1 to 1.3, and the ratio of 1.3 may thus be a crude estimate of just how different two species must be in order to coexist syntopically (Hutchinson, 1959; Schoener, 1965). Clearly the preceding argument applies only to saturated communities, and ecological overlap, or a greater similarity between syntopic species, can presumably be greater in unsaturated habitats.

The third type of evidence for competition comes from studies on so-called "incomplete" biotas, such as islands, where all the usual species are not present (see also Chapter 8). Those species which do invade such areas often expand their niches and exploit new habitats and resources that are normally exploited by other species on areas with more complete faunas. On the island of Bermuda, for example, there are considerably fewer species of land birds than there are on the mainland, with the three most abundant being the cardinal, catbird, and white-eyed vireo. Crowell (1962) found that, compared with the mainland, these three species are much more abundant on Bermuda and that they occur in a wider range of habitats and microhabitats. In addition, all three have somewhat different feeding habits on the island, and one species at least (the vireo) employs a greater variety of foraging techniques.

Mountain tops represent islands on the terrestrial landscape just as surely as Bermuda is an island in the ocean and often show similar phenomena. For example, two congeneric species of salamanders, *Plethodon jordani* and *P. glutinosus* occur in sympatry on mountains in the eastern United States. In sympatry, the two species are altitudinally separated, with *glutinosus* occurring at lower elevations than *jordani*; vertical overlap between the two species never exceeded 70 meters (Hairston, 1951). On mountain tops where *jordani* occurs, *glutinosus* is restricted to lower elevations, whereas on adjacent mountains that lack *jordani, glutinosus* is found at higher elevations, often right up to the peak.

Niche expansion under reduced interspecific competition has been termed "ecological release." Further evidence of competition stems from a corollary of ecological release; when mainland forms are introduced onto islands, native species are frequently driven to extinction, presumably via competitive exclusion. Thus many birds that once occurred only on Hawaii (called *endemic* Hawaiian species) became extinct shortly after the introduction of mainland birds, such as the English sparrow and the starling. Similar extinctions have apparently occurred in the Australian marsupial fauna (e.g., the Tasmanian wolf) with the introduction of placental mammal species (e.g., the dingo dog and the fox). Of course, fossil history is replete with cases of natural invasions and subsequent extinctions. The simplest and most plausible explanation for many of these observations is that the surviving species were superior competitors and that niche overlap was too great for coexistence. Before natural selection could produce character displacement and niche separation, one species had become extinct. Elton (1958) discusses many other examples of ecological invasions among both plants and animals.

A final observation, involving the taxonomic composition of communities, can be used to assess whether or not competition has been an important force in nature. Because closely related species should be strong competitors, one might predict that fewer pairs of congeneric species will occur within any given natural community than would be found in a completely random sample from the various species and genera occurring over a broader geographic area. Such a paucity of congeners, if observed, would suggest that competitive exclusion occurs more often among congeneric species than it does in more distantly related ones. This test was applied to many communities by Elton (1946), who was well aware of the problem of defining a "community." Frequently, two abutting communities each supports its own member of a congeneric pair, and such pairs must be excluded wherever possible. Despite this potential bias toward an increased proportion of congeneric species pairs, Elton found fewer congeners than expected on a strictly random basis. Elton's analysis has since been shown to be incorrect by Williams (1964), who gave a corrected statistical approach to the problem. Using this correct technique, Terborgh and Weske (1969) calculated the expected number of congeneric species pairs of Peruvian birds in seven habitats (Figure 5.13). They found that the four habitats richest in total number of species had greater than the expected number of congeneric pairs, thus refuting any increased incidence of competitive exclusion among

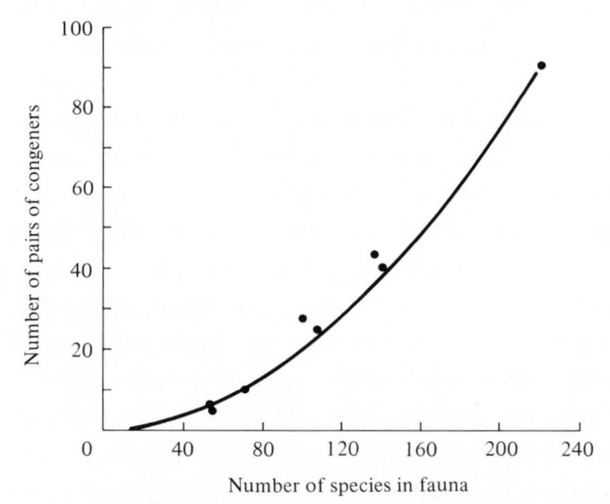

FIGURE 5.13 Observed (points) and expected (curve) numbers of pairs of congeneric bird species in seven Peruvian habitats. The combined avifauna of all seven areas contained 92 different pairs of congeners among a total of 221 species (plotted in the uppermost right-hand corner). Random samples of this total fauna would contain the numbers of congeneric pairs indicated by the curve. If competitive exclusion occurred more frequently among congeners, the observed number of such pairs (points) would fall below the expected curve. [From Terborgh and Weske (1969). By permission of Duke University Press.]

congeners in this particular avifauna. Pianka (1973) also failed to find any consistent impoverishment in the number of congeneric species pairs in a number of lizard communities. More analyses of this sort on a broad range of taxa in different communities are desirable.

Predation

Predation is readily observed and easily studied, and neither its existence nor its importance in nature are doubted. It is directional in the sense that one member of the pair (the predator) benefits from the association, while the other (the prey) is affected adversely. This is in contrast to competition, which is a symmetric process in that both species are affected adversely and, where possible, each tends to evolve mechanisms whereby the relationship with the other is avoided.

Individual predators that are better able to capture prey should have more resources at their disposal and should therefore normally be more fit than those which are less proficient at capturing prey. Hence, natural selection acting on the predator population tends to increase the predator's efficiency at finding, capturing, and eating its prey. However, members of the prey population that are better able to escape predators should normally be at a selective advantage within the prey population. Thus, selection on the prey population favors new adaptations that allow prey individuals to avoid being found, caught, and eaten. Obviously, these two selective forces oppose one another; as the prey become more adept at escaping from their predators, the predators in turn evolve more efficient mechanisms of capturing them. Hence, in the evolution of a prey–predator relationship, the prey evolves so as to dissociate itself from the interaction, while the predator continually maintains the relationship. Long-term evolutionary escalations of this sort have resulted in some rather intricate and often exceedingly complex adaptations. Consider, for instance, the complex social hunting behavior of lions and wolves, the long sticky tongues and accurate aim of some fish, toads, and certain lizards, the folding fangs and venom-injection apparatus of viperine snakes, spiders and their webs, the deep-sea angler fish, or snakes such as boas that suffocate their prey by constriction. Other examples would include the rapid and very accurate strikes of predators as diverse as praying mantids, dragonflies, fish, lizards, snakes, mammals, and birds. Prey have equally elaborate predator escape mechanisms, such as the posting of sentinels, predator alarm calls, background color matching, and thorns (see also section on predator escape devices, pp. 169–174). Many prey organisms recognize their predators at some distance and employ appropriate avoidance tactics well before the predator gets close enough to make a kill; this behavior, in turn, has forced many predators to hunt by ambush.

One of my favorite examples of the joint adaptations of a predator and its prey is provided by the starling and the peregrine falcon. The peregrine is a magnificent bird hawk whose hunting behavior must be seen to be fully appreciated. These falcons take other birds as large as themselves; nearly all prey is captured on the wing and in the air. Peregrines have exceedingly keen vision; foraging individuals climb high up into the sky and move across country. When a potential prey is sighted flying along below, the peregrine closes its wings and dives or "stoops." In order to make its ambush most effective, the falcon often "comes out of the sun" at its prey. Diving peregrines have been estimated to reach speeds of over 300 kilometers per hour (nearly 100 meters a second!). Most prey are killed instantly by the sudden jolt of the peregrine's talons. Large prey are allowed to fall to the ground and eaten there, but smaller items may be carried away in the air. (Little wonder that falconers and their dogs find it extremely difficult to get small birds to fly when a peregrine is "waiting on" overhead! On occasion, game birds even allow themselves to be overtaken on the ground by dogs in preference to taking to the air and risking the falcon's deadly stoop.) Starlings normally fly in loose flocks, but when a peregrine is sighted, often at some considerable distance, they quickly assume a very tight formation. This tight flocking is a response specific to the peregrine, and it is not employed with other hawks. Falcons are much less likely to attack a tight flock than a single bird; indeed, "stragglers" slightly out of the starling flock formation are often taken by peregrines. Presumably the falcon itself could be injured if it were to stoop into a tight flock. Thus, even a predator as effective as the peregine falcon has had its hunting efficiency impaired by appropriate behavioral responses of its prey.

There is an important difference between predation on animals and predation on plants. In most animals, predation is an all-or-none proposition in that the predator kills the prey outright and consumes most or all of it; however, when plants are eaten, usually only a portion of the plant is consumed by its predator. Hence predation on plants (herbivory) is more like parasitism among animals. But even partial predation must often reduce a prey individual's ability to survive and/or to reproduce. Because of this fundamental difference, though, selective pressures on animals to avoid being eaten may be stronger than they are on plants. Plants have evolved elaborate antipredator devices, some of which are taken up later (pp. 169–175).

Theories: Predator–Prey Oscillations

The theory of predation has, in many ways, lagged behind that for competition; perhaps its asymmetry makes it more difficult to model. Lotka (1925) and Volterra (1926, 1931) wrote the following simple pair of predation equations:

$$\frac{dN_1}{dt} = r_1 N_1 - p_1 N_1 N_2 \tag{6}$$

$$\frac{dN_2}{dt} = p_2 N_1 N_2 - d_2 N_2 \tag{7}$$

where N_1 is the prey population density, N_2 is the population density of the predator, r_1 is the instantaneous rate of increase of the prey population (per head), d_2 is the death rate of the predator population (per head), and p_1 and p_2 are predation constants. Each population is limited by the other and there are no self-limiting density effects (that is, no second order N_1^2 or N_2^2 terms). Thus, in the absence of the predator, the prey population expands exponentially, and the rate of increase of the prey population is potentially unlimited. The product of the densities of the two species, $N_1 N_2$, reflects the number of contacts between them; after multiplication by the constant, p_2, this term becomes the maximal rate of *increase* of the predator population ($p_2N_1N_2$). The same term multiplied by the constant, p_1, appears with a negative sign in the prey equation, and acts to *decrease* the rate of growth of the prey population. This pair of differential equations has a periodic solution, with the population densities of both prey and predator changing cyclically and out of phase over time (see Wilson and Bossert, 1971).

Indeed, in some ways the Lotka-Volterra competition equations themselves are an improvement over the above pair of prey–predator equations. If the potential rate of increase of the predator (r_2) is taken as *very small* (near, but not quite zero) and α_{21} is assumed to be negative (α_{12} is positive as in competition), the equations can be written:

$$\frac{dN_1}{dt} = r_1N_1 - \frac{r_1N_1^{\,2}}{K_1} - \frac{r_1N_1\alpha_{12}N_2}{K_1} \quad (8)$$

$$\frac{dN_2}{dt} = r_2N_2 - \frac{r_2N_2^{\,2}}{K_2} + \frac{r_2N_2\alpha_{21}N_1}{K_2} \quad (9)$$

These equations can be rewritten in a considerably simpler form by consolidating constants, as follows:

$$\frac{dN_1}{dt} = r_1N_1 - z_1N_1^{\,2} - \beta_{12}N_1N_2 \quad (10)$$

$$\frac{dN_2}{dt} = r_2N_2 - z_2N_2^{\,2} + \beta_{21}N_1N_2 \quad (11)$$

where z_1 and z_2 are equal to r_1/K_1 and r_2/K_2, respectively; similarly β_{12} is $z_1\alpha_{12}$ and β_{21} is $z_2\alpha_{21}$. Because α_{21} is negative and α_{12} is positive, contacts between the predator and its prey cause the predator population to increase but these same contacts cause the prey population to decrease. Both populations are self-limited as indicated by the second term in each equation. However, even this pair of equations has several serious and fundamental flaws. The predator population can increase, albeit slowly, in the absence of any prey. There is no way in which prey density directly affects the self-limitation of the predator population; that is, there is no provision for competition among the predators for prey individuals. Moreover, the predator population has a carrying capacity, K_2, even in the total absence of prey.

Another pair of equations which model the prey-predator relationship reasonably well are:

$$\frac{dN_1}{dt} = r_1N_1 - z_1N_1^{\,2} - z_1\alpha_{12}N_1N_2 \quad (12)$$

$$\frac{dN_2}{dt} = \alpha_{21}N_1N_2 - \beta_2\frac{N_2^{\,2}}{N_1} \tag{13}$$

The prey equation is the simple Lotka-Volterra competition Equation (10), but the predator equation combines aspects of both Equations (7) and (9), and, in addition, has a new twist in that competitive inhibition of the predator population is now a function of the relative densities of predator and prey. Thus, inhibition of the predator population increases both with increased predator density and with decreased prey density. Notice that the predator population cannot increase unless there are some prey. However, even though this pair of equations overcome many of the faults of previous pairs, they are still unrealistic in at least one important way. Imagine a situation in which there are more prey than the predator population can possibly exploit; in such a case growth rate of the predator cannot be simply proportional to the product of the two densities as in (13), but some sort of threshold effects must be taken into account.

Equations like the above omit entirely many important subtleties from the prey-predator interaction. For instance, Solomon (1949) distinguished two separate components to the way in which predators respond to changes in prey density. First, individual predators capture and eat more prey per unit time as prey density increases until some satiation threshold is reached, above which the number of prey taken per predator is more or less constant (Figure 5.14*a*); second, increased prey density raises the predator's population size and a greater number of predators eat an increased number of prey (Figure 5.14*b*). Solomon termed the former the functional response and the latter the numerical response of the predator. Using the "systems" approach (see also Chapter 9) that relies upon continued feedback between observation and model, Holling (1959a, 1959b, 1966) developed elaborate models of predation incorporating both the functional and the numerical responses, as well as other parameters including various time lags and hunger level. These models are more realistic and descriptive than the others (above), but are also more complex and restricted. Obviously, a realistic model of prey–predator relationships *must* be quite complex!

A simple graphical model of the prey–predator interaction was developed by Rosenzweig and MacArthur (1963), who reasoned somewhat as follows. In the absence of any predators, the maximum equilibrium population density of the prey is K_1, the prey's carrying capacity. Similarly, some lower limit on prey density must exist, below which contacts between individuals are too rare to ensure reproduction and the prey population thus decreases to extinction. Likewise, at any given density of prey, there must be some

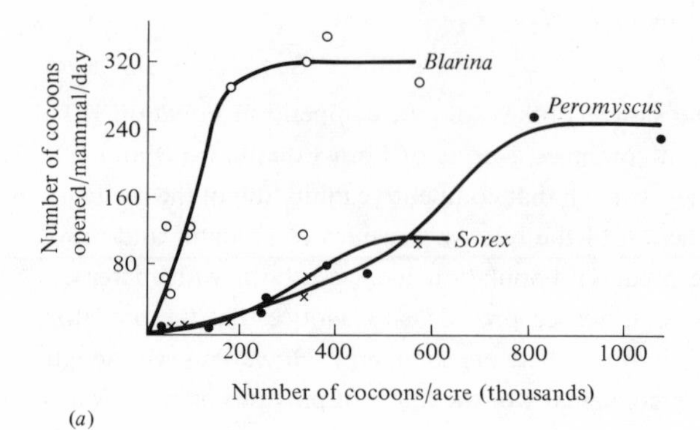

(*a*)

(*b*)

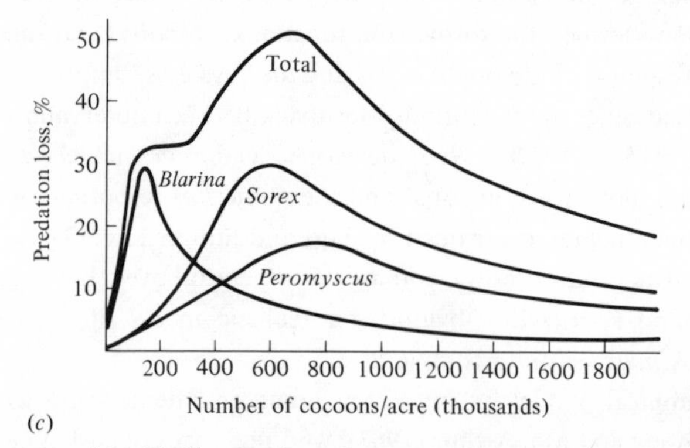

(*c*)

FIGURE 5.14 (*a*) The number of cocoons (prey) eaten per mammal per day by three small mammals is plotted against the density of their prey (the so-called functional response). (*b*) The density of each of the three mammal predators plotted against prey density (the so-called numerical response of the predators). (*c*) Combined functional and numerical responses of each predator species and the total, which represents the overall intensity of predation on the prey population, as a function of prey density. [From Holling (1959a).]

maximal predator density that can just be supported without either an increase or decrease in the prey population. Using the preceding arguments, a prey isocline ($dN_1/dt = 0$) can be drawn in the N_1–N_2 plane (Figure 5.15), similar to those drawn earlier in Figures 5.3 and 5.4. As long as the prey isocline has but a single peak, the exact shape of the curve is not important to the conclusions that can be derived from the model. Above this line, prey populations decrease; below it they increase. Next, consider the shape of the predator isocline ($dN_2/dt = 0$). Below some threshold prey density, individual predators cannot gather food enough to replace themselves and the predator population must decrease; above this threshold prey density, predators will increase. For simplicity, first assume (this assumption is relaxed below) that there is little interaction or competition between predators, as would occur when predators are limited by some factor other than availability of prey. Given this assumption the predator isocline should look somewhat like that shown in Figure 5.16*a*. If there is competition between predators, higher predator densities will require denser prey populations for maintenance and the predator isocline will slope somewhat as in Figure 5.16*b*. In both examples, the carrying capacity of the predator is assumed to be set by something other than prey density. Only one point in the N_1–N_2 plane represents a stable equilibrium for both species—namely, the point of intersection of the two isoclines (where dN_1/dt and dN_2/dt are both zero). Consider now the behavior of the two populations in each of the four quadrants marked *A*, *B*, *C*, and *D* in Figure 5.17. In quadrant *A*, both species are increasing; in *B*, the predator increases and the prey decreases; in *C*, both species decrease; and in *D*, the

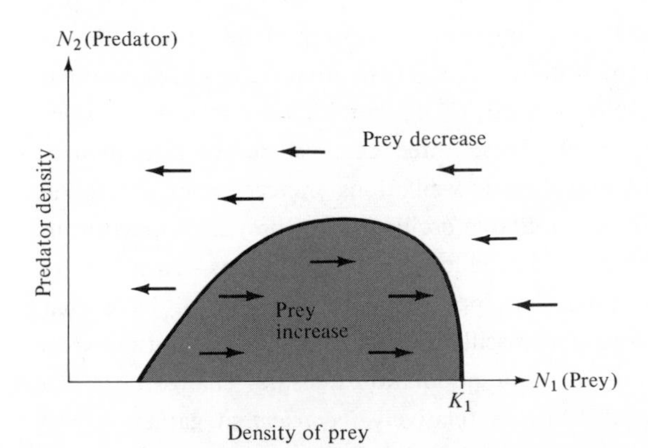

FIGURE 5.15 Hypothetical form of the isocline of a prey species ($dN/dt = 0$) plotted against the densities of prey and predator. Prey populations increase within the shaded region and decrease above the line enclosing it. Prey at intermediate densities have a higher turnover rate and will therefore support a higher density of predators without decreasing.

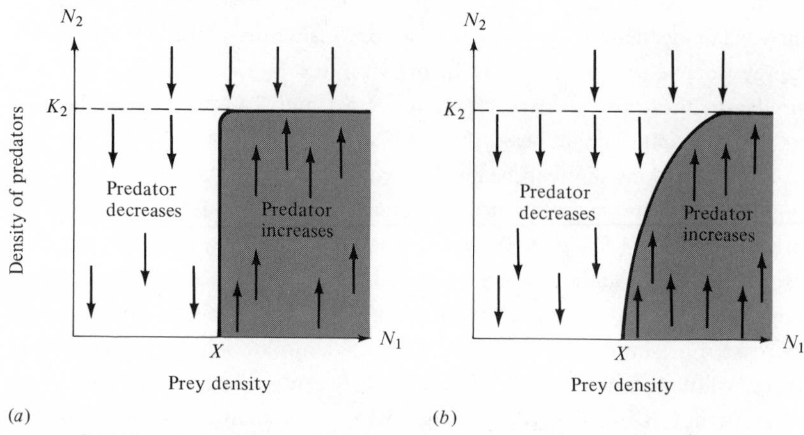

FIGURE 5.16 Two hypothetical predator isoclines. (*a*) Below some threshold prey density, X, individual predators cannot capture enough prey per unit time to replace themselves. To the left of this threshold prey density, predator populations decrease, while they increase to the right of it provided the predators are below their carrying capacity, K_2 (shaded area). So long as predators do not interfere with one another's efficiency of prey capture, the predator isocline rises vertically to the predator's carrying capacity as shown in (*a*). (*b*) Should competition between predators reduce their foraging efficiency at higher predator densities, the predator isocline might slope somewhat like the curve shown. More rapid learning of predator escape tactics by prey through increased numbers of encounters with predators would have a similar effect.

prey increases while the predator decreases. Arrows or vectors in Figure 5.17 depict the above changes in population densities.

Relative magnitudes of the changes in the population densities of prey and predator determine another important property of this model, that is, whether or not a stable equilibrium exists. There are three cases, corresponding to vectors that (1) spiral inward, (2) form a closed circle, or (3) spiral outward (Figure 5.17*a*, *b*, *c*). These three cases correspond to damped oscillations, stable oscillations, and oscillations increasing in amplitude, respectively (Figure 5.17*a*, *b*, *c*). Stable oscillations of prey and predator are one of the mechanisms that could produce population "cycles," such as those of lemmings and their predators (pp. 90–92 in Chapter 4). Note that, given time, the case with damped oscillations will reach its equilibrium value at which neither prey nor predator population densities change; this case corresponds to a predator which is relatively *inefficient* at gathering prey (the predator cannot even begin to exploit the prey population until prey are fairly near their own carrying capacity). Similarly, the case that produces oscillations of ever increasing amplitude corresponds to a very *efficient* predator, which can exploit the prey population nearly down to its limiting

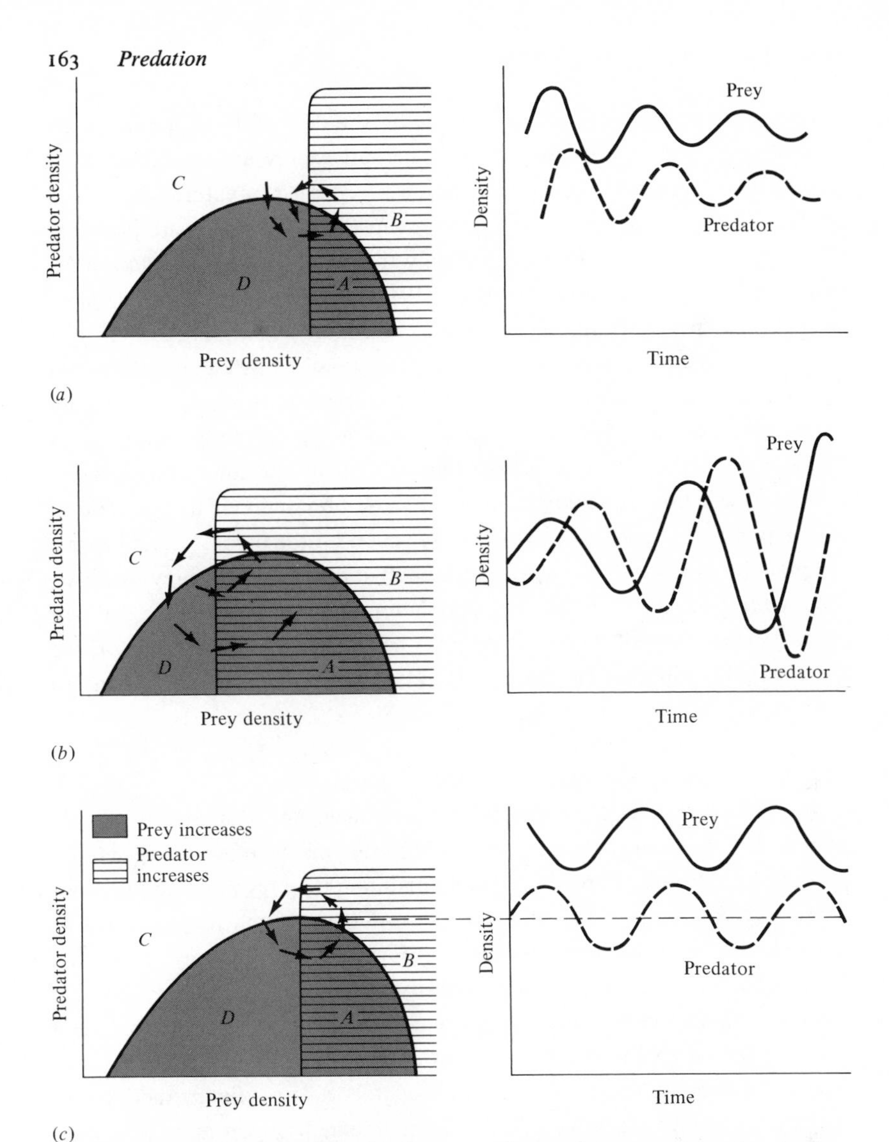

FIGURE 5.17 Prey and predator isoclines superimposed upon one another to show stability relationships. (*a*) An inefficient predator that cannot successfully exploit its prey until the prey population is near its carrying capacity. Vectors spiral inward, prey-predator population oscillations are damped, and the system moves to its joint stable equilibrium point (where the two isoclines cross). (*b*) An extremely efficient predator which can exploit very sparse prey populations near their limiting rareness. Vectors now spiral outward, the amplitude of population oscillations steadily increases, eventually leading to the extinction of either the predator or both the prey and the predator. (*c*) A moderately efficient predator which can begin to exploit its prey at some intermediate density. Vectors here form a closed circle, and populations of prey and predator oscillate in time; because the amplitude of the oscillations does not change, these stable oscillations persist indefinitely. [After MacArthur and Connell (1966).]

rareness. Such an overly efficient predator should rapidly exterminate its prey (and thus go extinct itself unless alternative prey are available); little wonder that increasing oscillations are never observed in nature! Because of predator-escape tactics of prey, many (or most) real predators are probably relatively inefficient, tending to crop only those prey present in excess of a substantial prey population (Errington, 1946); hence, the case with damped oscillation is probably the most realistic reflection of nature.

Individual predators that reproduce successfully at low prey densities will normally outcompete and replace less efficient individuals that require higher prey densities; hence, natural selection acting on the predator moves its isocline to the left and *reduces* the stability of the interacting system. However, selection operating in favor of those prey individuals best able to escape predators opposes the action of selection on the predators, and forces the predator isocline to the right (presumably it also raises the prey isocline); thus natural selection on the prey population tends to *increase* the stability of the system. Indeed, unless the prey is one step ahead of the predator, the latter can be expected to overeat its prey and take both populations to extinction.

"Prudent" Predation and Optimal Yield

It is sometimes suggested that an intelligent predator should crop its prey so as to maximize the prey's turnover rate and therefore the predator's yield. Such a "prudent" predator would maintain the prey population at that density which gives the maximum rate of production of new prey biomass. In terms of the prey isoclines depicted in Figure 5.17, this prey density of "optimal yield" corresponds to the density at the peak of the prey isocline. Man has the capacity to be such a prudent predator; indeed, optimal yield has long been a goal in the management of exploited populations in fisheries biology. But do other, less-intelligent predators also maximize their yield? A truly "prudent" predator should prey preferentially upon those prey individuals with low growth rates and low reproductive values, and leave those with rapid growth rates and higher reproductive values alone. In fact, predators often do take aging and decrepit prey individuals which are frequently easy to catch, while younger, more vigorous ones escape.

However, there is a potential flaw in the prudent predation interpretation, provided that several predator individuals encounter the same prey items. If, say, young juicy prey are less experienced and easier to catch, an individual predator who "cheated" and ate them would be likely to leave more genes than the prudent genotypes which did not exploit this food supply; as a result, nonprudence would become incorporated into the gene pool and spread. Exactly the same considerations apply to a competing species that is able to use the prey individuals in question. Hence we would expect prudence to evolve only in a situation where a single predator has exclusive use of a prey population; perhaps some feeding territories are examples.

Another, perhaps more likely, explanation for the occurrence of apparent "prudent" predation in nature can be made in terms of the prey organisms themselves. As was pointed out in Chapter 4, the intensity of natural selection is directly proportional to the expectation of future offspring (reproductive value); thus one might predict that individual prey with high reproductive values would have more to gain from escaping predators than would those with low reproductive values. After one's expectation of future offspring has dropped to zero, nothing further can be gained from being able to escape a predator, and predator avoidance cannot be evolved. Thus, many cases of apparent "prudent" predation may well be simply part and parcel of old age; the evolution of old age has wide significance! Viewed in this way, the susceptibility of prey to predation should be inversely related to their reproductive value.

Selected Experiments and Observations

Predation can be readily studied in the laboratory, and, under certain favorable circumstances, in the field. Gause (1934) studied a simple prey–predator system in the laboratory using two microscopic protozoans, *Paramecium caudatum* and *Didinium nasutum. Didinium* are voracious predators on *Paramecium.* He found that when both species were placed together in a test tube containing clear medium (which supports a culture of bacteria, the food for *Paramecium*), *Didinium* overate its food supply, exterminated it, and then starved to death itself (Figure 5.18*a*). When some sediment was added to the medium (making it "heterogeneous" rather than "homogeneous"), providing a refuge or "safe site" for the prey, *Didinium* went extinct but the *Paramecium* population recovered (Figure 5.18*b*). In a third experiment (Figure 5.18*c*), Gause introduced new individuals of each species at regular time intervals; such "immigrations" resulted in two complete cycles of prey and predator. In other experiments, using *Paramecium aurelia* as the predator and a yeast, *Saccharomyces exiguus*, as prey, Gause (1935) obtained nearly three complete cycles.

Huffaker (1958) performed very similar laboratory experiments on two species of mites, using oranges as the plant food for the system; one mite was herbivorous, eating the oranges, while the second mite was a predator on the herbivorous one. In simple systems with oranges close together and evenly spaced the predator simply overate its prey and both species became

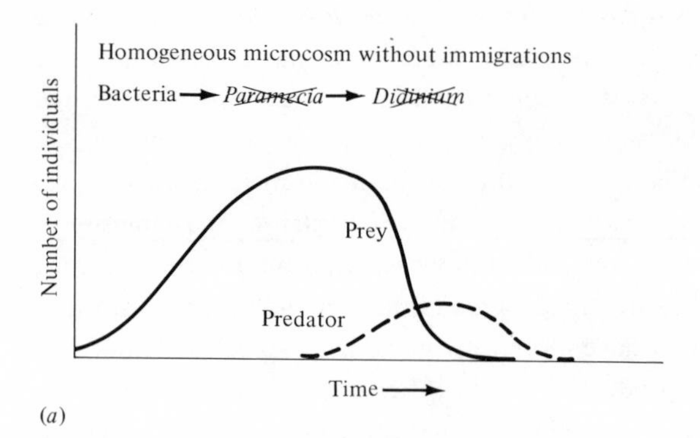

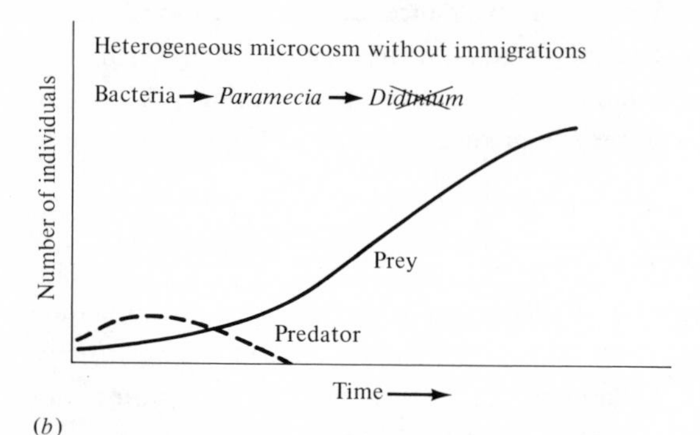

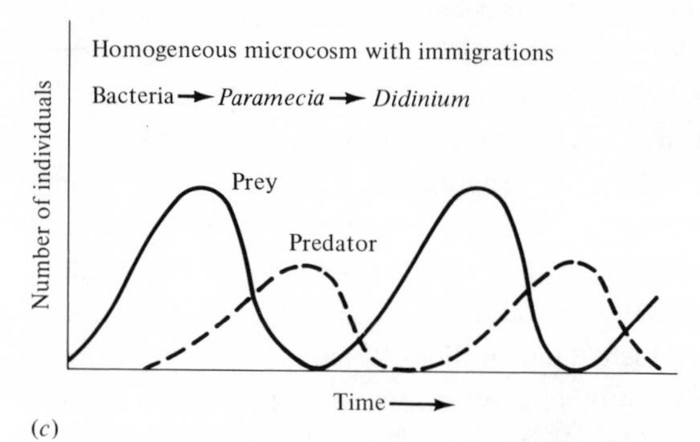

FIGURE 5.18 Three laboratory prey-predator experiments with protozoans. (*a*) In a simple homogeneous microcosm without immigration of new prey or predators, the predator quickly overeats its prey, exterminates the prey, and then all predators starve to death themselves. (*b*) In a more heterogeneous system, the predator goes extinct first and the prey population recovers to expand to its carrying capacity. (*c*) Even in a homogeneous microcosm, immigration of prey and predators results in both populations oscillating in time. [From Gause (1934). *The Struggle for Existence*. Hafner, New York.]

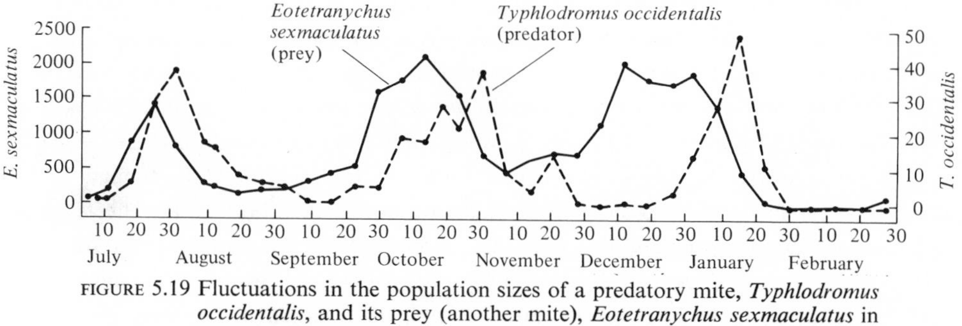

FIGURE 5.19 Fluctuations in the population sizes of a predatory mite, *Typhlodromus occidentalis*, and its prey (another mite), *Eotetranychus sexmaculatus* in a spatially heterogeneous environment. [After Huffaker (1958).]

extinct. Increasing the distances between oranges only lengthened the time required for extinction, but did not allow coexistence. However, by introducing barriers to dispersal and thus making the system still more complex, Huffaker obtained three complete prey–predator cycles (Figure 5.19). Thus, environmental heterogeneity increased the stability of the system of a predator and its prey. In addition, these experiments illustrate the existence of the stable oscillations predicted by theory as discussed earlier.

In a very revealing, although more complex, laboratory investigation, Utida (1957) examined both competition and predation simultaneously. His system was composed of three species: a beetle (*Callosobruchus chinensis*) as the prey, and two species of predatory wasps as competitors (*Neocatolaccus mamezophagus* and *Heterospilus prosopidis*). The prey beetle was provided with a continually renewed food supply. Both species of wasps, which have similar life histories, were dependent upon the beetle population as a common food source. Population densities of the three species fluctuated widely and erratically (Figure 5.20), but after 4 years, some 70 generations later, all

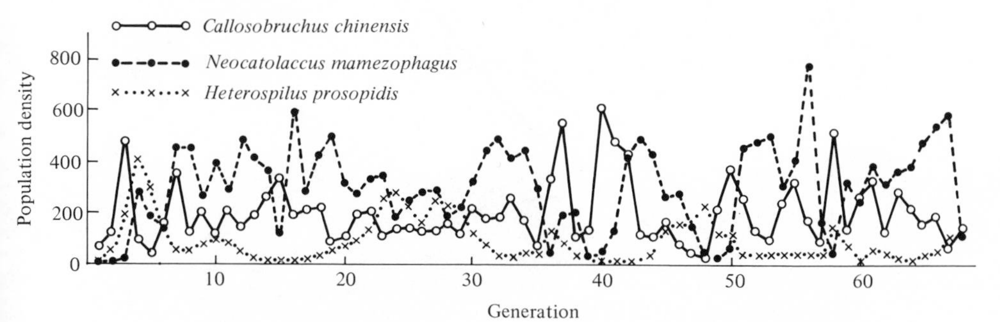

FIGURE 5.20 Changes in the population sizes of a beetle host (*Callosobruchus*) and two parasitic wasps over a 4-year period. Although the fluctuations appear erratic, all three species survived the entire period of some 70 generations. [After Utida (1957).]

three were still coexisting. Populations of the two predatory wasps tended to fluctuate out of phase with one another. Analysis showed that *Heterospilus* was more efficient at finding and exploiting the beetle population when it was at low densities but that *Neocatolaccus* was more efficient at high prey densities (Figure 5.21); thus the competitive advantage shifted between the two wasps as the density of the beetle prey changed in time. Utida thought that the fluctuations of the beetle population were caused both by the effects of the two wasp predators and by density-dependent changes in the rate of reproduction of the beetle itself. Thus, this system provides a neat example of coexistence of two competitors on a single resource due to changing abundance of that resource; the stability of the system is apparently a result of biotic interactions.

Perhaps the most potent technique for studying the effects of predation in nature is to exclude predators from an area and then to compare this experimental area with an adjacent "control" area, which is similar but unaltered with normal access for predators. Hopefully, resulting differences between the two areas can best be ascribed to predation. Paine (1966) performed just such a predator removal experiment along the rocky intertidal seacoast of the Olympic peninsula in Washington State. When the major top predator, the sea star *Pisaster ochraceus*, was removed, there was a drastic reduction in the number of species remaining. Control areas with *Pisaster* supported some 15 species of marine invertebrates, but the area without the starfish had only 8 species. The rocky intertidal is a space-limited system. In the absence of predation, more efficient occupiers of space, especially the

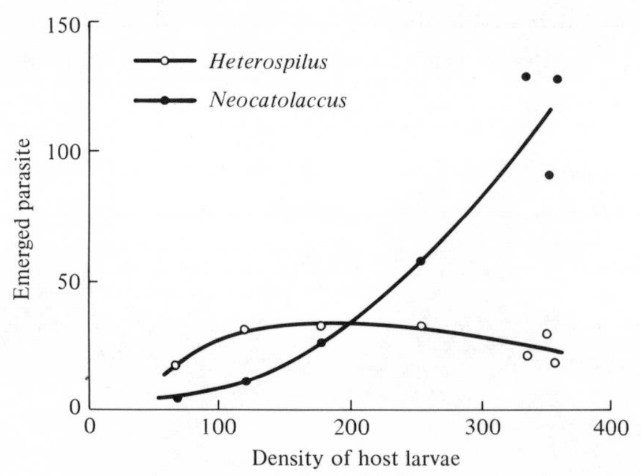

FIGURE 5.21 The number of parasitic wasps emerging from larvae of the host beetle plotted against the density of these larvae. *Neocatolaccus* is a more successful parasite at high host densities, whereas *Heterospilus* has the advantage at lower densities of host larvae. [After Utida (1957).]

bivalve *Mytilus californianus* (a mussel), dominated the area. Presumably by preying on *Mytilus*, *Pisaster* continually open up new spaces that are rapidly colonized by less efficient competitors for space. Thus, by reducing the level of competition among lower trophic levels, this predator allows the coexistence of otherwise competitively incompatible species.

Evolutionary Consequences of Predation: Predator Escape Devices

Generalized predators feeding on a variety of prey obviously must be adapted to cope with a wider variety of predator escape tactics than more specialized predators that need to deal with fewer types of prey. Indeed, by diverging in their "strategies" of predator avoidance over evolutionary time, two (or more) prey species can make it increasingly difficult for a single predator to capture and exploit both prey types efficiently; thus prey evolution can "force" a predator to restrict the range of foods it eats.

Antipredator devices are extremely varied; some such mechanisms for predator escape are quite simple and straightforward, but others may be exceedingly intricate and subtle. As an example of the former, many lizards dig special escape tunnels in their burrow systems that come up near the surface and allow the lizard to break out should a predator corner it underground.

Behavior and anatomy often make animals difficult to detect and/or to follow; such cryptic adaptations can involve sound, smell, color, pattern, form, posture, and/or movements. Concealing or cryptic coloration is widespread, and often depends upon appropriate behavior; in order to hide itself, an animal must select the proper background and orient itself correctly. Some moths, for instance, normally align themselves with the dark markings on their wings parallel to cracks and crevices of the tree bark substratum. Nearly all diurnal animals and some nocturnal ones are countershaded, with their dorsal (upper) parts darker than their ventral (lower) parts. Lighting from above casts shadows below; in a countershaded animal these balance dorsoventrally, reducing contrast and producing a neutral density—the net effect, of course, is to make the animal more difficult to see. Countershading occurs in most insects, fish, amphibians, lizards, snakes, birds, and many mammals. A counterexample, which proves the point, is provided by a few animals that are normally upside down in nature, such as the "upside down" catfish *Synodontis nigriventris* and certain moth larvae; these animals are darker ventrally than they are dorsally! Note that, because a good predator must also be inconspicuous in order to catch its prey, crypticity is equally as important to predators as it is to prey.

Many insects resemble parts of the plants on which they live, especially leaves, twigs, thorns, or bark; leaf butterflies and walking sticks are familiar examples. Both a green and a brown color phase often occur in such cryptically shaped animals. For example, females of two southern grasshoppers, *Syrbula admirabilis* and *Chortophaga viridifasciata*, have green and brown color phases (strangely enough, males are almost always brown!). Green females predominate in wetter, greener, habitats; but in immediately adjacent drier and browner areas the brown form is most prevalent (Otte and Williams, 1972). Determination of a female's color is not under strict genetic control, but rather is developmentally flexible in response to local conditions.

Actual demonstrations that such coloration differences and background color matching have selective value are unfortunately rather scarce. The best documented example is that of the moth, *Biston betularia*, in England. This moth, along with several hundred other species, has evolved rapidly during the last century in response to human modification of its habitat. In the 1800s, *Biston* were pale colored, spending their daytime hours on pale, lichen-covered tree trunks. However, with the build up of industry and concomitant air pollution, the lichens have died and tree trunks in some areas have been covered with a layer of soot and grime, becoming quite dark. In early collections, black moths (melanics) were very rare, but they have become increasingly more common until now these melanistic varieties comprise the vast majority of moth populations in polluted areas. This phenomenon of directional selection, termed *industrial melanism*, has also taken place in the United States and in Europe. In an elegant series of experiments, Kettlewell (1956) made reciprocal transfers of pale moths from a non-polluted woods with melanic moths from a polluted area (Table 5.5). These moths, along with resident moths occurring at each locality, were marked with a tiny inconspicuous paint spot beneath their wings, and attempts were

TABLE 5.5 Numbers of Typical and Melanic Marked Moths (*Biston betularia*) Released and Recaptured in a Polluted Woods Near Birmingham and an Unpolluted Woods Near Dorset[a]

	Polluted Woods	Unpolluted Woods
Numbers of Marked Moths Released		
Typical	64	496
Melanic	154	473
Number Recaptured		
Typical	16 (25%)	62 (12.5%)
Melanic	82 (53%)	30 (6.34%)

Source: From data of Kettlewell (1956).

[a] The wild population in the polluted woods was 87 percent melanic.

made to recapture them on later days. As expected, pale moths had lower survivorship in the polluted woods and melanic moths had lower survivorship in clean, lichen-covered forests. Moreover, Kettlewell actually observed foraging birds catching mismatched moths!

A black lava flow on white desert sands in New Mexico provides a "natural" experiment, which strongly suggests that background color matching has evolved and is adaptive (Benson, 1933). This lava flow is completely surrounded by white sandy areas and has presumably been stocked mostly with animals derived from those that live on the white sands. Two closely related pocket mice live in the area: one, *Perognathus intermedius ater*, is nearly pitch black and occurs only on the lava; the other, *P. apache gypsi*, is a pale white and lives only on the white sands.

Some animals are covered with blotches of different colors which tend to break up their shape and to make them more difficult to see. Good examples are rattlesnakes and boa constrictors; often these snakes look so much like a pile of dead leaves that they may go undetected. This type of coloration, which is especially prevalent in larger animals that have difficulty in finding hiding places, such as leopards, tigers, and giraffe, has been termed *disruptive coloration.*

A form of adaptive coloration which is not necessarily cryptic is so-called *flash coloration.* Many inconspicuous insects, including some butterflies and grasshoppers, are extremely cryptic when at rest; but when disturbed they fly away and reveal brightly colored underwings (often red, yellow, or orange). These insects thus suddenly become extremely visible and conspicuous, catching the predator's eye. When they land, they close their wings and quickly move away from the spot they landed. As anyone who has chased grasshoppers knows, it is very difficult to keep track of the position of such an animal. (Squid and octopi employ a remotely similar strategy when they squirt out their "ink," leaving a dense cloud in the water: typically the animal immediately changes both color and course, becoming pale and swimming at right angles to its original direction of flight, evading the predator.) A rather derived kind of flash coloration occurs in some butterflies and moths, which have large owl-like eyes on their underwings. These eyes are normally hidden by the upper pair of wings. When a small bird approaches, the upper wings are suddenly twitched aside, revealing an owl-like face beneath. It is thought that many small birds are so startled that they fly away leaving the insect alone!

Another, smaller, type of eyespot actually invites the attack of a predator. Many predators instinctively go for the eyes of their prey since eyes are usually one of the most vulnerable parts of an animal and their loss readily incapacitates them (lions and wolves have found another "Achilles heel" on large ungulates—they simply hamstring the animal). Many species of butterflies possess small "fake" eyespots along the periphery of their wings that may actually invite attack. By painting eyespots on such butterflies in places without eyespots and releasing and recapturing the animals, it has been demonstrated that these spots are damaged, apparently being pecked at by birds (Sheppard, 1959). Thus, the butterfly obtains a second chance at escape by luring the predator's attack away from its own eyes. Behavioral adaptations may sometimes serve similar functions; certain snakes raise their tails and wave them around in a very headlike manner, occasionally actually making short lunges with their tails at the threatening predator! Should the unwary predator grab the snake by this "head," the serpent still has its real head free to bite back.

Many birds and mammals have various sorts of signals, which, when given, warn other animals that a predator is in the immediate vicinity. For instance, beaver warn one another of danger by slapping their tails loudly against the surface of the water. Similarly, rabbits in a warren (a colony using the same burrow system) often "thump" with their hind feet to signal the approach of a predator. Prairie dogs, many primates, and many foraging bird flocks, such as crows, frequently post sentinels (see also Chapter 4, p. 128), which watch for predators from a good vantage point, and warn the group should one appear in the distance. The white underside of the tail of some rabbits and ungulates (such as the white-tailed deer) is raised when the animal flees from a predator and it is thought that this may serve as a warning signal to nearby animals. Among birds, alarm calls in response to the presence of a bird-eating hawk are especially prevalent. Typically these faint shrill whistles are extremely difficult for a vertebrate predator to locate; often they are similar in widely different bird species, presumably having converged over evolutionary time.

Because different species may benefit from the call of one individual, hawk alarm calls appear to be somewhat "altruistic." However, warning calls are normally used only during the nesting season, and are therefore best interpreted as having arisen through kin selection (see Chapter 4). The ventriloquial nature of the call, coupled with the fact that the caller has already seen the predator and therefore knows exactly where the danger lies, ensures that the risk to the bird giving the alarm is slight. Likewise, the fact that banded birds often return to breed in the same area where they themselves were raised (Chapter 4) ensures that the birds in any given area will be related and share many genes, which in turn ensures that the gain to the many relatives of the pseudo-altruist will frequently be quite large. If hawk alarm calls have evolved via kin selection, one would predict that the frequency of use of such calls should be inversely related to the distance the caller is from his own nest and immediate relatives (no one has, however, yet demonstrated this empirically). The fact that these calls work across species lines is easily explained, because individuals that recognize warning signals of other sympatric species as alarms should be at a relative advantage within their own population. The convergence of hawk alarm calls can be explained by either or both of two mechanisms. The first concerns the ventriloquial properties of the call. The number of ways in which an alarm call can be loud enough to function as a warning, and yet still be difficult to locate, are decidedly limited. Thus convergence may be a simple result of limitation inherent in the system. A second, equally likely, possibility is that natural selection has favored convergence because it facilitates interspecific recognition of alarm calls; thus individuals tending to produce call variants more like those of another species would benefit because they would also be more likely to recognize the calls of the other species (as such, call convergence would be very similar to Müllerian mimicry, discussed below).

Yet another evolutionary consequence of predation is *warning coloration*; unpalatable or poisonous animals have often evolved bright colors that advertise their distastefulness. Such animals are usually colored with the same conspicuous colors men use for signs along highways: reds, yellows, and blacks and whites. Examples of animals with warning coloration are bees and wasps, monarch butterflies, coral snakes, skunks, and certain brightly colored poisonous frogs and salamanders. Indeed, signals that warn potential predators need not be only in colors, but may also often involve pattern, posture, smell, or sound. Thus, a rattlesnake's buzzing presumably serves to warn other larger animals such as the American bison not to come too near (unfortunately for the rattler, however, this warning only attracts human attention, which usually results in the snake's demise). It has been shown experimentally that avian predators learn to avoid distasteful prey. In the case of poisonous prey, this learning may actually be incorporated into the gene pool, and manifested as an "instinctual" avoidance.

Mimicry is an interesting sidelight of warning coloration, which nicely demonstrates the power of natural selection. An organism that commonly occurs in a community along with a poisonous or distasteful species can benefit from a resemblance to the latter warningly colored species, even though the "mimic" itself is nonpoisonous and/or quite palatable. Such false warning coloration is termed Batesian mimicry after its discoverer. Many species of harmless snakes mimic poisonous snakes; in Central America some harmless snakes are so similar to poisonous coral snakes that it takes an expert to distinguish the mimic from the "model." Similarly, harmless flies and clearwing moths often mimic bees and wasps, and the palatable viceroy butterfly mimics the distasteful monarch. Batesian mimicry is disadvantageous to the model, because some predators will encounter palatable or harmless mimics and thereby take longer to learn to avoid the model. The greater the proportion of mimics to models, the longer the time required for predator learning and the greater the number of model casualties. In fact, if mimics became more abundant than models, predators might not learn to avoid the prey item at all, but rather might actively search out model and mimic alike. For this reason Batesian mimics are usually much less abundant than their models; also these mimics are frequently polymorphic and mimic several different model species.

A different kind of mimicry occurs when two species, both distasteful or dangerous, mimic each other; this phenomenon is termed Müllerian mimicry.* Both bees and wasps, for example, are usually banded with yellows and blacks. Since potential predators encounter several species of mimics more frequently than a single species, they learn to avoid them faster, and the relationship is actually beneficial to both prey species (Benson, 1972). The resemblance need not necessarily be very exact as it must be under Batesian mimicry, because neither species actually deceives the predator, but rather each only reminds the predator of its dangerous or distasteful properties. Müllerian mimics can be equally common and are rarely polymorphic.

Plants, being sessile, cannot use many of the escape techniques of animals, but instead they are obviously and decidedly more limited as to the ways in which they can deter potential predators. A plant with a patchy or spotty distribution in time and/or space may escape some predation simply by virtue of its unpredictable availability; thus an annual that is here today but gone tomorrow may be more difficult for herbivores to find and use than an evergreen perennial that is always relatively available. Some plants, especially perennials, have evolved morphological adaptations, such as hairs, spines, and hooks (Gilbert, 1971), that discourage many herbivores quite directly. But by far the most widespread predator deterrent of plants is what might be termed chemical warfare. A great variety of secondary chemical substances occur in plants that are not known to serve any direct physiological function for their possessors. Many of these are evidently not the breakdown products of larger molecules due to metabolic processes and wastes, but rather are secondary substances produced by active synthesis from smaller molecular precursors. Such secondary chemical substances often

* Actually, the dichotomy between Batesian and Müllerian mimicry is somewhat artificial in that these two sorts of mimicry grade into one another.

contain nitrogen and other elements that are available to the plant in only limited supply; moreover, it takes energy to produce these chemicals. Clearly there are definite costs to the plant in the production of herbivore repellents. Nearly a century ago, the German botanist Stahl (1888) suggested that these secondary substances might reduce the plant's palatability to herbivores. Stahl's prediction has now been amply verified for many different plant–herbivore systems.

Agriculturalists and plant breeders have produced many strains that are highly resistant to normal herbivores. Genetic varieties, or morphs, of plants in nature have been shown to be differentially palatable to herbivores; thus Jones (1962, 1966) found that a number of herbivores ranging from insects and snails to *Microtus* (a mammal) preferred a "noncyanogenic" morph of the plant *Lotus corniculatus* over a "cyanogenic" one. Tannins have been implicated as the agent that repels some herbivores; oak leaf tannin significantly reduces the growth rate of larvae of the moth *Operophtera brumata* (Feeny, 1968). Frequently herbivores eat only relatively new growth and do not utilize older parts of plants, presumably because the latter contain tannins and other repellent chemicals (Feeny, 1970). Other chemical substances believed to protect plants from animals and fungi include essential oils and resins, alkaloids, terpenes, and turpenoids. The latter two classes of compounds have especially penetrating odors and tastes; sesquiterpenes are known to be fatal to sheep. Wild herbivores that have evolved alongside poisonous plants would be unlikely to eat them, whereas domestic sheep and cattle will eat many poisonous fodders.

Coevolution

In its broadest sense, *coevolution* refers to the joint evolution of two (or more) taxa that have close ecological relationships but do not exchange genes, and in which reciprocal selective pressures operate to make the evolution of either taxon partially dependent upon the evolution of the other (Ehrlich and Raven, 1964). Thus, coevolution includes most of the various forms of population interaction, from competition and predation to mutualism and protocooperation. The term coevolution is also often used in a more restricted sense, to refer primarily to the interdependent evolutionary interactions between plants and animals, especially herbivores. For example, a plant may evolve a secondary chemical substance that deters the vast majority of predators, but if a herbivore can in turn evolve a physiological means of coping with the chemical deterrent, it can thereby obtain an uncontested food supply. By this kind of coevolution, many herbivores have become strongly specialized on a single species or a few closely related species of plants. Thus, *Drosophila pachea* is the only species of fruit fly that can exploit the "senita" group of cacti; these plants produce an alkaloid that is fatal to the larvae of all other fruit flies, but *D. pachea* has apparently evolved a means of detoxifying this chemical (Kircher *et al.*, 1967).

In some cases plants have actually formed cooperative relationships with animals resulting in their protection from certain herbivore species. Thus, Janzen (1966) showed that some species of *Acacia* deprived of their normal epiphytic ant fauna are highly palatable to herbivorous insects, whereas other species that do not normally have ants for protection from herbivores are less palatable. Indeed, the acacias benefiting from ant protection actually produce nectaries and swollen thorns that attract and in turn benefit the ants! Thus, these plants are putting matter and energy into attracting ants which defend their leaves, rather than into more direct chemical warfare. This antiherbivore ploy is broad based, since the ants attack a wide range of herbivores.

Because most land plants cannot move, they often must rely upon animals both for pollination and for seed dispersal (some rely on wind as well). The seeds of many fruits pass unharmed through the intestines of herbivores and germinate to grow a new plant from the droppings of the animal-dispersing agent. Colorful flowers with nectar and brightly colored fruits can only be interpreted as having been evolved solely for the purpose of attracting appropriate animals. Here, as in plant–herbivore interactions, a high degree of plant–animal specificity has arisen. Animals that pollinate a particular plant are referred to as *pollinating vectors*. As an example, in Central America different species of male euglossine bees are highly specific to particular species of orchids; the males of these bees travel long distances between orchids. Different bee species are attracted by different orchid fragrances (Dressler, 1968), as can be shown by putting out "baits" of artificially synthesized orchid "perfumes." These bees are probably necessary for, and may have allowed the evolution of, the great diversity of tropical orchids, many of which are quite rare and far apart. Such specificity of pollinating vectors assures that the plant's pollen is transmitted to the ovules of its own species; the advantages of this specificity to male euglossine bees have not yet been determined. Whereas female euglossine bees are not as specific to the plant species they pollinate as males, individual females travel distances up to 23 km (Janzen, 1971a) and regularly move long distances between sparsely distributed plants in gathering nectar and pollen; thus they probably promote outcrossing among tropical plants at low densities. Indeed,

Janzen suggests that such "traplining" by these female bees may actually permit the very existence of plant species forced to very low densities by factors such as competition and predation on their seeds and seedlings.

Some pollinators, such as *Heliconius* butterflies (Gilbert, 1972), obtain amino acids from eating the pollen of the plants they pollinate. Because the production of nectar and pollen (and fruit) requires matter and energy, attracting animal pollinators (and seed dispersers) has its costs to the plant. Pollen-eating pollinators presumably cost a plant more than strict nectar feeders. In turn, the returns from visiting a flower (or eating a fruit) must be great enough to the animal pollinator or seed disperser to make it worthwhile, yet small enough that the animal must move the distances necessary to disperse the pollen or seeds. This intricate energetic interplay between plants and their pollinators is reviewed by Heinrich and Raven (1972).

Predation on seeds may often be heaviest where they occur in greatest concentrations (such as acorns underneath a parent oak) because seed predator populations will generally be largest where the most food is available (Janzen, 1971b). As a result, the probability of an individual seed surviving to establish itself as a plant may often vary inversely with seed density. In many trees, the majority of seeds fall to the ground near the parental tree, with a continually decreasing number of seeds ending up at distances farther from the parental tree (Figure 5.22). Janzen suggests that, as a result of these opposing processes, recruitment is maximized at some distance from the parent tree (see Figure 5.22); moreover, this model of seed predation and recruitment may explain the high species diversity of tropical trees, which

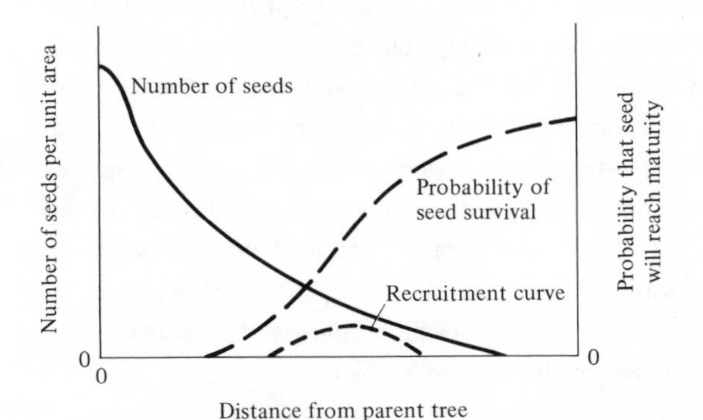

FIGURE 5.22 Hypothetical model of seed recruitment versus distance from the parent tree. Near the tree all seeds are eaten by seed predators, whereas at increasing distances from the tree the probability of seed survival increases as the density of seeds and their predators decreases. Although the number of seeds drops with distance from the parental tree, recruitment is highest at some distance from the tree. [After Janzen (1970).]

suffer heavy seed losses to specialized seed predators that eat the seeds of particular tree species.

The intricacies of coevolutionary relationships between pine squirrels (*Tamiasciurus*) and their coniferous food trees were studied in the Pacific northwest by C. Smith (1970). Conifer seeds constitute the staple food supply of these squirrels, which can effectively strip a tree of most of its cones. Trees reduce the effectiveness of squirrel predation in a number of different ways: (1) by producing cones that are difficult for the squirrels to reach, open, and/or carry; (2) by putting fewer seeds into each cone (squirrels eat only the seeds themselves and must "husk" the cones to get them); (3) by increasing the thickness of seed coats, requiring that the squirrels spend more time and energy extracting each seed; (4) or, similarly, by putting less energy into each seed (a drawback is that seedlings from smaller seeds have fewer resources at their disposal and are presumably poorer competitors than seedlings from larger seeds); (5) by shedding seeds from cones early, before the young squirrels of the year begin foraging; and (6) by periodic cone "failures" that decimate the squirrel population, reducing the intensity of predation during the next year. Thus squirrel predation has had profound evolutionary influences upon various reproductive characteristics of conifers, including details of cone anatomy and location, the number of seeds per cone (and the variability in the number per cone), the time at which the cones shed their seeds, the thickness of seed coats, as well as annual fluctuations in the size of cone crops. Evolution of these defense mechanisms by the conifers has, in turn, forced the squirrels to adapt in various ways, such as choosing cones carefully and stockpiling them.

Fluctuations in cone crops from year to year are pronounced and are best interpreted as in antisquirrel strategy, since they occur even when climatic conditions are favorable for the trees. Apparently, the conifers withhold their products of primary production and store them for later use. Cone crops and failures are often synchronized among different tree species in a given area, a further indication that the phenomenon is directed at the squirrels. Smith points out that different conifer species have *diverged* from one another in the evolution of cone anatomy, size, location, and time of shedding, but that these same conifer species have *converged* in their fluctuations in cone crops. Both reduce the squirrel's efficiency.

Symbiotic Relationships

Symbiosis means living together. Usually the term is used only to describe pairs of organisms that live together without harming one another, thus excluding parasitism (+, −) and amensalism (−, 0) in which one party is affected adversely (see introduction to this chapter and Table 5.1 for explanation of symbols). Hence symbiotic relationships include neutralism (0, 0), protocooperation (+, +, nonobligatory), mutualism (+, +, obligatory), and commensalism (+, 0). As pointed out earlier, true neutralism is doubtlessly very rare or even nonexistent in nature and therefore need not be considered. However, protocooperation, mutualism, and commensalism are fairly widespread, particularly in diverse communities. The ant–acacia interaction referred to above is an example of protocooperation. Numerous other cases of protocooperation are known, such as some small birds riding on the back of water buffalo (the bird obtains food while the mammal is freed of many insect pests) or other birds picking between the teeth of crocodiles (again, the birds obtain food while the reptile gets its teeth cleaned). A bird known as the honey guide has formed a unique alliance with the honey bear (a mammal); the honey guide locates a beehive and leads the honey bear to it, whereupon the mammal tears open the bee's nest and eats its fill of honey and bee larvae, after which the bird has its meal. The honey guide can find bee hives with relative ease but cannot open them, while the honey bear is in exactly the opposite situation; cooperation clearly increases the efficiency of both species.

Mutualism is less common than protocooperation, probably because both populations are completely dependent upon the relationship and neither can survive without the other. Thus, termites cannot themselves produce enzymes to digest the cellulose in wood, but, by harboring a population of protozoans in their intestines which can make such enzymes, the insects are able to exploit wood as a food source successfully. Neither termite nor protozoan could survive without the other. These intestinal symbionts are passed on from one generation of termites to the next through exchange of intestinal contents (see Chapter 4, pp. 127–128). Another putative example of mutualism is lichens, which are composed of a fungus and an alga; the fungus provides the supportive tissue while the alga performs photosynthesis. (However, algae of some lichens can be grown without the fungi, so that some lichens may not represent true mutualism but rather protocooperation.)

Commensalism occurs when one population is benefited but the other is unaffected (+, 0); it is probably uncommon. Small epiphytes such as bromeliads and orchids, which grow on the surfaces of large trees without obvious detriment to the tree, might be an example. A well documented case of commensalism is the association between cattle egrets and cattle (Heatwole, 1965). These egrets follow cattle that are grazing in the sun and capture prey (including crickets, grasshoppers, flies, beetles, lizards, and frogs) that move as cattle approach. Heatwole found that the number of cattle egrets associated with cattle was strongly dependent upon the activities of the cattle; thus, he observed fewer egrets than expected on a random basis near resting cattle, but nearly twice as many egrets as expected (if the association were entirely random) accompanied cattle that were actively grazing in the sun. Since the birds seldom take prey, such as ticks and other ectoparasites, directly from the bodies of the cattle, the mammals probably do not directly benefit from their relationship with egrets. Moreover, Heatwole demonstrated that egret feeding rates, energy expended during foraging, and feeding efficiency were all markedly higher when these birds were associated with cattle (Table 5.6).

A Complex Example of Population Interactions

The exceedingly great complexity that can develop in interactions between populations is nicely demonstrated by the case of a Panamanian cowbird and its hosts, studied by N. Smith (1968). The giant cowbird, *Scaphidura oryzivora*, which is a brood parasite, or an animal that puts its eggs into the nest of another species and lets the latter raise its young, has four hosts in this area, three species of oropendulas and one cacique (oriole-like birds). These oropendulas and caciques build long dangling oriole-like nests in mixed-species colonies, often with over a hundred pairs of birds nesting together. Some colonies are known to have been in existence for over 20 years. Two colonies, averaging 173 nests, fledged an average of 111 cowbirds per year or nearly one per nest. Local Panamanians said that such high rates of production of *Scaphidura* were of regular occurrence (indeed, these people called the cowbirds "black oropendulas"). However, still other colonies fledged many fewer *Scaphidura*. Smith cut down several thousand nests, examined and manipulated their contents, and replaced them. He found striking variation from colony to colony in the coloration and marking of cowbird eggs; at some colonies, *Scaphidura* produced mimetic (mimicking) eggs, which closely matched the host eggs, yet at other colonies the cowbirds laid nonmimetic eggs. Mimetic cowbirds never deposited more than a single egg in a host's nest and always laid their eggs in nests in which the hosts had already deposited eggs. Nonmimetic cowbirds, which Smith nicknamed "dumpers," however, deposited their eggs in empty nests as well as full ones, and often left 2 or 3 eggs in a given nest. Moreover, mimetic cowbirds were

TABLE 5.6 Various Aspects of the Association of Cattle Egrets with Cattle

Category	No. of Cattle	Percent Cattle	Number of Associated Egrets	
			expected	observed
Grazing in sun	735	39.1	239	439
Grazing in shade	55	2.9	18	21
Standing in sun	146	7.8	48	46
Standing in shade	257	13.7	84	17
Lying in sun	503	26.8	164	69
Lying in shade	143	7.6	47	17
Walking	39	2.1	13	3
Total	1878	100.0	612	

	Mean Number per Minute	No. of Times Count Was Higher Than for Opposite Egret	Percent of Times Count Was Higher Than for Opposite Egret
FEEDINGS, $N = 84$			
Associated	2.34	58	69
Nonassociated	1.71	26	31
STEPS, $N = 62$			
Associated	20.1	7	11
Nonassociated	32.1	55	89
FEEDINGS/STEP, $N = 59$			
Associated	0.129	52	88
Nonassociated	0.051	7	12

Source: From Heatwole (1965).

Note: Upper box shows the numbers of egrets associated with cattle engaged in different activities. Lower box shows feeding rates, steps taken per prey item (energy expended in foraging), and feeding efficiencies of egrets associated with and not associated with cattle.

very cryptic and cautious to avoid being seen by the host birds, whereas dumpers often laid their eggs conspicuously, in full view of the host. By using model eggs, Smith showed that hosts in colonies with mimetic cowbirds discriminated between eggs and threw mismatched ones out of their nests; in contrast, hosts in colonies with the nonmimetic dumpers did not discriminate, but readily accepted even mismatched *Scaphidura* eggs. Whereas 73 percent of the nondiscriminator nests contained cowbird chicks, only 28 percent of the discriminators had *Scaphidura* in their nests.

Oropendula and cacique colonies are often clustered near the nests of wasps or stingless, but biting, bees (presumably these insects provide effective protection against vertebrate nest predators). However, the major cause of nestling mortality was not vertebrates, but rather botflies of the genus *Philornis*, which lay their eggs on the baby birds. Upon hatching, botfly larvae (maggots) burrow into the chick's body and feed upon its tissues; chicks with more than seven maggots normally die. By examining chicks and old nests for botfly larvae and pupae, respectively, Smith found that colonies near large nests of bees or wasps had a substantially lower incidence of botfly parasitism than did colonies without bees or wasps. Even within wasp-protected colonies, nests farther than 7 meters away from the wasp nest were more likely to have botflies than those closer to the wasp nests. The precise mechanism of botfly protection provided by proximity to bees or wasps is unknown, but it seems likely that the latter insects actively capture the former. Colonies protected against *Philornis* by bees or wasps were invariably composed of discriminator hosts and mimetic cowbirds, whereas colonies without bees or wasps were always composed of nondiscriminator hosts and nonmimetic cowbirds of the "dumper" type. However, even in colonies unprotected by bees or wasps, host chicks in nests containing at least one cowbird chick were seldom infested with botfly maggots (8.5 percent), whereas those in nests without *Scaphidura* chicks were heavily parasitized (90 percent of all such chicks had botflies). *Scaphidura* chicks, which are precocial, actively preen their nest mates and remove botfly eggs and maggots. (Occasionally, cowbird chicks even capture adult botflies!) Thus, the cowbird chicks protect the host chicks from botfly infestation.

Smith kept careful records of the composition of 4807 nests and the fledging success of hosts and cowbirds (Figure 5.23), which clearly show that

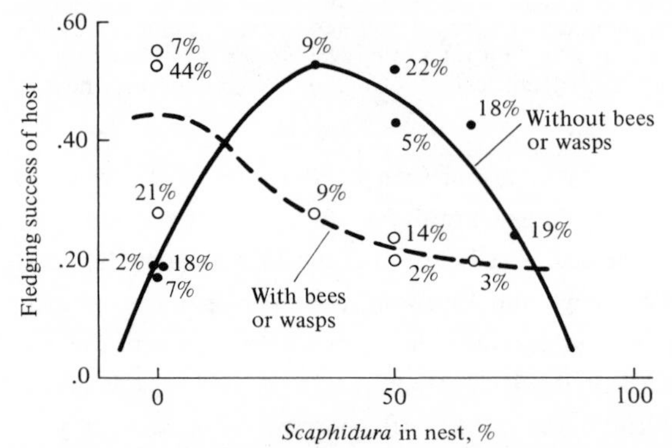

FIGURE 5.23 The fledging success of host oropendulas and caciques plotted against the percentage of cowbird chicks in the nest for broods with different compositions of host chicks and cowbird chicks (total brood size also varies). Colonies without bees or wasps depicted with closed circles and a solid line; those with bees or wasps with open circles and a dashed line. Percentages represent the fraction of host's nests observed in each of the various situations. [Data from N. Smith (1968).]

nondiscriminating hosts can gain an advantage from association with cowbirds, whereas reproductive success of discriminator hosts is reduced by *Scaphidura* parasitism. The average number of host chicks fledged per nest in colonies with and without bees or wasps was 0.39 and 0.43, respectively. Similar average numbers of young fledged per nest for cowbirds in the two types of colonies were 0.76 and 0.73. Hence reproductive success of both host and brood parasite is nearly the same under either situation, suggesting that a delicate balance has been reached (if one strategy were markedly better than the other, one would not expect to observe birds doing both). This example thus illustrates an intricate interplay between parasitism (mimetic cowbird and discriminator host, botfly and oropendula or cacique chicks), protocooperation (nonmimetic cowbird and nondiscriminator hosts), and commensalism (oropendulas–caciques and bees or wasps).

Selected References

Introduction

Krebs (1972); MacArthur (1972); MacArthur and Connell (1966); MacArthur and Wilson (1967); Odum (1959, 1971).

Competition

Birch (1957); Crombie (1947); Elton (1949); Hazen (1964, 1970); Miller (1967); Milne (1961); Milthorpe (1961).

Lotka-Volterra Equations and Competition Theory
Andrewartha and Birch (1953); Bartlett (1960); Levins (1966, 1968); Lotka (1925); MacArthur (1968, 1972); Pielou (1969); Slobodkin (1962); Vandermeer (1970, 1972); Volterra (1926a, 1926b, 1931); Wangersky and Cunningham (1956); Wilson and Bossert (1971).

Competitive Exclusion
Cole (1960); DeBach (1966); Gause (1934); Hardin (1960); MacArthur and Connell (1966); Patten (1961); Pianka (1972).

The Balance Between Intraspecific and Interspecific Competition
Connell (1961a, 1961b); Fretwell (1972); MacArthur (1972); MacArthur, Diamond, and Karr (1972).

Evolutionary Consequences of Competition
Collier *et al.* (1973); Connell (1961a, 1961b); MacArthur (1972); MacArthur and Wilson (1967); Orians (1962); Pianka (1973); Ricklefs and Cox (1972).

Laboratory Experiments
Gause (1934, 1935); Gill (1972); Harper (1961a, 1961b); Lerner and Ho (1961); Neill (1972); Neyman, Park, and Scott (1956); Park (1948, 1954, 1962); Park, Leslie, and Mertz (1964); Wilbur (1972).

Evidence from Nature
Beauchamp and Ullyott (1932); Brown and Wilson (1956); Cody (1968); Connell (1961a, 1961b); Crowell (1962); Dayton (1971); Elton (1946, 1949, 1958); Gadgil and Solbrig (1972); Hairston (1951); Hutchinson (1959); Kohn (1959, 1968); MacArthur (1958); Orians and Willson (1964); Pianka (1969, 1973); Pittendrigh (1961); Schoener (1965); Terborgh and Weske (1969); Vaurie (1951); Williams (1964).

Predation

Errington (1946); Janzen (1971b); MacArthur (1972); MacArthur and Connell (1966); Wilson and Bossert (1971).

Theories: Predator-Prey Oscillations
Elton (1942); Errington (1946); Gause (1934, 1935); Holling (1959a, 1959b, 1966); Keith (1963); Levins (1966); Lotka (1925); Pielou (1969); Rosenzweig (1971, 1973a, 1973b); Rosenzweig and MacArthur (1963); Solomon (1949); Volterra (1926a, 1926b, 1931); Wangersky and Cunningham (1956); Wilson and Bossert (1971).

"Prudent" Predation and Optimal Yield
Beverton and Holt (1957); MacArthur (1960b, 1961); Slobodkin (1968).

Selected Experiments and Observations
Connell (1970); Errington (1946, 1956, 1963); Force (1972); Gause (1934, 1935); Holling (1959a, 1965); Huffaker (1958); Menge (1972); Murdoch (1969); Neill (1972); Paine (1966); Salt (1967); Utida (1957); Wilbur (1972).

Evolutionary Consequences of Predation: Predator Escape Devices
Benson (1933); Benson (1972); Cott (1940); Feeny (1968, 1970); Fisher (1958b); Gordon (1961); Janzen (1966, 1967, 1970, 1971b); Jones (1962, 1966); Kettlewell (1956, 1958); Otte and Williams (1972); Sheppard (1959); Stahl (1888); Whittaker and Feeny (1971).

Coevolution

Brower (1969); Brower and Brower (1964); Caswell *et al.* (1973); Chambers (1970); Dressler (1968); Ehrlich and Raven (1964); Faegri and van der Pijl (1971); Gilbert (1971, 1972); Gordon (1961); Heinrich and Raven (1972); Janzen (1966, 1967, 1971a, 1971b); Kircher and Heed (1970); Kircher *et al.* (1967); C. Smith (1970).

Symbiotic Relationships

Allee (1951); Allee *et al.* (1949); Heatwole (1965); N. Smith (1968).

6 The Ecological Niche

The concept of the niche pervades all of ecology; indeed, were it not for the fact that the ecological niche has been used in so many different ways, ecology might almost be defined as the study of niches. Many aspects of the niche have been considered earlier (others are examined in Chapters 7 and 8); this chapter is especially intricately interrelated to both Chapters 4 and 5.

History and Definitions

Udvardy (1959) ascribed the first use of the term niche to Grinnell (1917, 1924, 1928). Grinnell's conception of niche was that it constituted the functional role and position of the organism in the community. As such, he considered it essentially a behavioral unit, although he also stressed the niche as the ultimate distributional unit, thereby including spatial features of the physical environment in his meaning of niche. Later, Elton (1927) defined the niche of an animal as "its place in the biotic environment, *its relations to food and enemies*" (his italics), and also as "the status of an organism in its community." He further stated that "the niche of an animal can be defined to a large extent by its size and food habits." Others, such as Dice (1952), have used the term to refer to a subdivision of the habitat; thus Dice states "the term (niche) does not include, except indirectly, any consideration of the function the species serves in the community." These two separate meanings in use for the term niche were clearly distinguished by Clarke (1954), who referred to them as the "functional niche" and the "place niche." Clarke noted that different species of animals and plants fulfill different functions in the ecological complex, and that the same functional niche may be filled by quite different species in different geographical regions. The idea of such "ecological equivalents" was first stressed by Grinnell in 1924 (see also Chapter 7, pp. 244–246).

Without doubt the most influential modern treatment of niche is that of Hutchinson (1957a). Using set theory he treats the niche somewhat more formally and defines it as the total range of conditions under which the individual (or population) lives and replaces itself (see also The Hypervolume Model, p. 190). Hutchinson's examples for niche coordinates are nonbehavioral, and have thus emphasized the niche as a place in space, rather like microhabitat, or the "habitat niche" of Allee *et al.* (1949). This emphasis is unfortunate to the extent that it tends to exclude the "behavioral niche" from consideration. Hutchinson's distinction between the fundamental and the realized niche is one of the most explicit statements that an animal's potential niche is seldom fully utilized at a given moment in time or place in space. This distinction has proven useful in clarifying the role of other species, both competitors and predators, in determining the niche of an organism.

More recently, Odum (1959) defined the ecological niche as "the position or status of an organism within its community and ecosystem resulting from the organism's structural adaptations, physiological responses, and specific behavior (inherited and/or learned)." He emphasized that "the ecological niche of an organism depends not only on where it lives but also on what it does." The place an organism lives, or where one would go to find it, is its habitat. Odum uses the analogy that the habitat is the organism's "address," whereas the niche is its "profession."

It has been suggested that the concept of niche is so broad as to be vague and ambiguous. Because of this, Weatherley (1963) suggested that the definition of niche be restricted to "the nutritional role of the animal in its ecosystem, that is, its relations to all the foods available to it." Some other ecologists prefer, however, to leave the term niche defined rather broadly and, when necessary, to subdivide it into components, such as the "food niche" and the "place niche."

Because concepts of the ecological niche have taken on so many different forms, it is often difficult to be sure of exactly what a particular ecologist means when he invokes this entity. Some therefore avoid using the word altogether and insist that we can get along perfectly well without it. However, no one denies that there is a broad zone of interaction between the traditional entities of "environment" and "organismic unit"; the major problem is to specify precisely in any given case just what subset of this enormous subject matter should be considered the "ecological niche." In this chapter a definition of niche is developed and various aspects, such as niche breadth and niche overlap, are treated in detail.

Adaptation and Deterioration of the Environment

Organisms are adapted to their environments in that, in order to survive, they must meet their environment's conditions for existence. Adaptation can be simply defined as the *conformity between organism and environment.* Plants and animals have adapted to their environments in innumerable ways, many of which have already been considered; thus, they can adapt either genetically or by means of physiological and/or developmental flexibility (the former would include instinctual behavior, and the latter includes learning).

Adaptation has many dimensions in that most organisms conform simultaneously to numerous different aspects of their environments. Thus, an organism must be adapted to cope with, all at the same time, intraspecific competition, interspecific competition, predation, its physical environment, etc. Conflicting demands of these various environmental components often require that an organism compromise in its adaptations to each. Conformity to any given component takes a certain amount of energy that is then no longer available for other adaptations. As an example, the presence of predators may require that an animal be wary, which in turn is likely to reduce its foraging efficiency and, hence, its competitive ability.

It is relatively easy for organisms to conform to and cope with highly predictable environments as long as they are not too extreme, even when they change in a regular way. It may usually be much more difficult to adapt to an unpredictable environment, and extremely unpredictable environments may prove impossible. Many organisms have evolved dormant stages in order to survive unfavorable periods, both predictable and unpredictable. Annual plants everywhere and brine shrimp in deserts are good examples. Brine shrimp eggs survive for years in the salty crust of dry desert lakes; when a rare desert rain fills one of these lakes, they hatch, grow rapidly to adults, and produce many eggs. Some seeds known to be hundreds of years old are still viable and have been germinated. Changes in the environment which reduce overall adaptation are collectively termed the "deterioration of environment"; such changes cause directional selection resulting in accommodation to the new environment.

A simple model of adaptation and undirected environmental deterioration was developed by Fisher (1930). He reasoned that no organism is "perfectly adapted," but that all fail to conform to their environments in some ways and to differing degrees. However, a hypothetical, perfectly adapted organism can be imagined, against which existing organisms may be compared. Fisher's mathematical argument is phrased in terms of an infinite number of "dimensions" for adaptation, but only three are used here for ease of illustration. One can envision an adaptational space of three coordinates representing, say, the competitive, predatory, and physical environments, respectively (Figure 6.1*a*). An ideal "perfectly adapted" organism lies at a particular point (say *A*) in this space, but any given organism is at another point (say *B*), some distance, *d*, away from this point of perfect adaptation. Changes in the position of point *A* correspond to environmental changes making the optimally adapted organism different, whereas changes in point *B* represent changes in the organism concerned, such as mutations. The distance between the two points, *d*, represents the degree of conformity between organism and environment or the level of adaptation. Fisher noted that very small *undirected* changes in either organism or environment have a 50 : 50 chance of being to the organism's advantage (that is, of reducing the distance between points *A* and *B*). The probability of such improvement is inversely related to the magnitude of the change as shown in Figure 6.1*b*. Very great changes in either organism or environment are always maladaptive, because even if they are in the right direction, they "overshoot" points of closer adaptation (of course, it is remotely possible that such major environmental changes or "macromutations" could put an organism into a completely new adaptive realm and thereby improve its overall level of adaptation). Fisher makes an analogy with focusing a microscope: very fine changes are as likely as not to improve focus, but grosser changes will almost invariably throw the machine further out of focus. Organisms may be thought of as "tracking" their environments in both ecological and evolutionary time, changing as they change; thus, as point *A* shifts because of daily, seasonal, and long-term environmental fluctuations, point *B* follows it.

Individual organisms with narrow tolerance limits, such as highly adapted specialists, generally suffer greater losses in fitness due to a unit of environmental deterioration than generalized organisms with more flexible requirements, all else being equal. Thus, more specialized organisms and/or those with restricted homeostatic abilities are able to tolerate less environmental change than generalists or organisms with well developed homeostasis (Figure 6.1*b*). That is, organisms that are either better adapted (smaller *d*) and/or those that are adapted to their environments in a greater number of

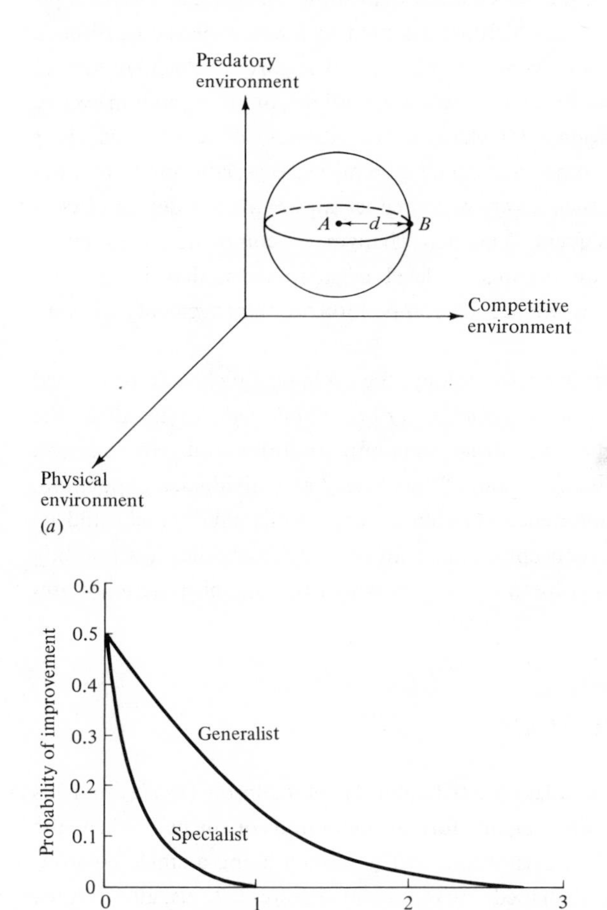

FIGURE 6.1 Fisher's model of adaptation and deterioration of the environment. (*a*) Point *A* represents a hypothetical "perfectly adapted" organism; an actual organism (point *B*) is never perfectly adapted and thus lies at some distance, *d*, away from point *A*. The surface of the sphere represents all possible points with a level of adaptation equal to the organism under consideration. Very small undirected changes in either organism (point *B*) or environment (point *A*) are equally likely to increase or to decrease the level of adaptation, *d*. (*b*) The probability of improvement of the level of adaptation (that is, of reducing *d*) plotted against the magnitude of an undirected change when the number of dimensions is large. Two hypothetical organisms are shown, one highly adapted such as a specialist with narrow tolerance limits and one less highly adapted such as a generalist with broader tolerance limits and/or a greater number of niche dimensions. A random change of a given magnitude is more likely to improve the level of adaptation of the generalist than the specialist. [Partially adapted from Fisher (1958a).]

dimensions can tolerate greater changes than those which are less well adapted (greater d) and/or those which are adapted to fewer dimensions. Fisher's model applies only to *nondirected* changes in either party of the adaptational complex, such as mutations and climatic fluctuations, or other random events. However, many environmental changes are probably directed. Thus, it is likely that changes in other associated organisms, especially predators, are invariably directed in such a way as to reduce an organism's degree of conformity to its environment. Directed changes in competitors can either increase or decrease an organism's level of adaptation, depending upon whether they involve avoidance of competition or improvements in competitive ability, per se.

Following the above terminology, the ecological niche can be defined as *the sum total of the adaptations of an organismic unit*, or as all of the various ways in which a given organismic unit conforms to its environment. As with environment, we can speak of the niche of an individual, a population, or of a species. The difference between an organism's environment and its niche is that the latter concept includes an organism's abilities at exploiting its environment and involves the ways in which the organism actually *uses* its environment.

The Hypervolume Model

Building on the law of tolerance (Chapter 1), Hutchinson (1957a) and his students constructed an elegant formal definition of niche. When the tolerance or fitness of an organismic unit is plotted along a single environmental gradient, a bell-shaped curve, as in Figure 6.2, usually results. Tolerances for two different environmental variables can be plotted simultaneously as in Figure 6.3. In Figure 6.4, hypothetical tolerances for three

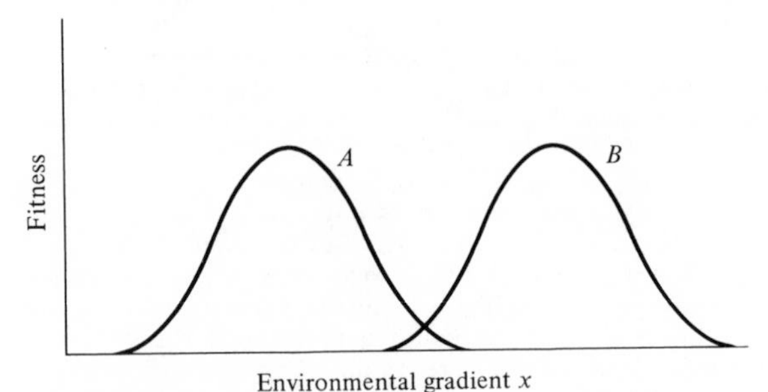

FIGURE 6.2 Fitness plotted along a single environmental gradient, x, for two organismic units, A and B.

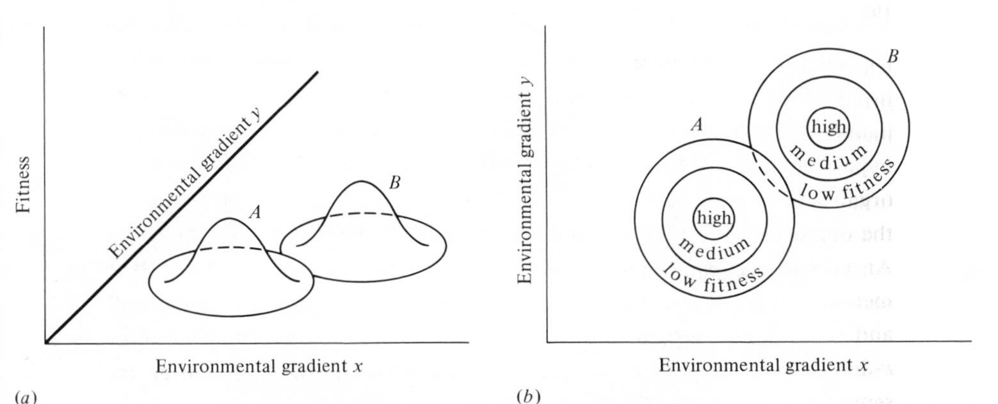

FIGURE 6.3 Two plots of the fitnesses of two organismic units, *A* and *B*, versus their position along two environmental gradients, *x* and *y*. (*a*) A three-dimensional plot analogous to Figure 6.2. (*b*) A two-dimensional plot with the fitness axis omitted; low, medium, and high fitness represented by contour lines.

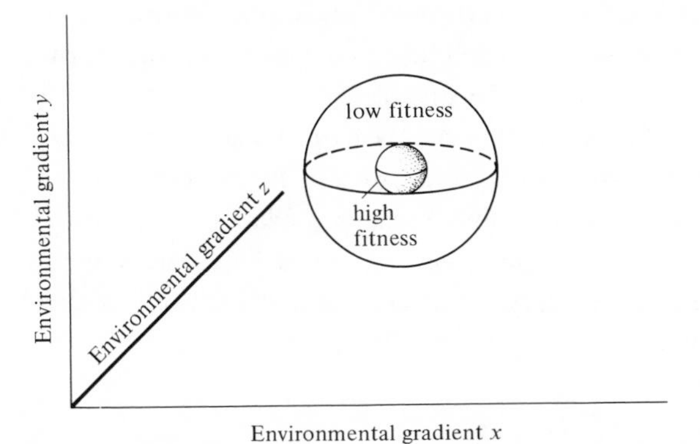

FIGURE 6.4 A plot like Figure 6.3*b* of fitness along three different environmental gradients, *x*, *v*, and *z*, showing zones of low and high fitness. A four-dimensional plot with a fitness axis analogous to Figure 6.3*a* is implicit in this graph.

different variables are plotted in a three-dimensional space. Adding each new environmental variable simply adds one more axis and increases the number of dimensions of the plot by one. Actually, Figure 6.4 represents a four-dimensional space, with the three axes shown representing the three environmental variables, while the fourth axis (this one cannot be shown directly but is implicit in the figure in the same way that Figure 6.3*b* is implicit in Figure 6.3*a*) represents reproductive success or some other convenient measure of performance, which we will call *fitness density*. Parts of this space with high fitness density are relatively optimal for the organism concerned, whereas those with low fitness density are suboptimal. Conceptually, this process can be extended to any number of axes, using *n*-dimensional geometry. Thus, Hutchinson defines an organism's niche as an *n*-dimensional hypervolume enclosing the complete range of conditions under which that organism can successfully replace itself. All variables relevant to the life of the organism must be included, and all must be independent of each other. An immediate difficulty with this model of the niche is that not all environmental variables can be nicely ordered linearly. In order to avoid this problem and to make the entire model more workable, Hutchinson translated the *n*-dimensional hypervolume formulation into a set theoretic mode of representation. Unfortunately, the fitness density attributes of the *n*-dimensional model are lost in the conversion to a set theory model.

Hutchinson designates the entire set of optimal conditions under which a given organismic unit can live and replace itself as its *fundamental niche*, which can then be represented as a set of points in environmental space. The fundamental niche thus refers to a hypothetical idealized niche in which the organism encounters no "enemies" such as competitors and predators and in which its physical environment is optimal. In contrast, the actual set of conditions under which an organism exists, which is always less than or equal to the fundamental niche, is termed the *realized niche*. As such, the realized niche takes into account various forces restricting an organismic unit, such as competition and perhaps predation. The fundamental niche has also been called the *precompetitive* or *virtual* niche, whereas the realized niche is the *postcompetitive* or *actual* niche. These two concepts are thus somewhat analogous to the notions of r_{max} and r_{actual}, as discussed in Chapter 4 (pp. 79–87).

Niche Overlap and Competition

Niche overlap occurs when two organismic units use the same resources or other environmental variables. In Hutchinson's terminology, each *n*-dimensional hypervolume includes part of the other, or some points in the two sets that constitute their realized niches are identical. Overlap is complete when two organismic units have identical niches, and there is no overlap if two niches are completely disparate. More usually, niches overlap only partially, with some resources being shared and others being used exclusively by each organismic unit.

Hutchinson (1957a) treats niche overlap in a simplistic way, assuming that the environment is fully saturated, that niche overlap cannot be tolerated for any period of time, and that competitive exclusion must occur in the overlapping parts of any two niches. Thus, competition is assumed to be intense and results in only a single species surviving in contested niche space. While this oversimplified approach has its shortcomings, it is useful to examine each of the logically possible cases (Figure 6.5) before considering niche overlap and competition in a more realistic way. First, two fundamental niches could be identical, corresponding exactly to one another, although such ecologic identity is infinitely unlikely. In this very improbable event, the organismic unit that is competitively superior excludes the other. Second, one fundamental niche might be completely included within another (Figure 6.5*a*); given this situation, the outcome of competition depends upon the relative competitive abilities of the two organismic units. If the one with the included niche is competitively inferior, it is exterminated and the other occupies the entire niche space, whereas if the former organismic unit is competitively superior, it eliminates the latter from the contested niche space. The two organismic units then coexist with the competitively superior one occupying a niche included within the niche of the other. Third, two fundamental niches may overlap only partially, with some niche space being shared and some used exclusively by each organismic unit (Figure 6.5*b* and *c*). In this case each organismic unit has a "refuge" of uncontested niche space and coexistence is inevitable, with the superior competitor occupying the contested (overlapping) niche space. Fourth, fundamental niches might abut against one another as in Figure 6.5*d*; although no direct competition can occur, such a niche relationship may reflect the avoidance of competition. Finally (Figure 6.5*e*), if two fundamental niches are entirely disjunct (no overlap), there can be no competition and both organismic units occupy their entire fundamental niche. Figure 6.6 illustrates the distinction between the fundamental and the realized niche for an organismic unit with six competitors.

Of course, a major shortcoming of the preceding discussion is that, in nature, niches often do overlap yet competitive exclusion does not take place. Niche overlap, in itself, obviously need not necessitate competition. Should resources not be in short supply, two organismic units can share them without detriment to one another. In fact, extensive niche overlap may often be correlated with *reduced* competition, just as disjunct niches may frequently indicate avoidance of competition in situations where it could potentially be severe. For these reasons, the ratio of demand to supply, or the degree of saturation, is of vital concern in the relationship between ecological overlap and competition. Indeed, much current research is designed to clarify, both theoretically and empirically, the relationship between competition and niche overlap; as pointed out in Chapter 5, modern ecologists

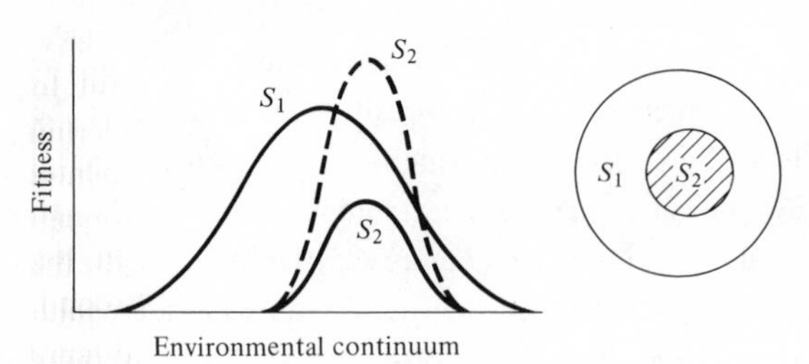

(*a*) Included

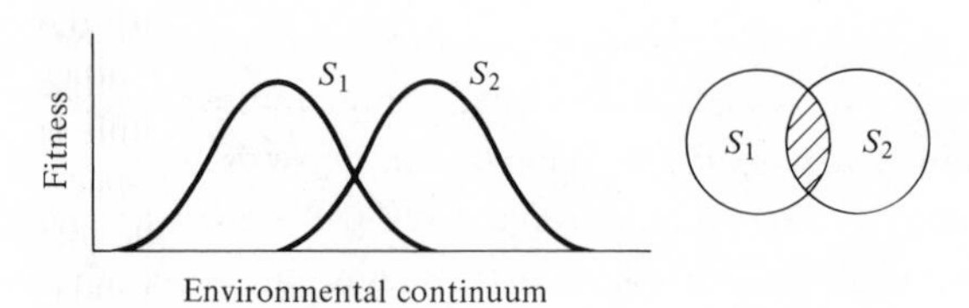

(*b*) Equal overlap

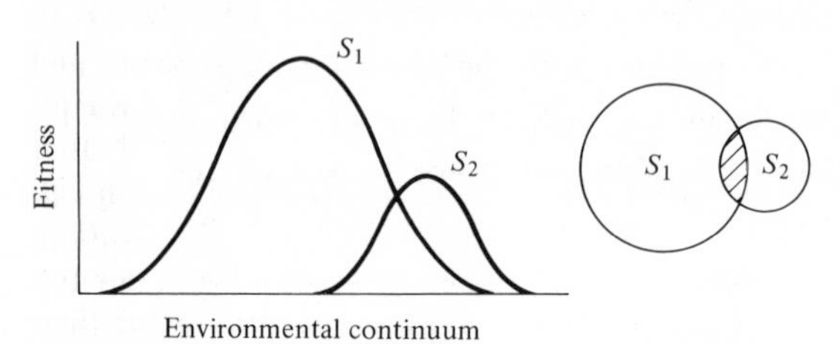

(*c*) Unequal overlap

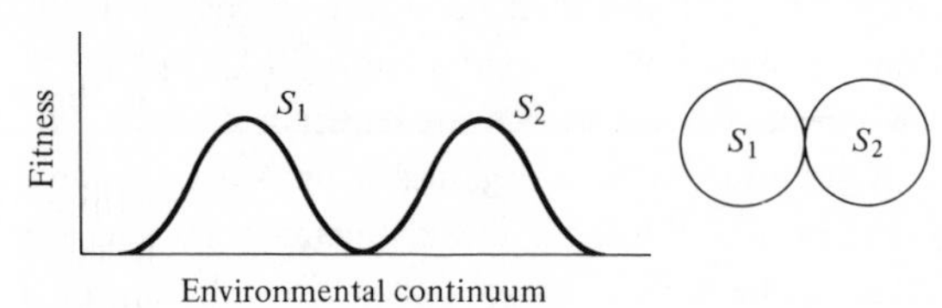

(*d*) Abutting

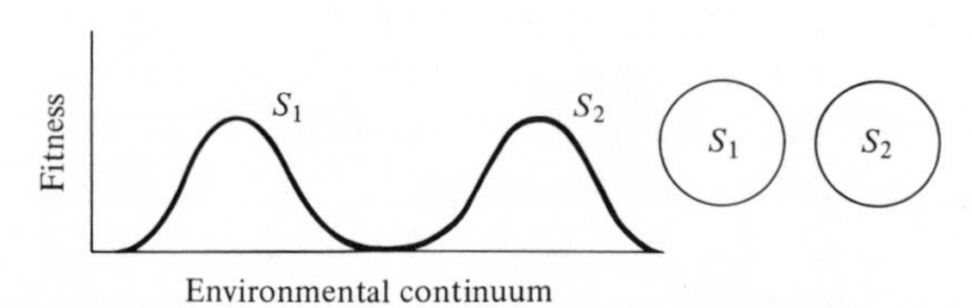

(*e*) Disjunct

FIGURE 6.5 Various possible niche relationships, with fitness density models on the left and set-theoretic ones on the right. (*a*) An included niche. The niche of species 2 is entirely contained within the niche of species 1. Two possible outcomes of competition are possible: (1) If species 2 is superior (dashed line) it persists and species 1 reduces its utilization of the shared resources. (2) If species 1 is superior (solid lines), species 2 is excluded and species 1 uses the entire resource gradient. (*b*) Overlapping niches of

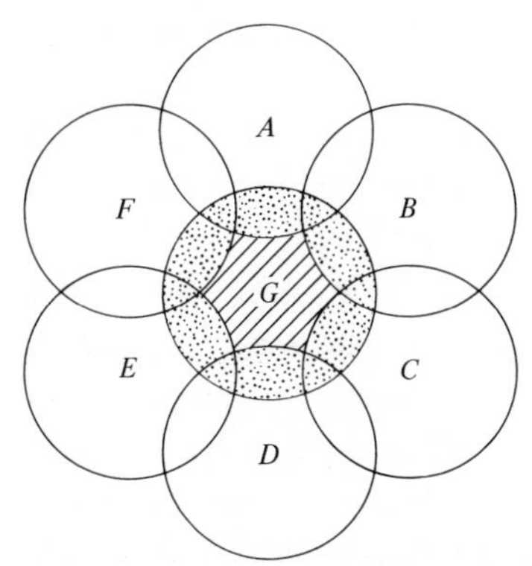

FIGURE 6.6 Set theoretic mode of representation of the fundamental niche (stippled and cross-hatched) of species *G*, and its realized niche (cross-hatched), which is a subset of the fundamental niche, after competition and complete competitive displacement due to six superior competitors, species *A*, *B*, *C*, *D*, *E*, and *F*.

are asking questions such as "how much niche overlap can coexisting species tolerate?" and "how does this maximal tolerable niche overlap vary with the degree of saturation?" Because the differential between supply and demand along any given resource spectrum should be approximately constant (see Figure 5.6 and pp. 142–143), however, the amount of niche overlap along any given gradient should at least be *proportional* to the intensity of competition.

Niche Dynamics

Realized niches of most organisms change both in time and from place to place as physical and biotic environments vary. Temporal niche changes can be considered at two levels: (1) on a short-term basis (that is, in an ecological time scale), usually during the life of a single individual, or, at most, a few generations, and (2) on a long-term basis (in an evolutionary time scale), that is, over geological time and many generations (see section on Evolution of Niches). Thus, the realized niche can be thought of as an ever

---

equal breadth. Competition is equal and opposite. (*c*) Overlapping niches of unequal breadth. Competition is not equal and opposite because species 2 shares more of its niche space than does species 1. (*d*) Abutting niches. No direct competition is possible, but such a niche relationship may result from competition in the past and be indicative of the avoidance of competition. (*e*) Disjunct niches. Competition cannot occur and, indeed, it is not even implicit in this case.

changing subset of the fundamental niche, or, in the *n*-dimensional hypervolume model, as a pulsing hypervolume bounded by the hypervolume corresponding to the fundamental niche.

Some organisms, such as insects, have entirely disjunct, non-overlapping, niches at different times in their life histories: thus larval mosquitos (commonly referred to as "wriggleworms") are aquatic herbivores whereas the familiar adults are terrestrial flying carnivores. Many other examples could be cited, such as caterpillars and butterflies, maggots and flies, tadpoles and toads, and planktonic larval and sessile adult barnacles. In all these cases, drastic and major modification of an animal's body plan at metamorphosis allows a pronounced niche shift. Niches of other organisms change more gradually and continuously during their lifetimes. Thus juvenile lizards eat smaller prey than adults, and are often active earlier at lower environmental temperatures (their smaller size and greater surface to volume ratio facilitates faster warming). Incomplete metamorphosis, as in grasshoppers, in which an animal changes rather gradually, is similar.

An organism's immediate neighbors in niche space, or its potential competitors, can, but need not, exert strong influences upon its ecological niche. While realized niches of relatively *r*-selected organisms are determined primarily by their physical environments, realized niches of more *K*-selected organisms are perhaps more strongly influenced by their biotic environments. Selective pressures, and niches, may vary during an individual's life. Thus, in temperate zones, early spring is a time when annual plants are relatively *r* selected, whereas later in the season, they become progressively more *K* selected (Gadgil and Solbrig, 1972). Similarly, even within a given species, some organisms may be more *r* selected than others, such as populations in different parts of a species' geographic range or individuals at different positions in the rocky intertidal.

Theoretically, reduced interspecific competition should often allow niche expansion. In an attempt to observe this, Crowell (1962) examined and compared ecologies of three species of birds on Bermuda with those of mainland populations. There are many fewer species of land birds on the island than on the mainland, with the three most abundant being the cardinal, catbird, and white-eyed vireo. On Bermuda, these three species have very dense populations; however, in mainland habitats where there are many more species of birds, and thus a greater variety of interspecific competitors, populations of these same three species are usually considerably smaller. Although there are, of course, inevitable differences between the habitats of Bermuda and mainland North America, the striking difference between avifaunas should nevertheless have a major effect on the ecologies of the birds. Although the catbird and cardinal have generally more restricted place and foraging niches on the island (perhaps the available niches are more restricted), the white-eyed vireo has expanded both its place niche and foraging niche. All three species nest at a wider variety of heights on Bermuda (see also Chapter 5, p. 153).

The effects of interspecific competition on niche breadth are complex, and, under different conditions, may actually favor niche contraction or niche expansion. Thus, a competitor that reduces food availability in some microhabitats, but leaves prey densities in other microhabitats unaltered, effectively reduces the expectation of yield in some patches but not others. As a result, a competitor that is an optimal forager should restrict its patch utilization to those with higher expectation of yield, thus decreasing the breadth of its place niche (see also pp. 108–109 and pp. 264–266). Conversely, a competitor that reduces food availability more or less equally in all microhabitats, by reducing the overall level of prey availability, can force its competitor to expand the range of resources used, and thus increase the breadth of its food niche. The reason for this is that, in a food-sparse environment, an optimal forager simply cannot afford to bypass as many potential prey items as it can in a food-dense environment; as a result, more suboptimal prey are eaten in the former type of habitat (see also pp. 201–210 and pp. 264–265). Reduced interspecific competition is often accompanied by an increase in the range of habitats a species uses, but marked changes in the variety of foods eaten with changes in interspecific competition seem to be much less common (MacArthur, 1972).

Niche Dimensionality

Although the *n*-dimensional hypervolume model of the niche is extremely powerful conceptually, it is too abstract to be of much practical value and is rather difficult to apply to the real world. Indeed, in order to actually construct such a hypervolume, we would have to know essentially everything about the organism concerned. Certainly, because we can never know *all* the factors impinging upon an organismic unit, the fundamental niche must remain an abstraction. Even the realized niches of the majority of organisms have so many dimensions that they defy quantification. When considering relatively *K*-selected organisms, the number of niche dimensions can be limited to those dimensions in which competition is effectively reduced. Competition is often avoided by differences in the place niche, time niche, and/or the food niche (Chapter 5), and so the effective number of niche

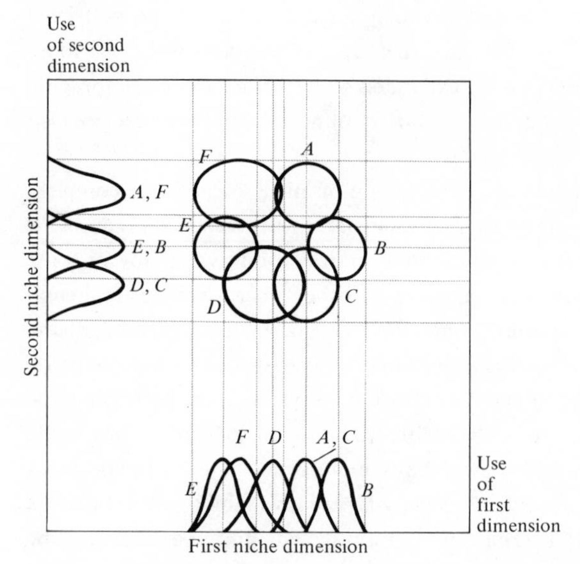

FIGURE 6.7 Illustration of hypothetical niches of six species differing along two different, independent, niche dimensions. Although overlap is broad or complete along either single dimension alone (bell-shaped curves), niches overlap minimally or not at all when both niche dimensions are considered (circles and ellipses).

dimensions can be reduced to only three: place, time, and food. We can think of a saturated community as occupying some volume in a space of these three dimensions; thus, a community is something like a three-dimensional jigsaw puzzle, with each piece being one species that occupies only a part of the overall volume.

Niche dimensionality, or the number of ways in which niches differ, however, is important for the following reasons. A greater number of niche dimensions means that any given niche has the potential for more immediate neighbors in niche space. In a one-dimensional niche space (Figure 6.2), any given niche can be bounded only on two sides, whereas there can be many more neighbors in a two-dimensional niche space (Figure 6.3), and still more in three or more dimensions. Another feature of increased niche dimensionality is that niches may be overlapping or identical along one axis and yet separated or even disjunct along another (Figure 6.7). Thus an observer oblivious to the first niche dimension in Figure 6.7 would consider species pair *A* and *F*, pair *B* and *E*, and pair *C* and *D* as broadly or completely overlapping, when in fact they are partially or entirely separated along the unknown dimension!

Specialization versus Generalization

Some organisms have smaller niches than others. Niche breadth, also called "niche width" and "niche size," can be thought of as the extent of the hypervolume representing the realized niche of an organismic unit. Thus a koala bear, *Phascolarctos cinerus*, which eats only leaves of certain species of *Eucalyptus*, has a more specialized food niche than the Virginia opossum, *Didelphis virginianus*, which is a true omnivore that eats nearly anything. Statements about niche breadths must invariably be comparative, and we can only say that a given organismic unit has a niche that is narrower or broader than that of some other organismic unit. Highly specialized organisms, like the koala, usually, though not always, have narrow tolerance limits along one or more of their niche dimensions. Often such specialists have very specific habitat requirements and as a result they may not be very abundant. In contrast, organisms with broad tolerances are typically more generalized, with more flexible habitat requirements, and are usually much more common. In other words, specialists are often rare while generalists are more abundant. Rare organisms may, however, frequently occur in clumps so that their local density need not necessarily be low.

The only currency of natural selection is differential reproductive success. This fact raises a question: If specialization involves becoming less abundant, why have organisms become specialized at all? Since generalized organisms can usually exploit more food types, occupy more habitats, and build up larger populations, they might be expected, by their very numbers, to outreproduce slightly more specialized competing members of their own population and swamp the population gene pool. The answer to this apparent dilemma lies in the old adage that a jack-of-all-trades is a master-of-none. More specialized individuals are more efficient on their own grounds than generalists.

Under what conditions will a jack-of-all-trades win in a competition with more specialized species? MacArthur and Levins (1964, 1967) considered this question and developed the following model. First, imagine an ant-eating lizard in an environment that contains only a single food resource type, say colonies of ants 3-mm in length, and a variable population of ant-eating lizards that are exploiting the ant populations. Assume that ants are eaten whole and that lizards differ only in the size of their jaws and form a fairly continuous phenotypic spectrum. Some phenotypes will be best adapted to use the 3-mm ants and very effective at harvesting them, whereas others will be less efficient, either because their jaws are too large or too small. Next, consider another "pure" environment, this one composed solely of colonies of 5-mm ants, and the same phenotypic spectrum of lizards. Almost certainly, the best adapted phenotype will be different from that in the 3-mm ant environment (Figure 6.8*a*), and the phenotype most efficient at using 5-mm ants will be one with a larger mouth. Now consider a *mixed* ant colony with equal numbers of 3-mm and 5-mm ants (say, two castes) in a homogeneous

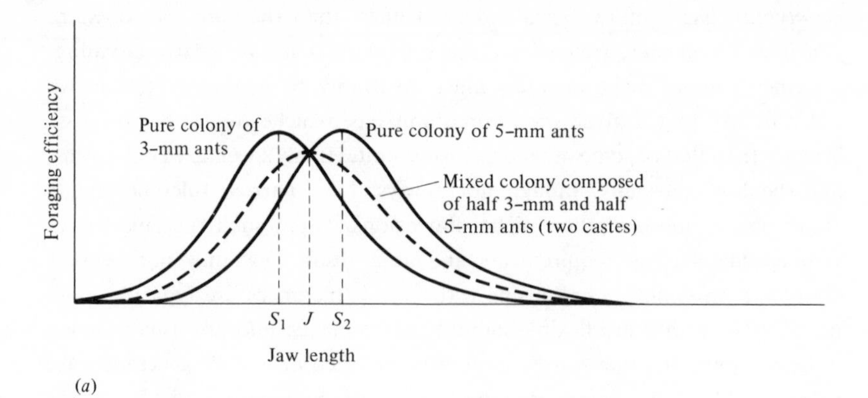

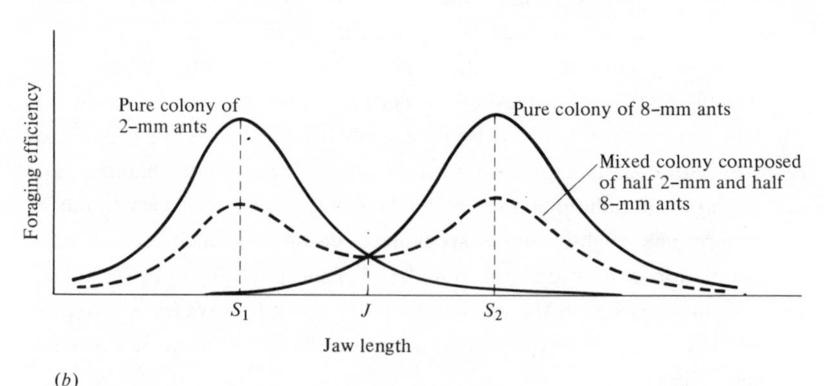

FIGURE 6.8 Example of the conditions under which a jack-of-two-trades outperforms two specialists. Foraging efficiency, measured in useful calories gathered per unit of foraging time, is plotted against a phenotypic spectrum of jaw lengths for a hypothetical population of ant-eating lizards. (*a*) Solid curves: performance of various lizard phenotypes in "pure" environments containing only 3-mm and 5-mm ants, respectively Dashed line: performance of various phenotypes in an environment containing a fine-grained mixture of equal numbers of both 3-mm and 5-mm ants (say, two castes). The jack-of-both-trades, *J*, has a higher foraging efficiency than either specialist, S_1 or S_2, which are the most efficient foragers in the respective "pure" environments. (*b*) A similar plot, but showing foraging efficiencies of the various lizard phenotypes when the *difference* between the two ant types is greater, say, 2-mm versus 8-mm ants. In this case, performance on a fine-grained mixture is bimodal, and the two specialists outperform the jack-of-both-trades. [After MacArthur and Connell (1966).]

mixture. Which phenotype will be optimal in this new mixed environment? Assuming that the lizards encounter and use the two ant sizes in exactly equal proportions, the relationship between phenotype and harvesting effectiveness must be exactly intermediate between the two similar relationships in "pure" environments (Figure 6.8*a*). The dashed line can thus be drawn in Figure 6.8*a* midway between the first two (if the two resources were not in exactly equal proportions, this new line would simply be closer to one or the other of the original lines); depending upon the shapes of the curves and the distance between them, this new line can take either a unimodal (Figure 6.8*a*) or a bimodal shape (Figure 6.8*b*). In the former case, the phenotype of highest harvesting efficiency is intermediate between the best "pure 3-mm ant eater" and the best "pure 5-mm ant eater," and this "jack-of-both-trades" (probably the phenotype that could best exploit 4-mm ants) is competitively superior. In the latter case, because the two types of ants are very different in size, the two phenotypes with high harvesting effectiveness are separated from one another by intermediate phenotypes with *lower* efficiencies at exploiting a mixture of 2-mm and 8-mm ants, and the two specialists will eliminate the jack-of-both-trades.

Time, Matter, and Energy Budgets

Any organism has a certain limited amount of time, matter, and energy available to devote to foraging, growth, maintenance, and reproduction. The way in which an organism allocates, or budgets, these resources among various conflicting demands is of fundamental interest. This apportionment determines the ways in which the organism can conform to many aspects of its environment and thus indicates a great deal about its ecological niche. Time and energy budgets vary widely among organisms; for example, relatively *r*-selected organisms allot more time and energy to reproduction than more *K*-selected ones (Chapter 4). Varying time and energy budgeting is a potent means of coping with a changing environment and yet retaining some degree of adaptation to it. Thus, a male marsh wren expends a great deal of energy on territorial defense during the breeding season but little at other times of the year. Similarly, in animals with parental care, an increasing amount of energy is spent on offspring as they grow until some point when progeny begin to become independent of their parents, whereupon the amount of time and energy devoted to them decreases. Indeed, adult female red squirrels, *Tamiasciurus*, at the height of lactation consume an average of 323 kcal of food per day whereas the average daily energy consumption of an adult male is only about 117 kcal (C. Smith, 1968); time budgets of these squirrels also vary markedly with the seasons. In a bad dry year, many annual plants "go to seed" while still very small, whereas in a good wet year, these plants grow to a much larger size before becoming reproductive; presumably more seeds are produced in good years, but perhaps none (or very few) would be produced in a bad year if individuals grew to the sizes they reach in good years.

Small animals, which have relatively high ratios of body surface to body volume, generally have higher energy requirements *per unit of body weight* than do larger ones. However, the *total amount of energy* needed for maintenance (and, therefore, the overall energy budget of the organism) increases with increasing body weight. Body temperatures of cold-blooded animals (*poikilotherms*) approximate the temperature of the environment and may vary widely from time to time, whereas warm-blooded mammals and birds (*homoiotherms*) maintain relatively constant body temperatures. Because energy is required to maintain a constant body temperature, homoiotherms have higher metabolic rates, and higher energy needs (and budgets), than poikilotherms of the same body weight. Moreover, because a high body surface to volume ratio allows a greater loss of heat from the body surface, small homoiotherms must expend relatively more energy maintaining a constant body temperature than larger ones. Hence there is a distinct lower limit on body size for homoiotherms (about 2 or 3 grams), which is the size of a small hummingbird or shrew. Indeed, both hummingbirds and shrews have very precarious energy budgets; they are dependent upon continual supplies of energy-rich foods. (A small hummingbird would starve to death during the night if it did not allow its body temperature to drop and go into a state of torpor!)

In the remainder of this section we investigate some of the ways in which time and energy budgets are influenced by a variety of ecological factors, including body size, mode of foraging, vagility, trophic level, prey size, resource density, environmental heterogeneity, rarefaction, competition, and predation.

Foraging and Feeding Efficiency

Foraging has both profits and costs. Its profits are the matter and energy gathered, which can be used in growth, maintenance, and reproduction. The costs of foraging are perhaps often more elusive, but a foraging animal must expose itself to potential predators and much of the time it spends foraging is unavailable for other activities, including reproduction. Natural selection should favor foraging behaviors that maximize the difference between foraging profits and their costs (see also Chapter 4, pp. 108–109).

Two extreme ways in which carnivorous animals forage are (1) the "sit-and-wait strategy," in which a predator waits in one place until a moving prey item comes by and then "ambushes" the prey, and (2) the "widely-foraging strategy," in which the predator actively searches out its prey (Pianka, 1966b; Schoener, 1969a, 1969b). Obviously the second strategy normally requires a greater energy expenditure than the first. The success of the sit-and-wait tactic usually depends upon one or more of the following conditions: a fairly high prey density, high prey mobility, and/or low predator energy requirements. The success of the widely foraging tactic is also a function of the density and mobility of prey and of the predator's energy needs, but here the distribution of the prey in space and the searching abilities of the predator assume paramount importance. Although these two tactics are actually pure forms of a spectrum of possible foraging strategies and hence are somewhat artificial, the foraging techniques actually employed by many organisms are rather strongly polarized. The dichotomy of sit-and-wait versus widely foraging therefore has substantial practical value. For example, among snakes, racers and cobras forage widely when compared with boas, pythons, and vipers, which are, relatively, sit-and-wait foragers. Similarly, among hawks, accipiters such as the Cooper's Hawk and the Goshawk often hunt by ambush using a sit-and-wait strategy, whereas buteos and falcons are relatively more widely foraging. Web-building spiders and sessile filter feeders such as barnacles are also good examples of animals that typically forage by sitting and waiting. Many spiders expend considerable amounts of energy and time building their webs rather than in moving about in search of prey; those which do not build webs forage much more widely.

Similar considerations can be applied to a comparison of herbivores and carnivores. Because the density of plant food almost always greatly exceeds the density of animal food, herbivores often need expend little energy, relative to carnivores, in finding their prey [of course, to the extent that the secondary chemical substances of plants, such as tannins, and other antiherbivore defenses of plants (pp. 174–176) reduce the palatability of plants or parts of plants, the *effective* supply of plant foods may be greatly reduced]. Because cellulose in plants is difficult to digest, herbivores must, however, expend considerable energy in extracting nutrient from their plant food. (Most herbivores have a large ratio of gut volume to body volume, harbor intestinal organisms that digest cellulose, and spend much of their time eating or ruminating—envision a cow chewing its cud.) By contrast, animal food, composed of readily available proteins, lipids, and carbohydrates, is more readily digested. Carnivores can afford to expend considerable effort in searching for their prey because of the large dividends obtained once they find it. As would be expected, efficiency of conversion of food into an animal's own tissues (*assimilation*) is considerably lower in herbivores than it is in carnivores.

Prey density can strongly influence an animal's time and energy budget. Gibb (1956) watched rock pipits, *Anthus spinoletta*, feeding in the intertidal along the English seacoast in wintertime during two consecutive winters. The first winter was relatively mild and the birds were observed to spend $6\frac{1}{2}$ hours feeding, $1\frac{3}{4}$ hours resting, and 45 minutes fighting in defense of their territories (total daylight slightly exceeded 9 hours). The next winter was much harsher and food was considerably scarcer; the birds spent $8\frac{1}{4}$ hours feeding, 39 minutes resting, and only 7 minutes on territorial defense! Apparently the combination of low food density and extreme cold (homoiotherms require more energy in colder weather) demanded that over 90 percent of the bird's waking hours be spent in feeding and no time remained for frivolities. This example also illustrates that food is less defendable at lower densities as indicated by reduced time spent on territorial defense. Obviously, food density in the second year was near the lower limit that would still allow the survival of rock pipits. When prey items are too sparse, encounters may be so infrequent that an individual cannot survive. Gibb (1960) calculated that, in order to balance their energy budget during the winter, in some places, English tits must find an insect on the average once every $2\frac{1}{2}$ seconds during the daylight hours.

Many carnivores have extremely efficient prey-capturing devices (Chapter 5); often the size of a prey object markedly influences this efficiency. Using simple geometry (Figure 6.9), Holling (1964) estimated the diameter of a prey item that should be optimal for a praying mantid of a particular size. He then offered five hungry mantids prey objects of various sizes and recorded the percentages attacked (Figure 6.10); the mantids were noticeably reluctant to attack prey that were either much larger or much smaller than the estimated optimum! Thus natural selection has resulted in efficient predators both by producing efficient prey-capturing devices and by programming animals in such a way that they are unlikely to attempt to capture decidedly suboptimal items. Larger predators tend to take larger prey than smaller ones (see Figure 5.12). It may, in fact, be better strategy for a large predator to overlook prey below some size and to spend the time that would have been spent in capturing and eating these small items in searching out larger prey. Similarly, the effort a predator will expend on any given prey item is proportional to the size of that item (and hence, expected returns). Thus a lizard sitting and waiting on a perch will not usually bother to go far for very small prey items, but will often move considerable distances in attempts to obtain larger prey.

Because small prey are generally much more abundant than large prey, most animals encounter and eat many more small prey items than large ones. Moreover, small animals that eat small prey items encounter prey of suitable size much more frequently than larger animals that rely upon larger prey items; as a result, larger animals tend to eat a wider range of prey sizes. Because of such increased food niche breadths of larger animals, the size

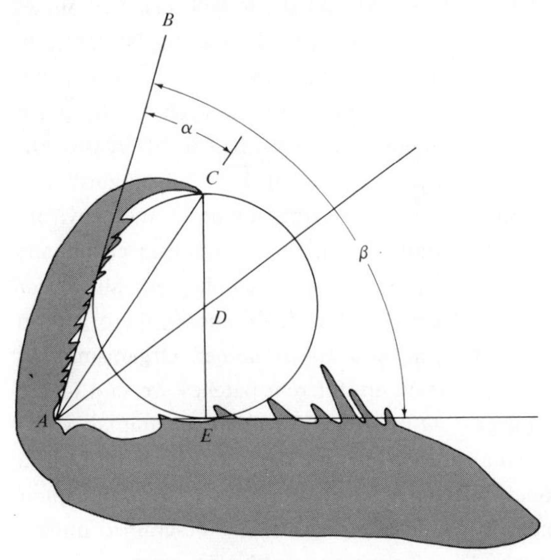

FIGURE 6.9 Diagram showing how Holling used geometry to calculate an estimated optimal size of a prey item from the anatomy of a mantid's foreleg. [From Holling (1964).]

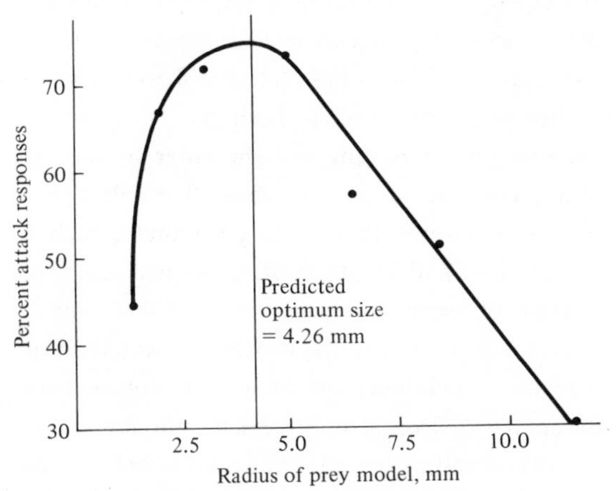

FIGURE 6.10 The percentage of prey items actually attacked by hungry mantids versus prey size. [From Holling (1964)].

differences between predators increase markedly with increasing predator size (MacArthur, 1972).

Optimal Use of a Patchy Environment

Environments that are decidedly discontinuous, that is, heterogeneous ones consisting of a patchwork of rather different resources are termed "patchy" environments, whereas those with similar or well mixed resources are homogeneous or uniform environments. Frequently, a particular organism in a heterogeneous environment exploits only a part of the environmental mosaic, and different organisms are specialized to use each of the patch types. However, as the degree of difference between patches decreases, the advantage of being generalized increases. Thus, again we distinguish two different extremes as to the way in which organisms can utilize their environments. When, in the course of its daily activities, an organism encounters and uses resources in the same proportions in which they actually occur (i.e., does not select resources or patches in the environment), the organism is said to utilize its environment in a *fine-grained* manner. Organisms that spend disproportionate amounts of time in different patches are said to use their environment in a *coarse-grained* manner. Note that an animal which encounters its prey in the actual proportions in which they occur (i.e., in a fine-grained manner), through selection of particular prey types, can exploit this fine-grained environment in a coarse-grained way. Fine-grained utilization of the environment may be forced upon an organism in situations where the size of the patches in the environmental mosaic (termed the *grain size* of the environment) is small compared to the size of the organism. Thus, larger animals, all else being equal, tend to encounter the world in a more fine-grained way than smaller ones. As an example, a meadow and an adjacent forest each has its own herbivorous mouse (often a *Microtus* and a *Peromyscus*, respectively), whereas a deer exploits both the meadow and the forest, using the former at night for feeding and the latter by day for sleeping and retreat. Small animals, such as insects, generally exploit their environments in a more coarse-grained way than do larger animals, such as vertebrates. Note that sessile animals and plants must, as individuals, tend toward coarse-grained utilization of space simply by virtue of their immobility. Because individual sessile organisms may be distributed widely in an environmental mosaic, however, populations of them may approximate fine-grained utilization of the spatial environment. Statements about environmental grain size or the degree of coarse-grained versus fine-grained utilization of resources are comparative; when one says that a given animal "lives in a coarse-grained environment," this must be relative to some other animal.

Consider now the possible ways in which insectivorous birds could forage in a mixed forest containing several different species of trees and a wide range of different types of insects (after MacArthur and MacArthur, 1961). Assume that no single bird has the capacity to exploit everything in the forest, but that some degree of coarse-grained utilization is necessary. A bird could forage only in one species of tree and fly from one such tree to another. Much time and energy would then be expended in flying between trees and over unsuitable ones. Alternatively, the bird could eat only one category of insect food, perhaps a particular spectrum of insect sizes, capturing these wherever they occur in any of the various tree species. In the course of searching out its particular insect prey, the bird would doubtlessly encounter a variety of other prey types (or sizes). The first strategy would presumably lead to different species of birds being specialized to each of the various tree species, whereas the latter would result in each bird species having its own particular range of prey types. A third way in which birds could exploit such a patchy environment would be to compromise in both of the above ways, but specialize as to exactly *where* in the trees they forage and as to *how* they feed. Thus, a bird might select a layer in the forest and forage through this layer, taking whatever prey are available, within broad limits, and feeding in different trees and on many different prey types. In this case, different species of birds would differ as to where and how they forage, with each species exploiting a "natural feeding route" in the forest. A compromise strategy such as this is usually a more efficient way for birds to exploit a variable environment than is specializing as to places or prey exploited; most birds have indeed developed their own unique patterns as to where and how they forage. [Thus kinglets and titmice tend to forage in the crowns of trees, while other species such as many warblers (see also Figure 5.8) forage in other parts of trees. Woodpeckers often exploit crevices in tree trunks in an ascending spiral up to some height, and then they fly or glide down to a low point on the trunk of a nearby tree and repeat the process.] Only in cases of extreme *concentrations* of food does it pay to specialize on a particular food type. Thus many parrots are quite specialized in the fruits and nectars they eat; when these foods are encountered, they are usually very dense and in superabundance. Such exceptionally rich energy sources are worth searching out because once one is located the dividends are relatively great.

Let us now consider a model for utilization of patchy environments, which is based upon optimization of the animal's time budget, although the model applies equally well to an energy budget (MacArthur and Pianka, 1966). Assume that environmental resources, say prey species, are encountered by a foraging animal in the same proportions in which they actually occur, or that the environment is fine-grained (this assumption is relaxed in the following paragraph), but the animal is able to select from the available array of prey types, or to *use* this fine-grained environment in a coarse-grained way. For convenience, we also assume that the animal either does or does not eat any given kind of prey, that is, no one type is eaten only part of the time it is encountered. What number of different kinds of prey will provide the animal with the maximal return per unit time (or expenditure)? Total foraging time, per item eaten, can be broken down into two components: time spent on search (search time) versus that spent on pursuit, capture, and eating (pursuit time). A fine-grained environment is searched for all types of food simultaneously, whereas prey are pursued, captured and eaten singly. Prey types are ranked from those providing the highest harvest per unit time (and/or energy) to those of lowest yield; that is, from the prey species whose capture requires the least expenditure per calorie assimilated to that requiring the most. The diet of course includes the most rewarding item; as an animal expands its diet to include progressively less rewarding kinds of prey, more and more acceptable items are encountered and search time (and/or energy), per prey item, *decreases*. However, as new, varied prey, often hard to catch or to swallow, are added to the diet, pursuit time (and/or energy) will usually *increase*. So long as the time or energy saved in reduced search is greater than the increase in time or energy expended in increased pursuit, the diet should be enlarged to include the next, less rewarding, kind of prey. At some point losses accompanying further enlargement balance or exceed gains, marking the optimal diet. Though no general statements about an animal's diet are possible from the model, some testable *comparative* predictions can be made. For instance, search time per item eaten should be less in a food-dense environment than in a food-sparse one. Pursuit time, which is a function of the relative abilities of predator and prey and of the variety of prey types and predator escape mechanisms, should however be little altered by changes in food density, per se. As a result, productive environments should be used in a more specialized way than less productive ones. Similarly, animals that spend little effort searching for their prey should be more specialized than those with higher ratios of time spent in search to that spent on pursuit. Should food be scarce, foraging animals are unlikely to bypass potential prey, whereas during times of (or on areas with) abundant food, individuals may be more selective and restrict their diets to only the better food types. Thus a low expectation of finding prey, or a high mean search time per item, demands generalization, whereas a higher expectation of locating prey items (short mean search time per item) allows some degree of specialization.

So far, the environment under consideration has been homogeneous and fine grained. Let us now extend consideration to patchy environments in which various patches contain different arrays of prey. In this case, types of patches are ranked in order of decreasing *expectation of yield*, or from the patch in which the most calories are likely to be obtained from prey per unit expenditure to that with the least return. The two components of the animal's time (and/or energy) budget are now the "hunting time," per item captured, spent *within* suitable patches (hunting time is equivalent to "foraging time" above, or to the sum of search time plus pursuit time), versus the "traveling time," per item caught, spent traveling *between* suitable patches. Obviously, the time spent traveling between patches decreases as an animal expands the number of different kinds of patches on its itinerary, while the time spent hunting within patches must increase as the itinerary is enlarged to include an increased variety of patches. The optimal use of a patchy environment therefore depends upon the rate of decrease in traveling time (per prey item) relative to the rate of increase in hunting time (per prey item) associated with expanding the number of different patch types exploited. Suppose that food density is suddenly increased in all patches; *both* traveling time and hunting time, per item, will then be reduced. Since only search time decreases with increased food density (see last paragraph), animals that expend greater amounts of energy on search (searchers) will have their hunting time reduced more than those which spend relatively more energy on pursuit (pursuers). Hence, under high food densities, pursuers should restrict the variety of patches they use more than searchers.

We next consider the effects of patch size on the optimal number of patch types exploited. Envision two environments differing only in the sizes of their patches and not at all in the proportions or qualities of the various patch types. Hunting time, per item, in identical but different sized patches is the same in both environments because the quality of the patches is unaltered. Traveling time (per item), however, *decreases* as patch size increases, because the distance between patches varies linearly with the linear dimension of a patch, while the hunting area or volume within a patch varies as its square or cube, depending upon whether the animal exploits space in two or three dimensions, respectively. Since larger patches offer smaller traveling time per unit of hunting time, they can be used in a more specialized way than can smaller patches. In the extreme, as patches become vanishingly small compared to the size of the organism, patch selection is impossible and completely fine-grained utilization must take place. Similarly, patches that are very large compared to an organism approximate a one-patch environment and an animal can spend all of its time in that one patch. Thus smaller and/or more mobile animals should use fewer different patches than larger and/or less mobile ones.

Competitors should normally reduce the density of some types of prey in some patches. Any prey item worth exploiting in the absence of competition is still worth eating if there is competition for it, but this is not true of patterns of patch exploitation. The decision of whether or not to forage within a given patch type depends upon an organism's expectation of yield in that patch. Should food within a given patch type become scarce, due to competitors or some other factor, inclusion of that patch type in the itinerary increases mean hunting time per item sharply and reduces efficiency. Thus the presence of competitors in some patches should cause an optimal predator to restrict the variety of patch types it exploits (see also The Compression Hypothesis, pp. 264–265).

Energetics, Vagility, and Home Range

Large animals require more matter and energy for their maintenance than small ones, and, in order to obtain it, they usually must range over larger geographical areas than smaller animals with otherwise similar food requirements. Food habits and vagility may also affect home range size.

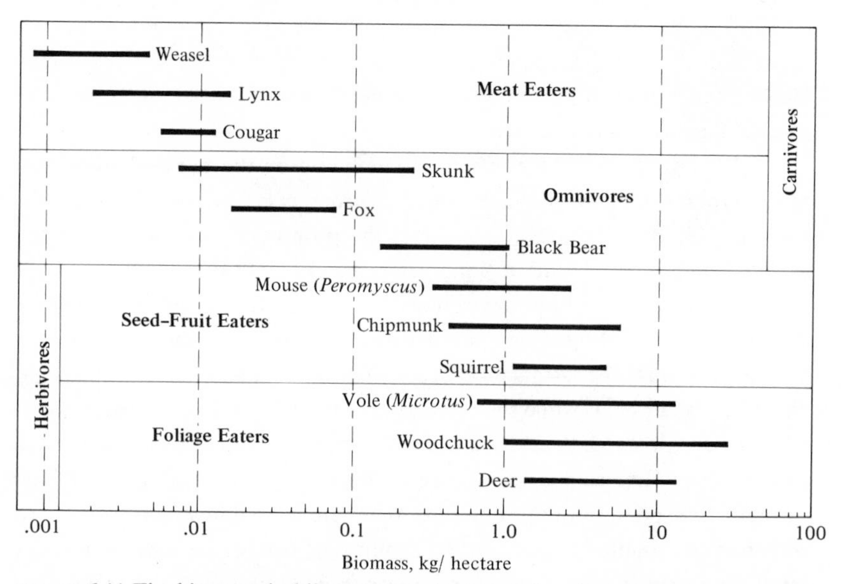

FIGURE 6.11 The biomass, in kilograms per hectare, of various mammals, arranged according to their food habits. Although mammals as a class vary over five orders of magnitude, the range within any given trophic niche is considerably less. Note that meat eaters and omnivores are much less dense than herbivores. [From Odum (1959) after Mohr (1940).]

Herbivorous animals that eat the green parts of plants (such as grazers that eat grasses and ground-level vegetation and browsers which eat tree leaves) normally do not spend much time searching for their food and therefore usually do not have very large home ranges. In contrast, carnivores and those herbivores that must search for their foods (such as granivores and frugivores, which eat seeds and fruits, respectively) frequently spend much of their foraging time and energy in search and have large home ranges. The first group of animals has been called "croppers" and the latter one "hunters" (McNab, 1963). Croppers generally exploit foods which occur in relatively dense supply, while hunters utilize less dense foods. Hunters are often territorial; croppers seldom defend territories. Obviously, croppers and hunters are not so discrete but in fact grade into each other (Figure 6.11). Thus a browser that eats only the leaves of a rare tree might be more of a hunter than a granivore that eats the seeds of a very common plant. However, such intermediates are uncommon enough that separation into two categories is adequate for many purposes. Figure 6.12 shows the correlation between home range size and body weight for a variety of mammalian species, here separated into croppers and hunters. Similar correlations have been obtained for birds (Schoener, 1968b) and lizards (Turner, Jennrich, and

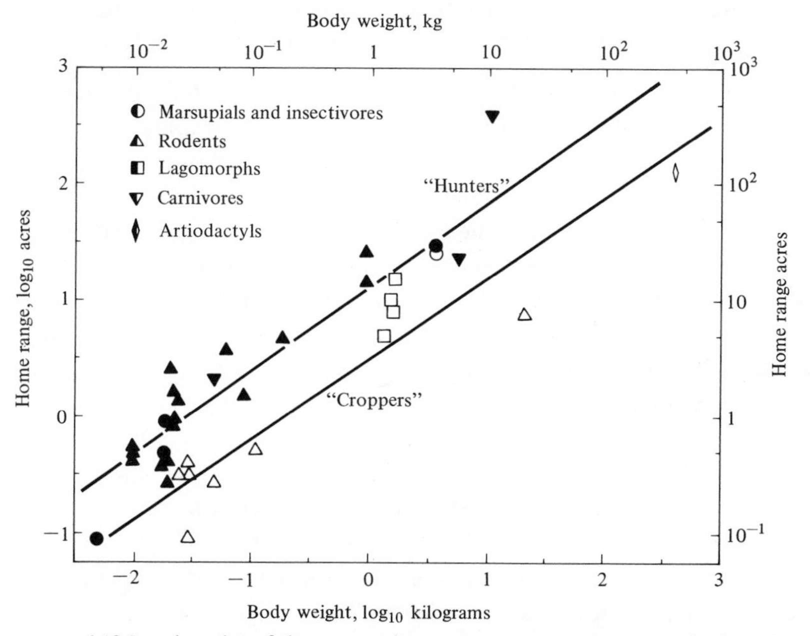

FIGURE 6.12 Log-log plot of the average home range area against mean body weight for a variety of mammals, separated into croppers (open symbols and lower regression line) and hunters (closed symbols and upper line). [After McNab (1963).]

Weintraub, 1969). Very mobile animals, such as birds, frequently have larger home ranges and territories than less vagile animals, such as lizards and mammals.

Home ranges in areas of low productivity, like deserts, may often be larger than they are in more productive regions. Large home ranges or territories usually result in low densities, which in turn markedly affect the evolution of sociality. Thus, McNab (1963) points out that complex social behavior has usually evolved only in croppers and/or exceptionally mobile hunters.

Reproductive Strategies

How much time and energy should an optimal organism allocate to reproduction at any particular time? Here again, R. A. Fisher (1930) considered this question years ago in the following penetrating and timeless statement (my italics):

> It would be instructive to know not only by what *physiological mechanism* a just apportionment is made between the nutriment devoted to the gonads and that devoted to the rest of the parental organism, but also what *circumstances in the life history and environment* would render profitable the diversion of a greater or lesser share of the available resources towards reproduction.

Fisher clearly distinguished between the proximate factor (physiological mechanism) and the ultimate factors (circumstances in the life history and environment) that determine the allocation of resources into reproductive versus nonreproductive tissues and activities.

Reproductive effort, or the proportion of the total resources available to an organism that is allocated to reproduction at any given time or over some interval of time, varies widely among organisms. In plants, energy expenditure on reproduction, integrated over a plant's lifetime, ranges from near zero to as much as 40 percent (Harper, Lovell, and Moore, 1970). Moreover annual plants tend to expend more energy on reproduction than most perennials (Harper *et al.* estimate these values at about 14–30 percent and 1–24 percent, respectively). Why are reproductive strategies so variable? Some of the reasons were considered in Chapter 4 (pp. 92–98 and 101–103); others are examined here.

Fisher's dichotomy for the apportionment of energy into reproductive versus nonreproductive (somatic) tissues, organs, and activities is a convenient starting point. Somatic tissues are clearly necessary for acquisition of matter and energy, but at the same time an organism's soma is of no selective value except inasmuch as it contributes to that organism's lifelong production of successful offspring. Increased reproductive effort clearly costs in reducing survivorship of the soma; this is easily seen in the extreme case of so-called "big bang" reproduction, as in salmon, annual plants, and many insects, in which the organism puts everything available into reproduction and then dies itself. Presumably, more subtle reductions in survivorship occur with minor increases in reproductive effort. How great a risk should an optimal organism take with its soma in any given act of reproduction? To answer this question, we need to know the *present value* in expectation of *future offspring* of an organism's soma at any instant. Williams (1966b) called this "residual reproductive value" and pointed out that it is simply the reproductive value at the next instant in time, or v_{x+1}. In order to maximize its overall lifelong contribution to future generations, an optimal organism should weigh its immediate prospects of reproductive success, measured by m_x, against its long term future prospects, as measured by v_{x+1}. Thus, an individual with a high probability of future reproductive success should be more hesitant to risk its soma in present reproductive activities than another individual with a lower probability of reproducing successfully in the future. One would predict that as an organism's reproductive value falls, due to aging or any factor that increases mortality, its reproductive effort should rise.

Heightened competition, both within and between species, can affect optimal time and energy budgets in several ways. First, by decreasing resource availability, competition may often reduce foraging efficiency and hence the total energy budget. In addition to reducing the total energy budget, an altered competitive milieu will usually require some reapportionment of the organism's time and energy budget. Recall from Chapter 4 (pp. 93–95) that optimal reproductive strategy is a compromise between conflicting demands for production of the largest possible total number of progeny versus production of offspring with the highest possible individual fitness. Since larger, better endowed, offspring should normally have higher survivorship and be generally better competitors than smaller ones, heightened competition should often increase the optimal expenditure per progeny. Thus, all else being equal (genetic background, etc.), one would expect a positive relationship between expenditure upon an individual offspring and that offspring's individual fitness. Gains in fitness per unit expenditure are likely to be greater at small expenditures per progeny than they are at higher expenditures (see also Figure 4.13); hence total fitness should peak at some optimal clutch size (Figure 4.14). Through such effects on optimal reproductive effort and expenditure per progeny, an organism's competitive environment should thus affect its optimal clutch or litter size, which is simply reproductive effort divided by average expenditure per progeny. Indeed, since heightened competition simultaneously reduces both the total energy budget and reproductive effort while it raises expenditure per individual progeny, the effects of competition on optimal clutch size should often be more pronounced than its effects on the other two parties of the triumvirate, reproductive effort and expenditure per progeny. Moreover, in order for expenditure per progeny to increase, the decrease in reproductive effort must be less than the corresponding decrease in clutch size.

Evolution of Niches

Niche changes over evolutionary time are rather difficult to document, although their occurrence cannot be disputed. As new species arise from the fission of existing ones through the process of speciation, new niches come into existence. Life on earth almost certainly arose in aquatic environments and early organisms were doubtlessly very small and simple. During the evolutionary history of life over geological time, organisms have become more and more complex and diversified, and earth has been filled with an overwhelming variety of plants and animals. Some taxonomic groups of organisms, such as the dinosaurs, have gone extinct and have been replaced by others. Major breakthroughs in the body plans of organisms periodically open up new adaptive zones and allow bursts of evolution of new and diverse species, termed *adaptive radiations*. One of the major forces that has led to niche separation and diversification is interspecific competition. Thus, the first terrestrial organisms found themselves in a wide open ecological and competitive vacuum, freed from competition with aquatic organisms, and they rapidly radiated into the many available new terrestrial niches. Similarly, the evolution of homoiothermy and aerial exploitation patterns have allowed major adaptive radiations; flight has evolved independently at least four times, in insects, reptiles, birds, and mammals. Often evolutionary interactions between two or more taxa have had reciprocal effects upon one another; thus the origin and radiation of flowering plants (angiosperms) in the Mesozoic presumably allowed insects to diversify widely, while species specificity of pollinating insects may well have, in turn, allowed considerable diversification of the plants. Indeed, Whittaker (1969) has suggested that organic diversity is self-augmenting.

The Periodic Table of Niches

In chemistry the urge of scientists to order and classify natural phenomena resulted in the well-known periodic table of the elements, which allowed chemists to predict new elements and their chemical properties, and led to our understanding of electron shells. Some ecologists wonder whether something like a "periodic table of niches" might be possible. Of course, nothing about niches is quite as simple and discrete as the number of electrons in the outer shell of a chemical element, but rather, most aspects of niches are more continuous. However, some patterns described earlier may be used to construct a very primitive periodic table of niches (Figure 6.13). Thus trophic niches repeat themselves in different sized organisms that are relatively more or less *r* and *K* selected. An aphid is more like a lemming, and a mantid more like a weasel, in their food niches, whereas in terms of body size and their position on the *r*→*K* selection continuum (Chapter 4), the aphid and mantid are relatively alike, as are the lemming and the weasel. Niche dimensions other than those used in Figure 6.13, such as diurnal and nocturnal time of activity, could also be used to construct similar, but different, periodic tables. A periodic table of niches for aquatic organisms would differ somewhat from that for terrestrial ones, especially in that there are very few relatively *K* selected aquatic primary producers. Perhaps one day, as the young science of ecology matures, we will be able to construct something analogous to the periodic table of the elements that will order niches, allow predictions, and improve our understanding of the elusive ecological niche.

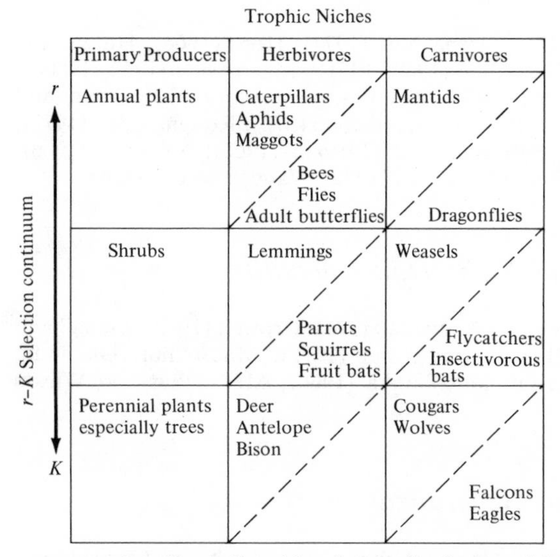

FIGURE 6.13 A "periodic table of niches" of terrestrial organisms, with examples. Dashed diagonal lines separate herbivores and carnivores which exploit space in two and three dimensions. Many other niche dimensions, such as time of activity, might be used in such classifications profitably.

Selected References

Hutchinson (1957a, 1965); Levins (1968, Chapter 3); MacArthur (1968, 1972); McNaughton and Wolf (1970); Odum (1959, 1971); Van Valen (1965); Whittaker, Levin, and Root (1973).

History and Definitions

Allee *et al.* (1949); Clarke (1954); Dice (1952); Elton (1927); Grinnell (1917, 1924, 1928); Hutchinson (1957a); Odum (1959); Parker and Turner (1961); Ross (1957, 1958); Savage (1958); Udvardy (1959); Vandermeer (1972b); Weatherley (1963); Whittaker, Levin, and Root (1973); Williamson (1971).

Adaptation and Deterioration of the Environment

Fisher (1930, 1958a, 1958b); Wallace (1973).

The Hypervolume Model

Green (1971); Hutchinson (1957a, 1965); Maguire (1967, 1973); Miller (1967); Shugart and Patten (1972); Vandermeer (1972b); Warburg (1965).

Niche Overlap and Competition

Colwell and Futuyma (1971); Hespenhide (1971); Horn (1966); Hutchinson (1957a); Klopfer and MacArthur (1960, 1961); MacArthur (1957, 1960a); MacArthur and Levins (1964, 1967); May and MacArthur (1972); Orians and Horn (1969); Pianka (1969, 1972, 1973); Pielou (1972); Roughgarden (1972); Schoener (1968a, 1970); Schoener and Gorman (1968); Selander (1966); Smouse (1971); Terborgh and Diamond (1970); Vandermeer (1972b).

Niche Dynamics and Niche Dimensionality

Cody (1968, 1973); Colwell and Futuyma (1971); Crowell (1962); Gadgil and Solbrig (1972); Levins (1968); MacArthur (1972); MacArthur, Diamond, and Karr (1972); MacArthur and Pianka (1966); MacArthur and Wilson (1967); Pianka (1973).

Specialization versus Generalization

King (1971); MacArthur and Connell (1966); MacArthur and Levins (1964, 1967); Roughgarden (1972); Van Valen (1965); Willson (1969).

Time, Matter, and Energy Budgets

Emlen (1966); Gadgil and Bossert (1970); Gibb (1960); Grodzinski and Gorecki (1967); Schoener (1969b, 1973); Slobodkin (1962); C. Smith (1968); Wiegert (1968); Zeuthen (1953).

Foraging and Feeding Efficiency

Cody (1968); Emlen (1966, 1968a); Gibb (1960); Holling (1964); MacArthur (1972); MacArthur and Pianka (1966); Morse (1971); Pianka (1966b); Rapport (1971); Royama (1970); Schoener (1969a, 1969b, 1971); Tullock (1970); Wolf and Hainsworth (1971); Wolf, Hainsworth and Stiles (1972).

Optimal Use of a Patchy Environment

Emlen (1966, 1968a); Hutchinson and MacArthur (1959); King (1971); Levins (1968); MacArthur (1972); MacArthur and Levins (1964); MacArthur and MacArthur (1961); MacArthur and Pianka (1966); Schoener (1969a, 1969b, 1971).

Energetics, Vagility, and Home Range

McNab (1963); Mohr (1940); Schoener (1968b); C. Smith (1968); Turner, Jennrich, and Weintraub (1969).

Reproductive Strategies

Fisher (1930); Gadgil and Bossert (1970); Gadgil and Solbrig (1972); Harper, Lovell, and Moore (1970); Harper and Ogden (1970); Istock (1967); Lewontin (1965); MacArthur and Wilson (1967); Pianka (1970, 1972); Schaffer and Tamarin (1973); Tinkle (1969); Tinkle, Wilbur, and Tilley (1970); Williams (1966a, 1966b).

Evolution of Niches

Hutchinson (1965); MacArthur (1968, 1972); MacArthur and Levins (1964, 1967); Whittaker (1969, 1972).

7 | *Community Structure*

Except for the brief treatment of biomes in Chapter 3, we have considered the ecology of individuals and populations until now; here we examine community ecology. Just as populations have properties that transcend those of the individuals comprising them, communities have both structure and properties not possessed by their component populations. Thus communities have trophic structures, rates of energy fixation and flow, efficiencies, stabilities, diversities, successional stages, and so on. Moreover, these various attributes of communities have profound effects upon the organisms living in a given community. The extreme complexity of most ecosystems makes their study difficult, but, at the same time, quite challenging. The concept of a community is itself an abstraction; communities are seldom clear cut and distinct, but almost always grade into one another. However, by considering ecological systems as "open" rather than "closed," and by allowing for continual inflow and outflow of materials, energy, and organisms, this difficulty can be partially overcome and the community concept can be useful. Thus communities change both in space and in time, and the picture developed in this chapter is essentially an instantaneous view of a fairly localized portion of a larger community.

Food Webs and Trophic Levels

Any community can be represented by a *food web*, which is simply a diagram of all the trophic relationships among and between its component species. Such a food web is generally composed of many *food chains*, each of which represents a single pathway up the food web. The direction of flow of matter and energy between species can be shown with arrows, as in Figure 7.1. A really complete food web would also include the rates of flow of energy between the various populations comprising a community.

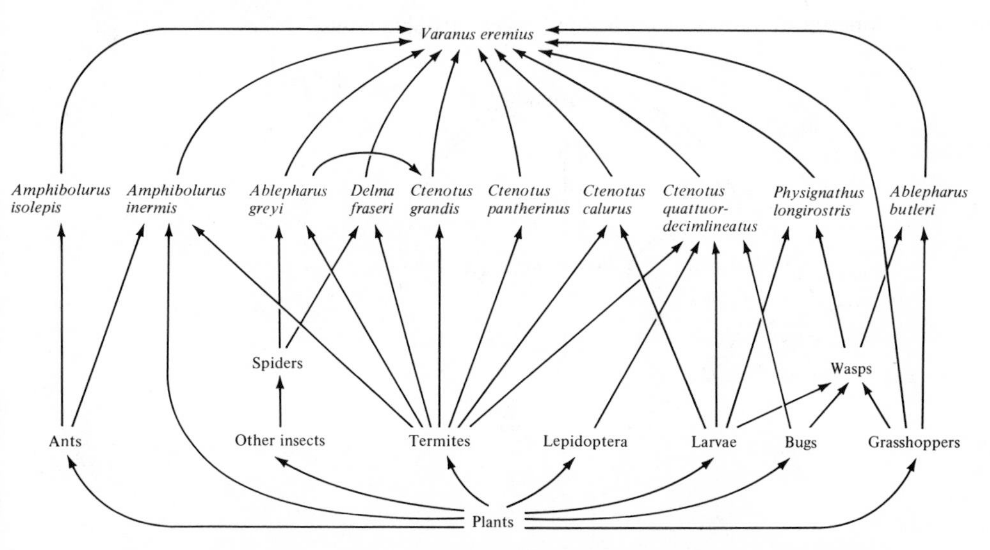

FIGURE 7.1 Part of the food web in an Australian sandy desert. The "top" predator, a monitor lizard, *Varanus eremius*, eats grasshoppers and 10 other species of lizards, which in turn have diets dominated by various sorts of arthropods or plants as indicated. A more detailed food web would separate all food types into species and would indicate the actual rate of flow of energy up each link in the food web.

Primary producers, or *autotrophs*, represent the first trophic level: they are the green plants that use solar energy to produce energy-rich chemicals. Primary producers are an essential part of a community in that practically all other organisms in the community are directly or indirectly dependent upon them for energy. Organisms other than the primary producers, or *heterotrophs*, include *consumers* and *decomposers*. Herbivores are primary consumers and represent the second trophic level. Carnivores that eat herbivores are *secondary consumers* or *primary carnivores* and are on the third trophic level; carnivores that eat primary carnivores, in turn, constitute the fourth trophic level and are termed *tertiary consumers* or *secondary carnivores*. Similarly those that prey upon secondary carnivores are *quaternary consumers* or *tertiary carnivores*, and so on. Because many animals, such as omnivores that eat both plant and animal matter, prey on several different trophic levels simultaneously, it is often impossible to assign them to a given trophic level. Nevertheless, such organisms can usually be assigned partial representation in different trophic levels in proportion to the composition of their diet. The trophic level concept has proven to be an extremely useful abstraction in the study of community structure; it facilitates examination of the flow of matter and energy through communities and underscores the differences between interactions that take place within trophic levels as opposed to those which operate between trophic levels.

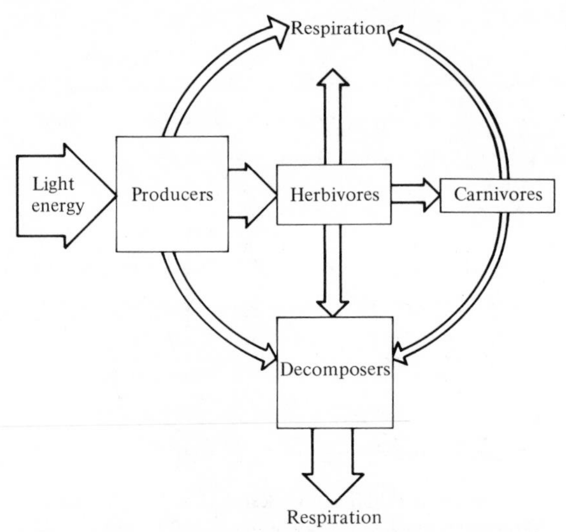

FIGURE 7.2 A trophic level "compartment" model of a hypothetical community, with arrows indicating the flow of energy through the system. The width of each arrow is intended to reflect the rate of flow of energy between particular parts of the system.

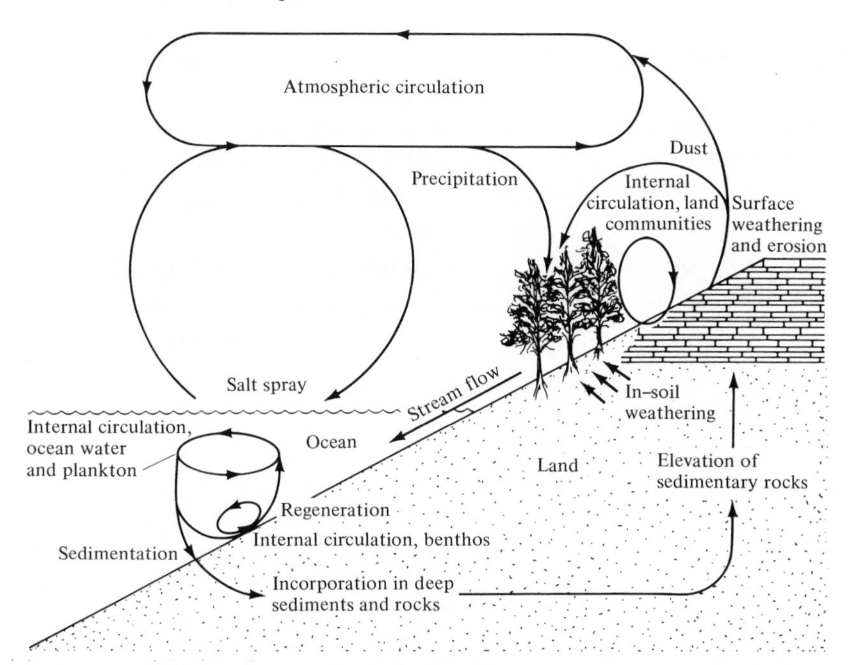

FIGURE 7.3 Diagrammatic representation of the major movements of calcium, termed a biogeochemical cycle. Many other materials circulate within and between ecosystems in a somewhat similar fashion. [From Whittaker (1970). *Communities and Ecosystems*. Reprinted with permission of Macmillan Publishing Co., Inc. © Copyright Robert H. Whittaker, 1970.]

Another way in which the major components of an ecosystem can be diagrammed conveniently is shown in Figure 7.2, where each trophic level is treated as a "compartment" and arrows again designate the direction of flow of matter and energy. Many materials, including calcium, carbon, nitrogen, and phosphorus, move from compartment to compartment as individual organisms are consumed by others at higher trophic levels, and eventually returned to the abiotic "nutrient pool," where they may be reused by primary producers; such movement of matter through ecosystems is termed *cycling* (Figure 7.3). An essential component of any ecosystem is its decomposers, or *reducers*, which function in returning materials to the nutrient pool. Whereas matter continually circulates through the various compartments of an ecosystem and is always being reused, energy can be used only once. All ecosystems are thus dependent upon a continual inflow of energy (see also Principles of Thermodynamics, pp. 223–225).

The Community Matrix

A table of numbers with rows and columns is known as a matrix. Building on the Lotka-Volterra competition equations, Levins (1968) formulated the concept of an alpha matrix, or more generally, a community matrix. For a community composed of n species, the community matrix is an $n \times n$ matrix giving the sign and degree of interaction between each pair of species. Figure 7.4 illustrates the community matrix of trophic relationships among 10 hypothetical species with a food web as shown. In pairs of competing species,

Secondary carnivore:

Primary carnivores:

Herbivores:

Producers:

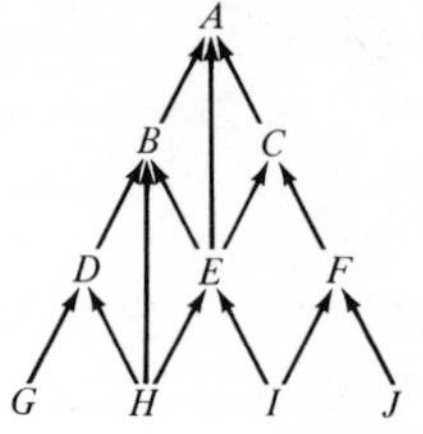

Species having the effect

Species affected	A	B	C	D	E	F	G	H	I	J
A	1	—	—	0	—	0	0	0	0	0
B	+	1	+	—	—	0	0	—	0	0
C	+	+	1	0	—	—	0	0	0	0
D	0	+	0	1	+	+	α_{DG}	—	0	0
E	+	+	+	+	1	+	—	—	—	0
F	0	0	+	+	+	1	0	0	—	—
G	0	0	0	α_{GD}	+	0	1	+	+	+
H	0	+	0	+	+	0	+	1	+	+
I	0	0	0	0	+	+	+	+	1	+
J	0	0	0	0	0	+	+	+	+	1

FIGURE 7.4 A community matrix for the hypothetical 10-species community shown. Only the signs of the alphas are given and only *direct* interactions between and within two trophic levels are considered. All members of a given trophic level are assumed to be in competition. Higher order interactions, such as the effects of a plant's antiherbivore defenses upon a carnivore that depends on the plant's herbivores for food, or the indirect effects of the predator on the plant through reducing herbivore densities, are ignored for simplicity. One pair of alphas is identified within the matrix, that between species D (predator) and species G (prey). Note that α_{DG} is negative, whereas α_{GD} is positive.

α_{ij} and α_{ji} values both have positive signs, indicating that each inhibits the other's population. The magnitude of the alpha values indicates the intensity of competitive inhibition, which need not be equal and opposite. A prey–predator relationship is indicated by α_{ij} and α_{ji} values with opposite signs, with the predator benefiting from the relationship and the prey suffering. Again, the degree of benefit or detriment is represented by the magnitude of the alpha values. Thus, if α_{ij} is negative and α_{ji} is positive, species *i* eats species *j*. Mutualistic relationships are represented by pairs of negative alphas. Mathematical manipulations of the community matrix, based on the Lotka-Volterra competition equations, have been developed by Levins (1968) and Vandermeer (1970, 1972a) that predict the maximal number of species in a stable community. These calculations, however, assume that alphas are constant and independent and that changes in community composition do not affect the various alpha values.

A virtue of the community matrix idea is that it facilitates abstraction and quantification of some of the interactions among the members of a community. The concept of the community matrix is independent of the Lotka-Volterra competition equations. Although one species can affect another simultaneously in both positive and negative ways, the appropriate alpha value in the community matrix represents the overall net effect. This abstract approach to a community is obviously a static one, which assumes that all relationships between species are constant. Interactions between real organisms must vary both in time and in space, and from individual to individual. Nevertheless, the concept of such an ever changing community matrix is a useful abstraction that helps us to visualize what presumably is actually happening in a real community.

Principles of Thermodynamics

An important facet of community ecology is the energy relationships between and among the members of a community. Before going into community energetics, we must first review briefly some fundamentals of thermodynamics.

All organisms require energy to persist and to replace themselves, and the ultimate source of practically all earth's energy is the sun. One can think of the sun as "feeding" earth via its radiant energy. But 99 percent or more of this incident solar radiation goes unused by organisms and is lost as heat and heat of evaporation. Thus only about 1 percent is actually captured by plants in photosynthesis and stored as chemical energy. Moreover, energy available from sunlight varies widely over earth's surface in both space and time (see Chapters 2 and 3).

Physics and chemistry have given us two basic laws of thermodynamics, obeyed by *all* forms of matter and energy, including living organisms. The first law is that of "conservation of matter and energy," which states that matter and energy cannot be created or destroyed. Matter and energy can be transformed, and energy can be converted from one form into another, but the total of the equivalent amounts of both must always remain constant. Light can be changed into heat, kinetic energy, and/or potential energy. Each time energy is converted from one form into another, some of it is given off as heat, which is the most random form of energy. Indeed, the only energy conversion that is 100 percent efficient is conversion to heat, or burning. Burning aliquots of dried organisms in "bomb calorimeters" is a common method of determining how much energy is stored in their tissues (Paine, 1971). Energy can be measured in a variety of different units such as ergs and joules, but the common denominator of all used in ecology is heat energy or calories.

The second law of thermodynamics states that energy of all sorts, whether it be light, potential, chemical, kinetic, or whatever, tends to change itself spontaneously into a more random, or less organized, form. This law is sometimes stated as "entropy increases," entropy being random, unavailable energy. As an example, suppose I heat up a skillet to cook an egg, and after finishing I leave it on the stove. At first, heat energy is concentrated near the skillet, which is, relative to the rest of the room, quite nonrandom. But by the next morning the skillet has cooled down to air temperature, and the heat energy has radiated throughout the room. Now that same heat energy is dispersed and unavailable for cooking; the system of the skillet, the room, and the heat has gone toward equilibrium, become more random, and entropy has increased. Unless an outside source of energy such as a stove, with fuel or electricity, is continually at work to maintain a nonequilibrium state, dispersion of heat results in a random equilibrium state. The same is true for all kinds of energy. According to the second law, our solar system and presumably the entire universe should theoretically become a completely random array of molecules and heat in the far distant future.

Life has sometimes been called "reverse entropy," because organisms maintain complex organized states compared to their surroundings. But they obey the second law just as any other system of matter and energy; all organisms must work continually to build and maintain nonrandom assemblages of matter and energy. This process requires energy, and organisms use the energy of the decaying sun (which of course also obeys the second law and tends toward increasing disorder) to "oppose" the second law within their own tissues by producing order out of increasing disorderliness. Wherever there is a live plant or animal, there must be an energy source. Without a continued influx of energy, no organism can survive for very long. Once again, this "reverse entropy" occurs only *within* each organism, and the overall energy relations of the entire solar system are in accord with the second law of thermodynamics, with the overall system continually becoming more and more random.

Pyramids of Energy, Numbers, and Biomass

Returning now to community energetics, the rate of flow of energy through a given trophic level decreases with increasing trophic level for several reasons. Because energy transfers are never 100 percent efficient, not all the energy contained in any given prey item is actually available to a predator; some of it is lost in converting prey tissue into predator tissue, and some is not even assimilated but rather passes through the predator's intestinal tract unchanged and is then decomposed by reducers. Moreover, the efficiency of transfer of matter and energy from prey to predator is often greatly reduced by predator-avoidance tactics of prey, such as the chemical defenses of plants. In addition, each organismic unit, and each trophic level, expends some of its available energy on its own activities, further reducing the amount of energy available at higher trophic levels. Ultimately, of course, at equilibrium all the energy captured by primary producers must be expended and dissipated back into space as heat, that is, the amount of energy entering the system must exactly balance that leaving it (see next section also).

The reduction in the rate of flow of energy from each trophic level to the next higher one determines many of a community's properties, including the total number of trophic levels as well as the proportion of predators to prey. Ecologists estimate that, after standardization per unit area and unit time, approximately 10 to 20 percent of the energy at any given trophic level is available to the next higher trophic level. Hence, if a thousand calories are available to primary producers, usually only a few calories of the thousand (from about 1 to 8 calories) are actually available to a secondary carnivore three trophic levels away. A result of this rapid reduction in the availability of energy is that animals at higher trophic levels are generally much rarer than those at lower ones. Moreover, decreasing availability of energy places a distinct upper limit on the number of trophic levels possible, with about five or six being the normal maximum.

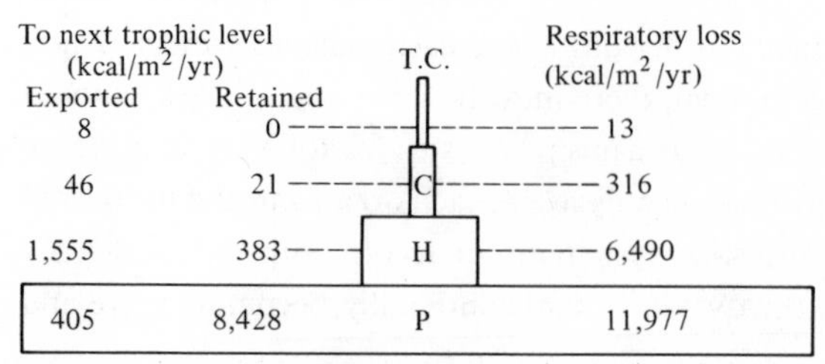

FIGURE 7.5 Pyramid of energy for Silver Springs, Florida. [From Phillipson (1966) after Odum.]

A convenient means of expressing the energetic structure of a community is the so-called *pyramid of energy*, which consists simply of the rates of energy flow through various trophic levels (Figure 7.5). The laws of thermodynamics and above considerations dictate that the pyramid of energy can never be inverted, that is, the flow of energy through each trophic level must always decrease with increasing trophic level.

Two other types of ecological "pyramids" are the so-called *pyramid of numbers* and the *pyramid of biomass*. These are instantaneous measures, rather than rates, having no time dimension (whereas the units of the pyramid of energy are calories/m²/year, the units of the pyramid of numbers are simply densities, number/m², and the units of the pyramid of biomass are grams/m²). The pyramid of numbers consists simply of a set of the densities of individuals in each trophic level, whereas the pyramid of biomass is the biomass (usually measured in grams dry weight) per square or cubic meter in each trophic level. Pyramids of numbers and biomass measure only the *standing crop*, or the amount present at an instant, of each trophic level, and not their turnover rate. Because they have no time dimension, these two pyramids may be inverted, with, for example, lower densities and/or smaller biomasses at lower trophic levels. Thus one tree may support many insects (an inverted pyramid of numbers); likewise, a rapid rate of turnover allows a small biomass of prey to support a larger biomass of predators with a slower turnover rate. Such an inverted pyramid of biomass often characterizes

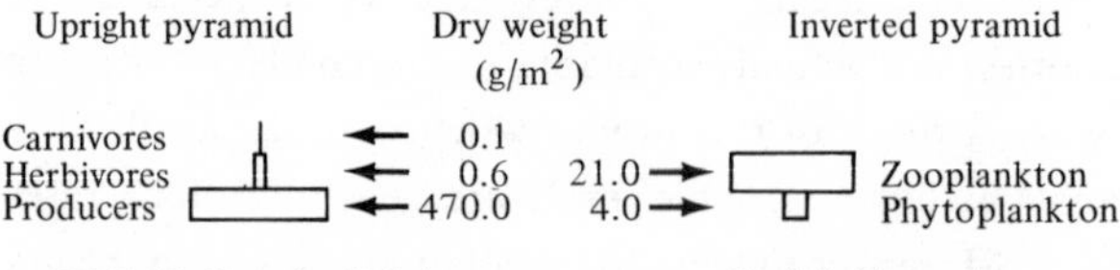

FIGURE 7.6 Two pyramids of biomass, one upright and one inverted. [From Phillipson (1966) after Odum.]

aquatic ecosystems, where the primary producers (phytoplanktonic algae) are small and rapidly dividing, while their zooplanktonic predators are larger and longer lived (Figure 7.6).

Energy Flow and Ecological Energetics

The energy content of a trophic level at any instant (that is, its standing crop in energy) is usually represented by the Greek letter capital lambda, Λ, subscripted to indicate the appropriate trophic level, as follows: Λ_1 = primary producers, Λ_2 = herbivores, Λ_3 = primary carnivores, and so on. Similarly the rate of flow of energy between trophic levels is designated by lower case lambdas, λ_{ij}, where the i and j subscripts indicate the two trophic levels involved, with i representing the level receiving, and j, that losing, the energy. Subscripts of zero denote the world external to the system, while subscripts of 1, 2, 3, etc., indicate trophic level as above.

Using these conventions, an ecosystem can be represented by a compartment model as in Figure 7.7. At equilibrium, the amount of energy contained in every compartment (trophic level) must be constant, which, in turn, requires that the rate of flow of energy into every compartment be exactly balanced by the flow of energy out of the compartment concerned. At equilibrium (that is, $d\Lambda_i/dt = 0$ for all i), the energy flow in the system portrayed in the figure may thus be represented by the following set of simple

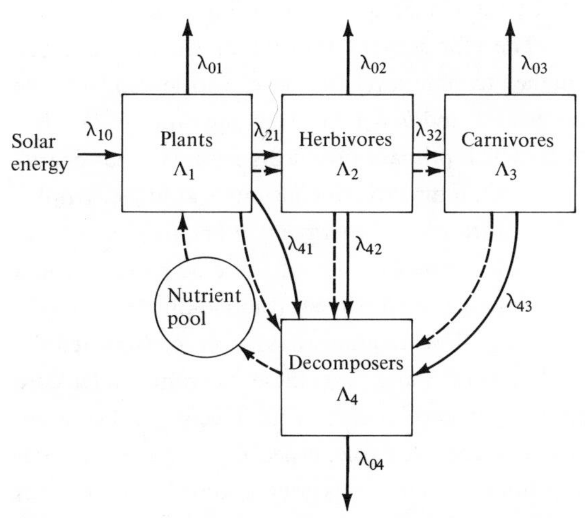

FIGURE 7.7 Another compartment model of an ecosystem. Energy flow is shown with solid arrows and the flow of matter with dashed ones. See text.

equations (with inputs on the left and rate of outflow to the right of the equals signs):

$$\lambda_{10} = \lambda_{01} + \lambda_{02} + \lambda_{03} + \lambda_{04}$$
$$\lambda_{10} = \lambda_{21} + \lambda_{01} + \lambda_{41}$$
$$\lambda_{21} = \lambda_{32} + \lambda_{02} + \lambda_{42}$$
$$\lambda_{32} = \lambda_{03} + \lambda_{43}$$
$$\lambda_{41} + \lambda_{42} + \lambda_{43} = \lambda_{04}$$

The rate at which energy is actually captured by plants (λ_{10}) is estimated at only about 1 percent of the total solar energy hitting the earth's surface. This rate of uptake of solar energy by primary producers, λ_{10}, is termed the gross productivity. It is usually given in *calories/m²/year*, which represents the *gross annual production* (GAP). Since plants use some of this energy in their own respiration (i.e., λ_{01}), only a part of the gross annual production is actually available to animals and decomposers; this fraction, λ_{21}, plus the energy used by decomposers, λ_{41}, is termed the net productivity, or on an annual areal basis, the *net annual production* (NAP). Net production may be considerably less than gross production; in some tropical rainforests plants use as much as 75 to 80 percent of their gross production in respiration. In temperate deciduous forests, respiration is usually from 50 to 75 percent of gross primary production, while in most other communities it is about 25 to 50 percent of the gross production. Whittaker (personal communication) estimates that perhaps 7 percent of the plant food on land is actually harvested by animal consumers, and the remainder of the net primary production is consumed by decomposers. The efficiency of transfer of energy from one trophic level to the next higher trophic level, say level i to level j, may be estimated as $\lambda_{ji}/\lambda_{ih}$, where $j = i+1$ and $h = i-1$. Thus the ratio $\lambda_{21}/\lambda_{10}$ is a measure of the efficiency with which primary producers pass the solar energy they capture on to herbivores, and, indirectly, to consumers at higher trophic levels as well. Such efficiencies of transfer of energy from one trophic level to the next are generally estimated to be between about 5 and 30 percent, and a reasonable average figure is about 10 to 15 percent (Slobodkin, 1960, 1962).

Energy budgets and energy flow diagrams have been constructed for some natural communities (Figures 7.8 and 7.9); these diagrams underscore the relatively minor energetic importance of carnivores in ecological systems. Decomposers often play a major energetic role, especially in terrestrial ecosystems, where much of the primary production is not consumed by herbivores but instead falls to the ground as dead leaves and other plant material. Indeed, as much as 90 percent of the net annual production in some communities may be consumed by their decomposers. Community ecologists are currently investigating energy flow and efficiency of transfer of energy in natural ecosystems and much remains to be learned about ecological energetics; such studies have obvious practical significance to human exploitation of ecological systems.

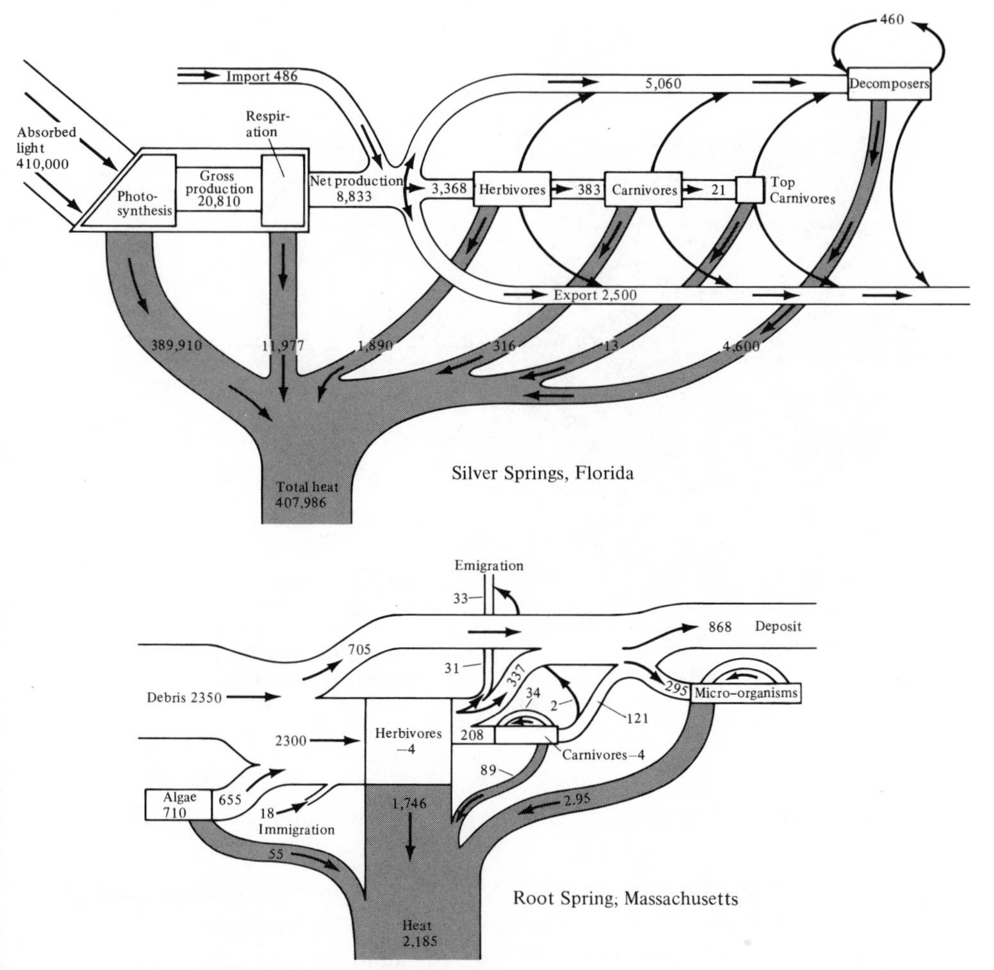

FIGURE 7.8 Two energy flow diagrams. Figures are in kcal/m^2/yr; numbers inside boxes in lower diagram represent changes in standing crops. [From Phillipson (1966) after Teal and Odum.]

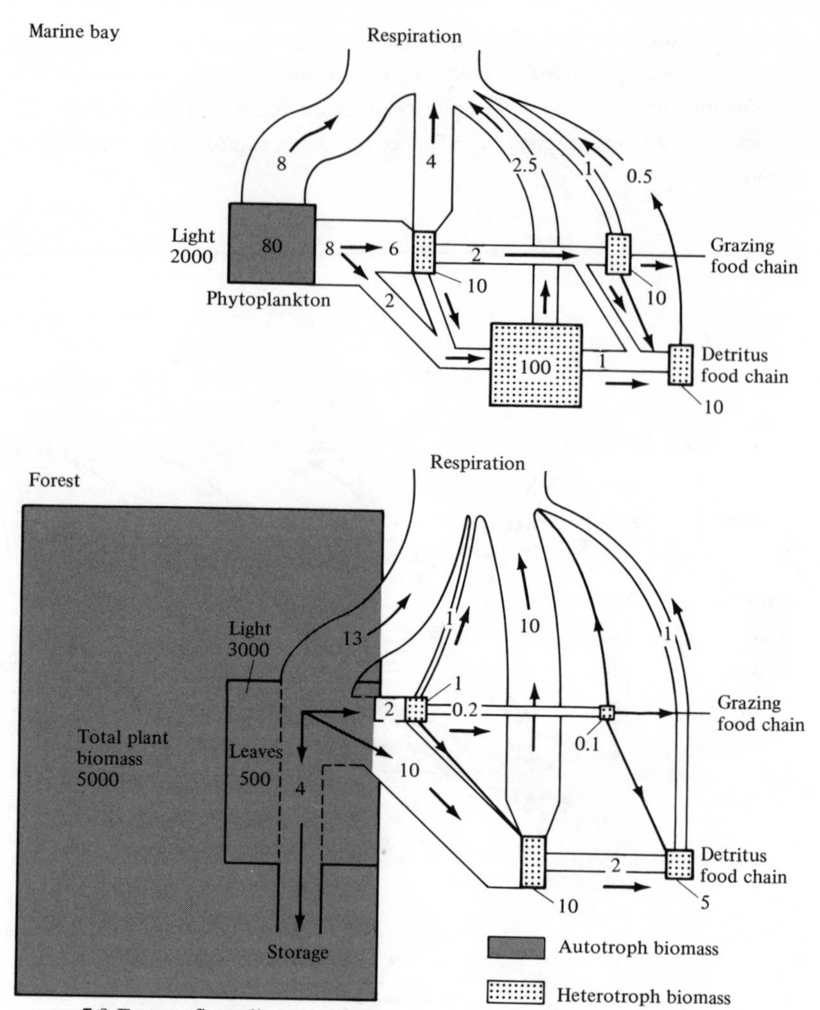

FIGURE 7.9 Energy flow diagrams for two ecosystems with very different standing crops, a marine bay (above) and a forest (below). Standing crop biomass measured in kcal/m^2; energy flow in kcal/m^2/day. [From Phillipson (1966) after Odum, Connell, and Davenport.]

Saturation with Individuals and with Species

In any closed ecosystem at equilibrium, all the energy of net production must be used up by consumers and decomposers in order for the system to have a balanced energy budget (Figure 7.7 and equations on page 228). Such an idealized system can be thought of as being saturated with individual organisms, because all the available energy is used and no more organisms could be supported. However, predators, by reducing the densities of organisms at lower trophic levels, can prevent their prey populations from reaching *otherwise maximal sustainable densities*, and thus, effectively preclude true saturation within that lower trophic level. If this is the case, only top predator populations reach a sort of "complete" saturation. Furthermore, anti-herbivore defenses of plants render much of the net primary productivity unusable by animal consumers and thus require that many plant tissues be routed directly through a community's decomposers.

Communities, or portions thereof, can be kept from reaching saturation with individuals in other ways, too. Real ecological systems are almost never truly closed, but rather they usually both receive materials and energy from other systems and lose them to others. A community or a portion of a community that is not closed may be rarefied by continual or sporadic removal of organisms. As a hypothetical example, consider a lake along a river; both the lake and river contain communities of phytoplankton and zooplankton, but the lake receives an inflow of river water containing no members of the lake community, while losing some water containing members of its community. Such a system can never become truly saturated because of the continual removal of organisms from it. Moreover, since the physical environment is changing continually (Chapters 2 and 3), and because it takes time to respond to these changes, populations and communities probably seldom actually reach equilibrium, although some K-selected organisms may occasionally approach it.

How much does the degree of saturation with individuals vary within and between communities? And, how does efficiency of transfer of energy change with degree of saturation with individuals? It is known, for instance (Chapter 5), that the turnover rate of prey is usually highest at intermediate prey densities; moreover, such prey populations often support higher predator densities than larger prey populations. The immense practical value of such knowledge is readily apparent.

Can communities be saturated with *species?* That is, is there a maximal possible number of *different* species that can exist within an ecological system? If so, a new species introduced into such a community must either go extinct or cause the extinction of another species which it replaces. Conversely, the successful invasion of a new species into a community without extermination of any existing species would imply that the original community was not saturated with species.

There is limited evidence that portions of some communities may indeed be saturated with species, at least within habitats. R. H. MacArthur and his colleagues have demonstrated that bird species diversity is strongly correlated with foliage height diversity (Figure 7.10), in a remarkably similar way on three continents: North America, South America, and Australia. Habitats with equal amounts of foliage (measured by leaf surface) in three layers (0–2, 2–25, and over 25 ft, respectively, above ground) are richer in bird species than are habitats with unequal proportions of foliage in the three layers. The diversity of bird species is lowest in habitats with only one of these layers of vegetation, such as a grassland. Interestingly enough, knowledge of plant species diversity does not allow an improvement in the prediction of bird species diversity (MacArthur and MacArthur, 1961), suggesting that the birds recognize the structure, rather than the type, of the vegetation. Despite the fact that avian niches are partitioned in a fundamentally different way in Australia, bird species diversity within a given habitat on that continent is very close to what it is in a habitat of similar structure in North America (Recher, 1969). In addition to illustrating that spatial heterogeneity regulates bird species diversity, the convergence of these data suggests that these

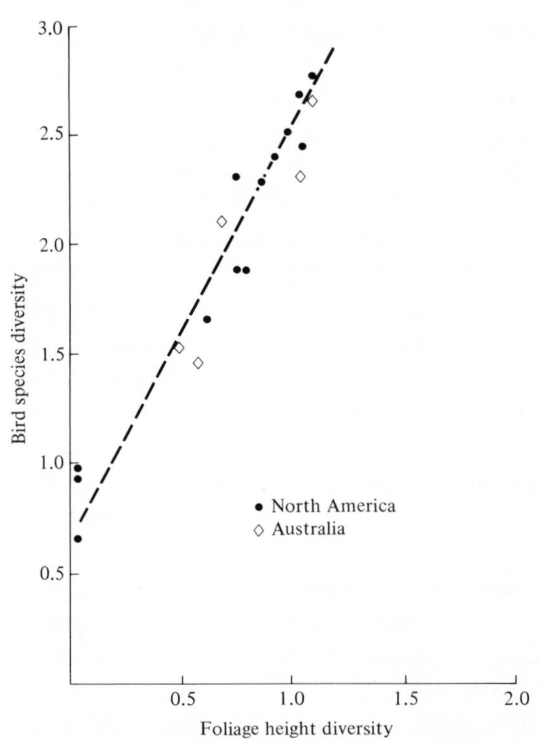

FIGURE 7.10 The observed correlation between foliage height diversity and bird species diversity. North American habitats shown as solid circles; Australian ones with diamonds. This is perhaps the best evidence that communities may sometimes be saturated with species. [After Recher (1969).]

avifaunas are saturated with species. However, such neat convergences in species densities of plants, insects, and desert lizards do not occur, which suggest that these groups may not always be saturated with species (Whittaker, 1969, 1970, 1972; Pianka, 1973).

The number of species that can coexist at any point in space may have a distinct upper limit as suggested above; however, there is no obvious limitation on the number of species that can occur in a given *area*, because horizontal replacement of species can allow coexistence of many more species than are actually sharing the use of a common point in space within that area. Indeed, MacArthur (1965) has suggested that the horizontal component of diversity ("between habitat" diversity) may be increasing continually during evolutionary time, whereas point diversities remain nearly constant. Some upper limit on horizontal turnover of species also seems likely.

Species Diversity

Why does one community contain more species than another? Some complex communities, such as tropical rainforests, consist of many thousands of different plant and animal species, whereas other communities, such as tundra communities, support considerably fewer species, perhaps only a few hundred. The number of species often varies greatly even at a local level; thus, a grassland habitat typically contains many fewer species of birds than does an adjacent forest. Indeed, different forest communities in the same general region usually vary in numbers as well as types of plant and animal species. The number of species is referred to as "species richness" or, more frequently, as "species density."

Communities with similar species densities often differ in yet another way: some contain a few very common species and many rare ones, whereas others support no very common species but many of intermediate abundance. Abundance is only one way of estimating the *relative importance* of various component species within a community; other measures frequently employed include both the biomass of, and the energy flow through, various species' populations. The relative importance of species varies within and between communities, and considerable effort has been expended in attempts to document such differences and to understand why they occur. The importances of different species within a community (or a portion of one) can be depicted conveniently by "species importance curves" as in Figure 7.11. Several hypothetical distributions of the relative importance of species within communities have been suggested, which generate different shaped species importance curves (see Figure 7.11 and Whittaker, 1970).

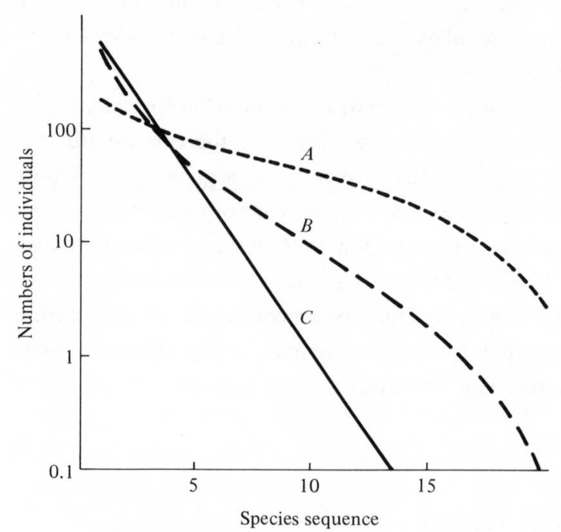

FIGURE 7.11 Species importance curves, with species ranked from most important to least important. Lines illustrate three different hypothetical curves based upon (*A*) random niche boundaries (MacArthur, 1957, 1960a), (*B*) multiple niche dimensions, which generate a lognormal distribution of species importances (Preston, 1948, 1960, 1962a, 1962b; Whittaker, 1970, 1972), and (*C*) niche preemption, which leads to a geometric series (Motomura, 1932; Whittaker, 1970, 1972). Data from various real communities fit each hypothetical line reasonably well. [After Whittaker (1972).]

Species density and relative importance have been combined in the concept of *species diversity*, which increases both with increasing species density and with increasing equality of importance among the members of a community. Species diversity is high when it is difficult to predict the species or the importance of a randomly chosen individual organism and low when an accurate prediction can be made. For example, an organism chosen at random from a cornfield would probably be a stalk of corn, whereas one would not venture to guess the probable identity of a randomly chosen organism from a tropical rainforest. We are currently trying to determine not only why different communities contain different numbers of species with differing relative importances, but also how such differences in species richness and importance affect other community properties, such as trophic structure and community stability.

Communities can differ in species diversity in several ways; more diverse communities may either (or both): (1) contain a greater range of available resources (that is, a larger total niche hypervolume space), and/or (2) their component species may, on the average, have smaller niche breadths (that is, each species might exploit a smaller fraction of the total niche hypervolume). The former corresponds roughly to "more niches" and the latter to "smaller niches." (3) Also, two communities with identical niche space and mean niche breadth can still differ in species diversity if they differ in the average degree of *niche overlap*, because greater niche overlap means that more species can exploit any particular resource (this situation is described as "smaller exclusive niches"). (4) Alternatively, of course, in communities that do not contain all the species they could conceivably support (i.e., those that are "unsaturated" with species), species diversity can vary with the extent to which all available resources are exploited by as many *different* species as possible (that is, with the degree of saturation with species, or with the number of "empty niches"). Resources are seldom if ever wasted, even in communities that do not contain their full quota of species, however, because those species that do occur in such communities generally expand their activities and exploit nearly all the available resources, although their efficiency of exploitation may be less than that of some better adapted species. (Thus most communities are probably effectively saturated with individuals even if they are not saturated with species.)

Because thorough understanding of community species diversity inevitably requires analysis of niche structure and diversification in component populations, investigations of species diversity usually go hand in hand with the study of niches. In practice, one is seldom able to study the species diversity of an entire community, and usually attention is focused on a portion of a community (an "assemblage"), such as trees, ants, lizards, or birds. Using the three major niche dimensions (Chapter 6), the total species diversity of an area can be partitioned into its spatial, temporal, and trophic components. Species replace one another along each of these three niche dimensions, and diversity is generated by separation along each.

I have studied species diversity and niche relations of desert lizards in certain deserts of North America (the Sonoran and Mojave deserts), southern Africa (the Kalahari desert), and Western Australia (the Great Victoria desert) (Pianka, 1973). Lizard niches in these deserts differ in the above three fundamental niche dimensions (place, time, and food). Moreover, the variety of resources actually used by lizards along each niche dimension, as well as the amount of niche overlap along them, differs markedly among desert systems. Food is a major dimension separating niches of North American lizards, whereas in the Kalahari food niche separation is slight and differences in the place and time niches are considerable. In the most diverse Australian lizard "communities," all three niche dimensions are important in separating niches and niche overlap is distinctly reduced. Differences between deserts in lizard species diversity are *not* accompanied by conspicuous differences in niche breadths, but rather they stem both from differences in the variety of resources used by lizards and from reduced niche overlap in the most diverse lizard communities of Australia (Pianka, 1973).

The spatial component of diversity is due to differential use of space by different populations. For convenience it can be broken down into horizontal and vertical components. At a gross geographic level, species replace one another horizontally as one moves from one habitat into another; this is the between-habitat component of overall diversity. Similar replacement of species occurs both horizontally and vertically *within* habitats. Birds, for instance, often tend to partition a given habitat by occupying different vertical strata, such as low bushes, tree trunks, lower foliage and high canopy. On the other hand, ground-dwelling mammals and lizards generally partition microhabitats horizontally, with some using open spaces between shrubs and others exploiting the ground beneath or near specific types of vegetation such as grasses, shrubs, and trees. Different populations, by occupying different microhabitats, are thus able to coexist within a given habitat and contribute to within-habitat diversity. The within-habitat component of diversity is most easily distinguished from the between-habitat component in relatively homogeneous communities [heterogeneous communities, such as edge communities and ecoclines (see Figure 3.13) include both components]. However, even a homogeneous community has an internal structure in that it consists of a mosaic of repeatable horizontal and vertical patches. Because communities and habitats frequently blend into one another, it is often difficult to distinguish "between-habitat" from "within-habitat" diversity. Where does one habitat "stop" and another "begin"? A sandridge gradually gives way to a sand plain and the intertidal grades into the deeper benthic zone. The problem of defining a habitat can be overcome by the use of "point diversities," which consist of the species diversity occurring at a point in space. Point diversities are difficult to estimate (one might have to wait a very long time to see *all* the species that use a particular point!). However, they should invariably be lower than any areal estimate of diversity because the different species in a community have each specialized somewhat as to the microhabitats they use.

Provided that different resources are utilized, temporal separation, both daily and seasonally, of species' populations can allow coexistence of more species and hence may also add to community diversity. Many instances of subtle differences in times of activity between populations are known, in addition to such conspicuous distinctions as that between nocturnal and diurnal animals.

Yet another means by which community diversity may be enhanced is by trophic differences. Here again, in addition to the conspicuous differences between trophic levels, such as herbivores, omnivores, and carnivores, there are more subtle but nevertheless important differences between species even within a given trophic level in the prey they eat. Thus different species of predators living in the same area tend to eat prey of different sizes and types, with the larger species taking larger prey items (this generalization applies to most fish, lizards, carnivorous mammals, and hawks). Moreover, the composition of the diet often varies markedly among potential competitors (see Table 5.4 and Lack, 1945). Finally, diversity of plant defensive chemicals doubtlessly creates numerous different potential food niches for herbivores, especially insects (Whittaker, 1969; Whittaker and Feeny, 1971), greatly facilitating trophic diversity at higher trophic levels.

A prevalent global pattern of species diversity is of some interest. The diversity of living organisms is usually high near the equator, and decreases rather gradually with increasing latitude, both to the north and south. Such "latitudinal gradients" in diversity are very widespread among different plant and animal groups and it is likely that a general explanation underlies these ubiquitous patterns. One reason species diversity is higher at lower latitudes than it is in the temperate zones is that there are often more habitats in the tropics. At high altitudes in the tropics, habitats similar to, but richer in species than, those in temperate zones often occur, whereas true tropical habitats are seldom found in temperate areas. However, the fact that there is a greater variety of species where there are more habitats is neither surprising nor theoretically very interesting. Diversity differences *within* a given habitat type are of much greater interest for they reflect the partitioning of available niche space within habitats. These two components of total species diversity are, respectively, the between-habitat and the within-habitat components.

There has been considerable speculation about the causes of both local and latitudinal patterns in species diversity, and many theories and hypotheses have been suggested for their explanation (Table 7.1), all of which probably operate in some situations (Pianka, 1966a). These various mechanisms for determination of diversity are clearly not independent and several may often act in concert or in series in any given case. Each hypothesis or theory is briefly outlined below and then some ways in which they could interact are considered.

These mechanisms can be classified as primary, secondary, or tertiary, depending upon whether they act mainly through, respectively, the physical environment alone, through a mixture of both the physical and the biotic

TABLE 7.1 Various Hypothetical Mechanisms for the Determination of Species Diversity and Their Proposed Modes of Action upon Ecological Niches

Level	Hypothesis or Theory	Mode of Action
Primary	1. Evolutionary time	Degree of unsaturation with species
Primary	2. Ecological time	Degree of unsaturation with species
Primary	3. Climatic stability	Mean niche breadth
Primary	4. Climatic predictability	Mean niche breadth
Primary or secondary	5. Spatial heterogeneity	Range of available resources
Secondary	6. Productivity	Especially mean niche breadth, but also range of available resources
Secondary	7. Stability of primary production	Mean niche breadth and range of available resources
Tertiary	8. Competition	Mean niche breadth
Primary, secondary, or tertiary	9. Rarefaction	Degree of allowable niche overlap and level of competition
Tertiary	10. Predation	Degree of allowable niche overlap and level of competition

environments, or through the biological environment alone (Poulson and Culver, 1969). Ultimately, thorough understanding of patterns in diversity requires knowledge of primary-level mechanisms.

1. *Evolutionary Time.* This theory assumes that diversity increases with the age of a community, although the validity of this assumption is still open to question. Temperate habitats are thus considered to be impoverished with species because their component species have not had time enough to adapt to, or to occupy completely, their environment since the recent glaciations and other geological disturbances. More "mature" tropical communities, however, are more diverse because there has been a longer period without major disturbances for organisms to speciate and diversify within them. The evolutionary time theory does not necessarily imply that temperate communities are unsaturated with individuals, because niche expansion may often allow nearly full utilization of available resources, even in a habitat that is impoverished with species.

2. *Ecological Time.* This theory is similar to the evolutionary time theory, but deals with a shorter, more recent, time span. Here we are concerned primarily with time available for dispersal, rather than with time for speciation and evolutionary adaptation. Newly opened or remote areas of suitable habitat, such as a patch of forest burned by lightning, an isolated

lake, or a patch of sand dunes, may not have their full complement of species because there has been inadequate time for dispersal into these areas. The dispersal powers of most organisms are good enough that this mechanism may be of relatively minor importance in most communities (see also Chapter 8).

3. *Climatic Stability.* A stable climate is one that does not change much with the seasons. Successful exploitation of environments with unstable climates often requires that organisms have broad tolerance limits to cope with the wide range of environmental conditions they encounter. Thus, by demanding generalization, such variable environments favor organisms with broad niches. Conversely, environments with more constant climates allow finer specialization and narrower niches. Plants and animals in the relatively constant tropics, for example, are often highly specialized in both the places they forage and the foods they eat. Obviously, in two habitats with the same range of available resources, the one whose component species each use a smaller fraction of these resources will support more species. By these means, the number of species should increase with climatic stability.

4. *Climatic Predictability.* Many aspects of climate, although temporally variable, are nevertheless highly predictable in that they repeat themselves fairly exactly, as for example, from day to day, or year after year. Such cyclical predictability can allow organisms to evolve some degree of dependence upon, as well as specialize on, particular environmental conditions and temporal patterns of resource availability, thus enhancing daily and/or seasonal replacement of species and the temporal component of total diversity. Deep freshwater lakes in the temperate zones, for example, typically have a consistent annual succession of primary producers; a major causal factor is nutrient availability, which changes markedly during the year, being greatest during the spring and fall turnovers (Chapter 3, pp. 61–64). Thus different species of phytoplankton have adapted to exploit lakes under particular environmental conditions that recur regularly every year. Annual plants in Arizona's Sonoran Desert have adapted to the marked bimodal annual precipitation pattern (Figure 2.10*b*) with distinct rainfall peaks in winter and summer: there are two distinct sets of species, one whose seeds germinate under wet and cooler conditions (the winter annuals) and another set that germinates when conditions are wet but warmer (the summer annuals that bloom in late summer after the flash floods).

5. *Spatial Heterogeneity.* A forest contains more different species of birds than a grassland, an arboreal desert generally supports more species of lizards than one without trees, and a tidal flat with a great variety of particle sizes and substrate types has more species of mud-dwelling invertebrates than a homogeneous mudflat. Structurally complex habitats obviously offer a

greater variety of different microhabitats than simple ones. Because there are more different ways of exploiting them, such spatially heterogeneous habitats usually support more species than homogeneous ones; thus species replace one another in space with greater frequency and the spatial component of diversity is higher. Correlations between the structural complexity of a habitat and the species diversity of its biota are widespread (one was discussed earlier in this chapter and is illustrated in Figure 7.10, p. 232).

6. *Productivity.* In habitats with little food, foraging animals cannot afford to bypass many potential prey items; where food is abundant, however, individuals can be more selective and confine their diets to better prey items (see also Chapter 6, pp. 205–209). Thus, more productive habitats, or those where food is dense, offer more prey choice, and hence allow greater dietary specialization than do less productive habitats. Because each species uses less of the total range of available foods, the same spectrum of food types

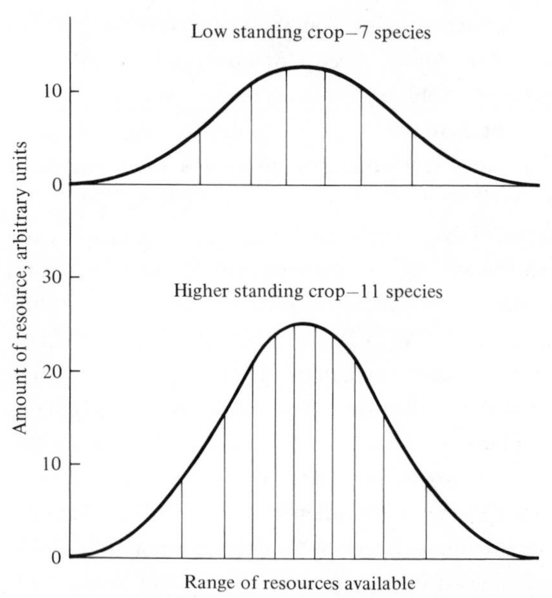

FIGURE 7.12 Graphic portrayal of the way in which more abundant resources can support a greater number of species. The horizontal axis represents the different kinds of available resources, ranked in any convenient order. The vertical axis is the amount of each resource type. Both curves cover the same range of resource types, but the lower curve is exactly twice the height of the upper one. All segments under both curves, except the tails, contain about the same area and therefore approximately the same amount of energy. With a low standing crop, component species must have broad niches and only seven species coexist; however, when standing crop is doubled, component species can have narrower niches and eleven species may exist in the community. [From Pianka (1971a).]

will support more species in a more productive environment (Figure 7.12). In addition, productive habitats may support more species than similar, less productive ones, by virtue of the fact that certain resources, too sparse to support a species in unproductive habitats, are dense enough to be successfully exploited in productive habitats (MacArthur, 1965). Thus an open desert with only one ant nest per hectare might not support a population of ant-eating lizard specialists, whereas another, richer, area with several ant nests per hectare would.

7. *Stability of Primary Production.* Just as more stable and more predictable climates support more species, areas with temporally stable and/or predictable patterns of production should allow coexistence of more species than would be possible in areas with more variable and/or erratic productivity. This secondary- or tertiary-level mechanism differs from those of climatic stability and climatic predictability in that plants themselves react to climatic conditions and hence they can alter temporal variability with their own homeostatic adaptations and storage capacities. Plants both buffer and enhance physical fluctuations by either releasing the products of primary production, that is, flowers or seeds, gradually and continuously or by expending them in temporally erratic blooms (see also Chapter 5, p. 178).

Mechanisms 3, 4, and 7 (climatic stability, climatic predictability, and stability of primary production) could be combined under a more inclusive heading of "temporal heterogeneity," to parallel "spatial heterogeneity."

8. *Competition.* In many diverse communities, such as tropical rainforests, populations are thought to be often near their maximal sizes (equilibrium populations of Chapter 4), with the result that intraspecific and interspecific competition are frequently keen. Selection for competitive ability (K selection) is therefore strong and most successful organisms in these communities have their own zone of competitive superiority. Thus, organisms that have specialized as to foods and/or habitats are at a competitive advantage (see Chapter 6, pp. 199–201), and the resulting small niches make high diversity possible. By way of contrast, populations in many less diverse communities, such as temperate and polar ones, are thought to be less stable and often well below their maximal sizes (opportunistic populations of Chapter 4). As a result, portions of the community are often unsaturated with individuals, and intraspecific and interspecific competition are frequently relatively lax. In such communities, it is often the physical world, rather than the biotic one, that demands adaptation. Selection for competitive ability is weak, whereas selection for rapid reproduction (r selection) is strong. Such r-selected populations typically have broad tolerance limits and relatively large niches.

9. *Rarefaction.* Rarefaction refers to the continual density-independent removal of organisms from a community (see pp. 230–231). The rarefaction hypothesis is essentially an alternative to the competition hypothesis (above), and these two mechanisms would seem to be mutually exclusive. In communities that are not fully saturated with individuals, competition is reduced and coexistence is possible *without competitive exclusion.* Thus, this hypothesis suggests that communities (or portions of communities) can, in some sense, actually be oversaturated with species in that more species coexist than would be possible if the system were allowed to become truly saturated with individuals. Rarefaction may operate through primary-, secondary-, or tertiary-level mechanisms (see also Predation). Catastrophic winter cold snaps and subsequent density-independent kills (see Table 4.3) illustrate primary-level rarefaction.

10. *Predation.* By either selective or random removal of individual prey organisms, predators can act as rarefying agents and effectively reduce the level of competition among their prey. Indeed, as described in Chapter 5 (pp. 168–169), predators can allow the local coexistence of species that are eliminated by competitive exclusion in the absence of the predator. Because many predators prey preferentially upon more abundant prey types, predation is often frequency dependent, which promotes prey diversity.

Clearly several of these mechanisms may often act together to determine the diversity of a given community, and the relative importance of each mechanism doubtlessly varies widely from community to community. A multitude of various possible ways in which these mechanisms could interact have been suggested; two are shown in Figure 7.13.

Community Stability

How much do populations in different communities fluctuate in size? And, how do a community's structure and the other members in a community influence such fluctuations in the size of its component populations and vice versa? Such questions are of great theoretical and practical interest, but their answers are essentially unknown.

MacArthur (1955) suggested that the stability of populations in a community should increase both with the number of different trophic links between species and with the equitability of energy flow up the various food chains. He reasoned that a community with many trophic links provides greater possibilities for checks and balances to operate between and among various species' populations. Should any one population begin to increase

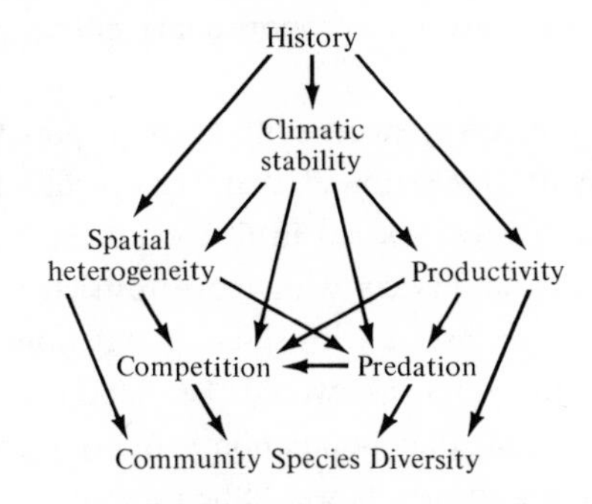

(*a*)

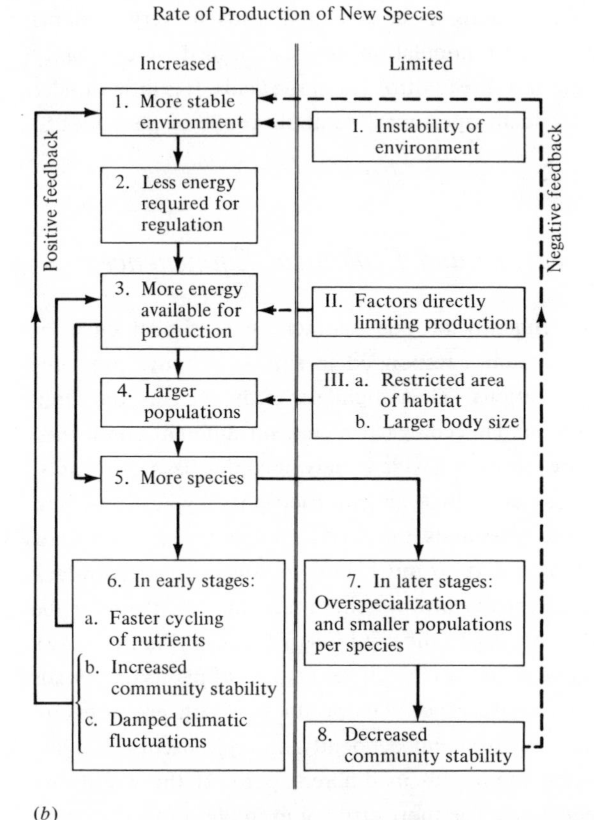

(*b*)

FIGURE 7.13 Two possible ways in which various mechanisms might interact to determine community diversity. (*a*) Simple schematic diagram showing some possible interactions of hypothetical processes. [From Pianka (1971a).] (*b*) A cybernetic model from Connell and Orias (1964) proposed for the production and regulation of diversity, which includes several of the 10 proposed mechanisms in Table 7.1.

markedly, its predators, by changing their diets and feeding selectively upon this abundant prey type, would thus exert disproportionate effects on its population growth.

Whether or not community diversity necessarily leads to population stability is still being debated. Watt (1968) found that herbivorous insect species that feed on a wide variety of tree species in Canada actually have *less* stable populations than do similar insects with more restricted diets. Watt does not indicate whether or not insects with stable populations have a greater variety of potential predators as would be expected from MacArthur's argument, but he did find that population stability increased with increasing numbers of competing species as would be expected.

Man's traditional method of agriculture, the monoculture of stands of a single species of plant, seems to be one that leads to very unstable ecological systems. Once a pest population gets established, it can easily grow at an exponential rate and spread throughout the fields. It is little wonder that we have become dependent upon massive applications of pesticides to control such pests!

Evolutionary Convergence and Ecological Equivalence

Organisms evolving independently of one another under similar environmental conditions have sometimes responded to similar selective pressures with nearly identical adaptations. Thus flightless birds such as the emu, ostrich, and rhea fill very similar ecological niches on different continents. Arid regions of South Africa support a wide variety of euphorbeaceous plants, some of which are strikingly close to American cacti phenotypically. A bird of some African prairies and grasslands, the African yellow-throated longclaw (*Macronix croceus*), looks and acts so much like an American meadowlark (*Sturnella magna*) that a competent bird watcher might mistake them for the same species; and yet they belong to different avian families (Figure 7.14*g*). Such convergent phenotypic responses by different stocks of plants or animals are known as evolutionary convergence. The products of convergent evolution, organisms that have evolved independently and yet occupy roughly similar niches in various communities in different parts of the world, are known as *ecological equivalents*. The more striking examples of evolutionary convergence (Figure 7.14) usually fall into either or both of two categories: they sometimes occur in relatively simple communities in which biotic interactions are highly predictable and the resulting number of different ways of exploiting the environment are few, and/or they occur under unusual conditions where selective forces for the achievement of a particular mode of existence are particularly strong. Examples of the latter include the independently evolved marsupial and placental "saber-toothed tigers" (Figure 7.14*f*), and the fusiform shapes of sharks, ichthyosaurs, and dolphins. Evolutionary convergence can easily be read into a situation by placing undue emphasis upon superficial similarities but failing to appreciate fully the inevitable dissimilarities between pairs of supposed ecological equivalents.

Often roughly similar ecological systems support relatively few conspicuous ecological equivalents, but instead are composed largely of distinctly different plant and animal types. For instance, although the bird species diversities of temperate forests in eastern North America and eastern Australia are similar [Recher (1969) and pp. 231–232], many avian niches appear to be fundamentally different on the two continents. Honeyeaters and parrots are conspicuous in Australia, while hummingbirds and woodpeckers are entirely absent. Apparently, different combinations of the various avian ecological activities are possible; thus an Australian honeyeater might combine aspects of the food and place niches exploited in North America by both warblers and hummingbirds. An analogy can be made by comparing the "total avian niche space" to a deck of cards. There are a limited number of ways this niche space can be exploited. Each bird population or species has its own ways of doing things, or its own "hand of cards," determined in part by what other species in the community are doing.

Ecotones, Vegetational Continua, and Succession

As indicated in the introduction of this chapter, communities are seldom discrete entities, but in fact usually grade into one another in both space and time. A localized "edge community" between two other reasonably distinct communities is termed an *ecotone*. Typically, such ecotonal communities are rich in species because they contain representatives from both parent communities and may also contain species distinctive of the ecotone itself. Often a series of communities grade into one another in an almost continuous fashion (see Figure 3.13); such a gradient community is called an *ecocline*. Ecoclines may occur either in space or in time.

Spatial ecoclines on a more local scale have led to so-called gradient analysis (Whittaker, 1967). The abundances and actual distributions of organisms along many environmental gradients have shown that the importances of various species along any given gradient typically form bell-shaped curves reminiscent of the tolerance curves discussed in Chapter 1. Moreover,

(a)
(b)
(c)
(d)
(e)
(f)
(g)
(h)

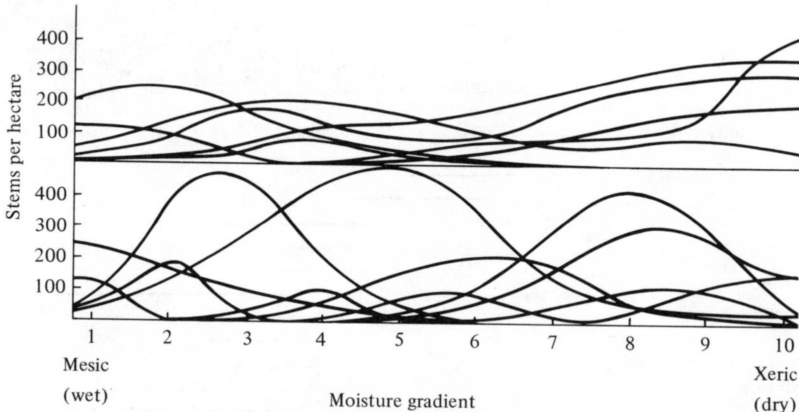

FIGURE 7.15 Actual distributions of some populations of plant species along moisture gradients from relatively wet ravines to dry southwest-facing slopes in the Siskiyou Mountains of northern California (above) and the Santa Catalina Mountains of Arizona (below). [After Whittaker (1967).]

these curves tend to vary independently of one another and often overlap broadly (Figure 7.15), indicating that each species' population has its own particular habitat requirements and width of habitat tolerance, and, hence, its own zone of maximal importance. Such a continuous replacement of plant species by one another along a habitat gradient is termed a *vegetational continuum* (Figure 7.15).

A temporal ecocline, or a change in the composition of a community in time, both by changes in the relative importances of component populations and by extinction of old species and invasion of new ones, is termed a

---

FIGURE 7.14 Some examples of convergent evolution in animals. Pairs of such independently evolved, but ecologically similar, species which occupy similar niches in different communities are known as "ecological equivalents." (*a*) A tropical Asian cyprinid fish, *Rasbora* sp. (above), and an African characin, *Neolobias* sp. (*b*) An Australian agamid lizard, *Amphibolurus cristatus* (left), and a North American iguanid, *Callisaurus draconoides*. (*c*) An African civet (left) and an American weasel. (*d*) Another Australian agamid lizard, *Moloch horridus* (left), and a North American iguanid, *Phrynosoma coronatum*. (*e*) An Australian marsupial, a wombat (left), with its skull and an American placental, a woodchuck, with its skull. (*f*) Skulls of two fossil (but not contemporary) saber-tooth carnivores, with the South American marsupial "cat," *Thylacosmilus*, on the left and the North and South American placental saber-toothed tiger, *Smilodon* on the right. (*g*) An American icterid, the Eastern Meadowlark, *Sturnella magna* (above), and an African motacillid, the yellow-throated longclaw, *Macronix croceus*. (*h*) A North American little auk (above) and a Magellan diving petrel, which belong to two different avian orders. [*a*, *c*, *e*, *f* after Salthe (1972) *Evolutionary Biology*. Copyright © 1972 by Holt, Rinehart and Winston, Inc. Reprinted by permission of Holt, Rinehart and Winston, Inc. *b*, *d* after Pianka (1971a). *g*, *h* after Fisher and Peterson (1964).]

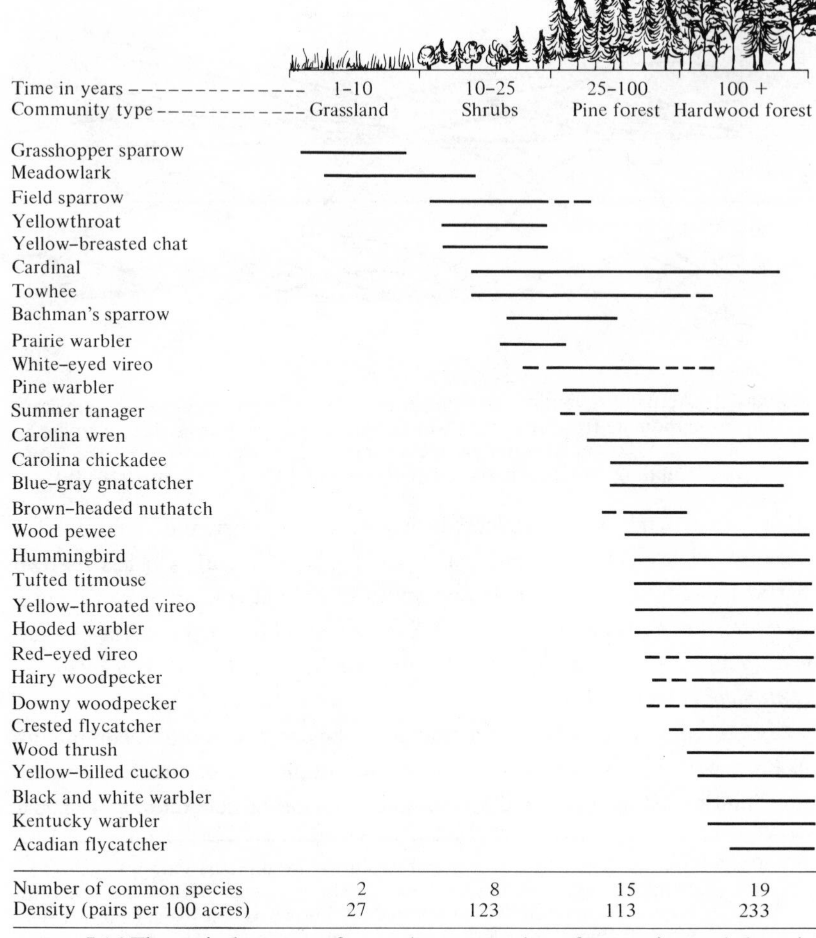

FIGURE 7.16 The typical pattern of secondary succession of vegetation and the avifauna on abandoned farmland in the southeastern United States. The number of bird species increases markedly with increased vertical structural complexity of the vegetation. [From MacArthur and Connell (1966) after Odum.]

succession. Primary succession refers to the development of communities from bare rock (see Chapter 3, p. 54), whereas secondary successions are changes that take place after destruction of the natural vegetation of an area with soil. In the southeastern United States, for instance, an old abandoned field allowed to change naturally is first invaded by annual weeds, such as crabgrass and asters, then by broomsedge, then by small perennials, then by various shrubby species of perennials, then by pine trees, and finally, by oak and hickory trees (Figure 7.16). The composition of the avifauna and bird species diversity also change drastically with these successional changes in the vegetation and are part of the community succession. The plants in each stage modify the environment, presumably making it more suitable for other species in following stages. Typically shade tolerance increases as succession proceeds. The entire process of secondary succession may take many years, over a hundred years in the above example. Only the oak–hickory forest is a stable community in a dynamic equilibrium that replaces itself; such a final stage in succession is termed its *climax*. In deserts, where the open vegetation alters microclimates very little and soil formation is virtually nonexistent, the first plants to invade are usually the climax species, and the succession, if one calls it such, is short. Earth's biomes (Chapter 3) represent the climax communities that prevail at different localities. Disturbances, both man made and natural (such as lightning, fires, droughts, landslides, and floods), are often frequent enough that extensive areas have not had time enough to reach their own climax state. An equilibrium is reached whereby the proportion of a habitat supporting early successional stages is determined by the frequency of disturbance. Largely undisturbed areas may be primarily in the climax state. During the course of succession, annual production exceeds annual respiration, and organic materials accumulate to form soils and, generally, an increasingly larger biomass of plants and animals. At climax, production equals respiration and organic materials cease to accumulate. An excellent discussion of ecoclines and succession may be found in Whittaker (1970).

Community Evolution

In the preceding section, we noted that many communities change during the lifetimes of the individuals that comprise them. In addition to such relatively short-term changes during ecological time, community characteristics are affected by evolution of and coevolution among the species' populations that are available to form the community over evolutionary time. At the same time, the community itself is a major determinant of the selective milieu of its component populations, and its characteristics presumably dictate many of their adaptations. The so-called taxon cycle (Chapter 8, pp. 273–275) is thought to be driven by biotic responses to competition and predation, or "counter adaptations" of the other species in a community (Ricklefs and Cox, 1972). Competition within, between, and among species results in the evolution of niche differences, which in turn assures that the resources of a given community, including plants and animals, are utilized more or less in proportion to their effective supply (Chapters 5 and 6). As pointed out earlier in this chapter, evolution of the species within a community has still other effects upon community structure. Evolution of prey reduces the efficiency of transfer of energy from one trophic level to the next, but increases stability, whereas evolution of predators acts to increase the efficiency of this transfer but reduces stability. The diversity of prey eaten by a predator as well as the predator's ability to alter his diet with changes in prey availability probably influences the stability of prey populations, and therefore of the community (see also Community Stability).

Can natural selection operate between entire communities? The notions of selection at the levels of communities and ecosystems (Dunbar, 1960, 1968; Lewontin, 1970) constitute apparent extremes of the idea of group selection (Chapter 1, pp. 12–13). Selection is quite unlikely to occur at these levels, in view of both the limited number of communities and ecosystems and their low rate of turnover. Most important, selection acts only by *differential reproduction* (Chapter 1, pp. 9–12), and it is most difficult to envision reproduction by a community or an ecosystem. The organisms comprising a community are not bound together by obligate relations, but instead, each evolves in a manner independent of, and often antagonistic to, other members of the community, such as its prey, competitors, and predators. Indeed, community stability may even be incompatible with efficient transfer of energy to higher trophic levels because of the antagonistic interactions between predators and their prey.

Selected References

Food Webs and Trophic Levels

Allee *et al.* (1949); Elton (1927, 1949, 1966); Gallopin (1972); Hairston, Smith, and Slobodkin (1960); Hubbell (1973a, 1973b); Kozlovsky (1968); Murdoch (1966a); Odum (1959, 1963, 1971); Paine (1966); Phillipson (1966).

The Community Matrix

Levins (1968); Parker and Turner (1961); Vandermeer (1970, 1972a, 1972b).

Principles of Thermodynamics

Bertalanffy (1957); Brody (1945); Gates (1965); Odum (1959, 1971); Paine (1971); Phillipson (1966); Wiegert (1968).

Pyramids of Energy, Numbers, and Biomass

Elton (1927); Kormondy (1969); Leigh (1965); Odum (1959, 1963, 1971); Phillipson (1966); Slobodkin (1962).

Energy Flow and Ecological Energetics

Bertalanffy (1969); Bormann and Likens (1967); Engelmann (1966); Gates (1965); Golley (1960); Hairston and Byers (1954); Hubbell (1971); Lindemann (1942); Mann (1969); Margalef (1963, 1969); Odum (1959, 1963, 1968, 1969, 1971); Paine (1966, 1971); Patten (1959); Phillipson (1966); Reichle (1970); Schultz (1969); Slobodkin (1960, 1962); Teal (1962).

Saturation with Individuals and with Species

Cody (1970, 1973); Levins (1968); MacArthur (1965, 1970, 1971, 1972); MacArthur and MacArthur (1961); Pianka (1966a, 1973); Recher (1969); Vandermeer (1972a); Whittaker (1969, 1972).

Species Diversity

Arnold (1972); Baker (1970); Connell and Orias (1964); Dobzhansky (1950); Fischer (1960); Fisher, Corbet, and Williams (1943); Futuyma (1973); Gleason (1922); Harper (1969); Hutchinson (1959); Janzen (1970, 1971a); Johnson, Mason, and Raven (1968); Klopfer (1962); Klopfer and MacArthur (1960, 1961); Lack (1945); Leigh (1965); Loucks (1970); MacArthur (1960a, 1965, 1972); MacArthur and MacArthur (1961); MacArthur, MacArthur, and Preer (1962); MacArthur, Recher, and Cody (1966); Margalef (1958a, 1958b, 1963, 1968); Odum (1969); Orians (1969a); Paine (1966); Patten (1962); Pianka (1966a, 1973); Poulson and Culver (1969); Preston (1948, 1960, 1962a, 1962b); Recher (1969); Richards (1952); Ricklefs (1966); Schoener (1968a); Schoener and Janzen (1968); Shannon (1948); Simpson (1949); Simpson (1969); F. E. Smith (1970a, 1970b, 1972); Tramer (1969); Vandermeer (1970); Watt (1973); Whiteside and Hainsworth (1967); Whittaker (1965, 1969, 1970, 1972); Whittaker and Feeny (1971); Williams (1944, 1953, 1964); Woodwell and Smith (1969).

Community Stability

Frank (1968); Futuyma (1973); Hairston *et al.* (1968); Harper (1969); Hurd *et al.* (1971); Leigh (1965); Lewontin (1969); Loucks (1970); MacArthur (1955, 1965); Margalef (1969); May (1973); Murdoch (1969); F. E. Smith (1972); Watt (1965, 1968, 1973); Whittaker (1972); Woodwell and Smith (1969).

Evolutionary Convergence and Ecological Equivalence

Grinnell (1924); MacArthur and Connell (1966); Raunkaier (1934); Recher (1969); Salthe (1972).

Ecotones, Vegetational Continua, and Succession

Clements (1920, 1949); Horn (1971); Kershaw (1964); Loucks (1970); Margalef (1958b); McIntosh (1967); Shimwell (1971); Terborgh (1971); Whittaker (1953, 1965, 1967, 1969, 1970, 1972).

Community Evolution

Darlington (1971); Dunbar (1960, 1968); Futuyma (1973); Kormondy (1969); Lewontin (1970); Odum (1969); Ricklefs and Cox (1972); Whittaker (1972); Whittaker and Woodwell (1971).

8 | *Biogeography*

Introduction

A major goal of ecology is to understand various factors influencing the distribution and abundance of animals and plants (Andrewartha and Birch, 1954; Krebs, 1972; MacArthur, 1972). Factors affecting abundances and microgeographic distributions (including habitat selection) were considered in Chapters 4, 5, and 6; here we examine grosser geographic distributions, namely the spatial distributions of plants and animals over large geographic areas, such as major land masses (continents and islands). The study of the gross geographical distributions of plants and animals, respectively, are termed phytogeography and zoogeography. Biogeography encompasses the geography of *all* organisms, and involves a search for patterns in the distributions of plants and animals and an attempt to explain how such patterns arose during the geological past. In addition to classifying present distributions, biogeographers seek to interpret and to understand the past movements of organisms. Ecology and biogeography are, of course, closely related and overlapping disciplines, and have profoundly affected one another.

Classical Biogeography

As early naturalists traveled to different parts of the world they discovered distinctly different assemblages of species; as data were gathered on these patterns, they came to recognize six major biogeographic "realms" or regions, three of which correspond roughly to the continents of Australia (Australian), North America north of the Mexican escarpment (Nearctic), and South America south of the Mexican escarpment (Neotropical). (The Neotropical region also includes the Antilles.) Africa south of the Sahara is known as the Ethiopian region. Eurasia is subdivided into two regions, the Palearctic north of the Himalayas, which also includes Africa north of the Sahara, and the Oriental south of the Himalayas (India, southern China, Indochina, the Phillipines, and Borneo, Java, and Sumatra, as well as other islands of Indonesia east to, and including, the Celebes). Each of the six biogeographic regions (Figure 8.1) is separated from the others by a major barrier to the dispersal of plants and animals, such as a narrow isthmus, high mountains, a desert, an ocean, or an oceanic strait. There is generally a high degree of floral and faunal consistency within regions and a marked shift in higher taxa such as genera and families in going from one region to another. Although biogeographers familiar with different plant and animal groups often disagree on the exact boundaries between regions (Figure 8.2), there is broad agreement on the usefulness of recognizing these six major regions.

High species diversities in the tropics (Chapter 7, pp. 237–242), among other things, have led to the notion that speciation rates in these areas must be extremely high and that such regions are often "source areas" for the production of new species, many of which then migrate into less hospitable areas such as the temperate zones. Thus, Darlington (1957, 1959) proposed the "area–climate hypothesis," which states that the majority of dominant animal species have arisen in geographically extensive and climatically favorable areas; he considers the Old World Tropics, which includes the tropical portions of the Ethiopian and Oriental regions, the major source

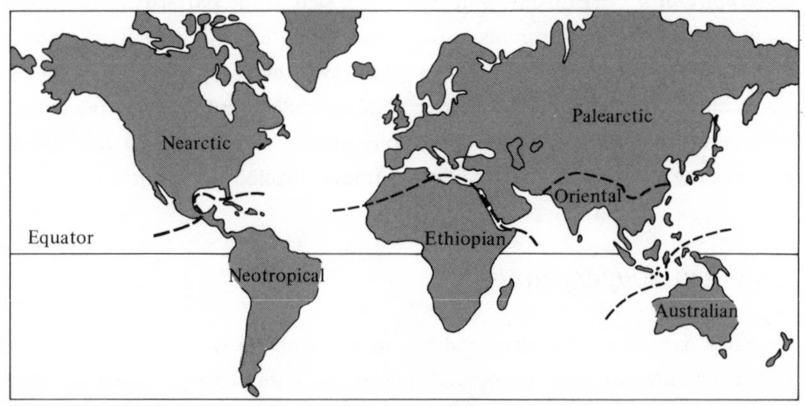

FIGURE 8.1 The six major biogeographic regions of the world.

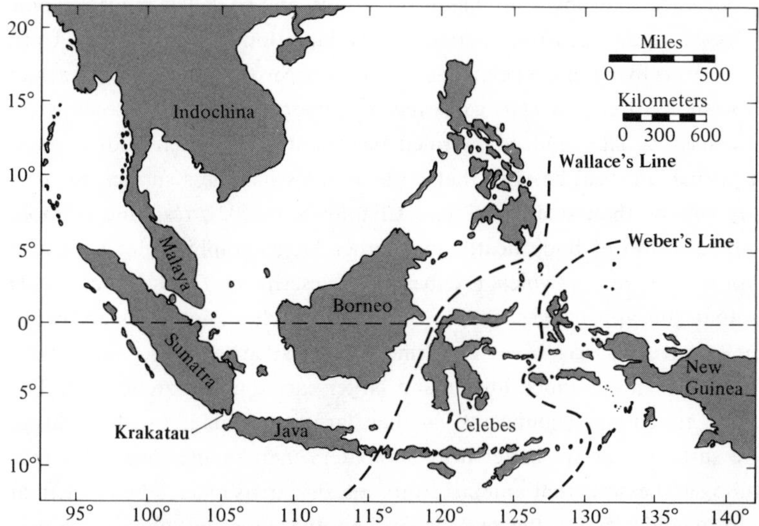

FIGURE 8.2 Wallace's and Weber's "lines" in southeast Asia, which separate the Oriental from the Australian regions. The position of the volcanic island of Krakatau between Sumatra and Java is marked with an arrow (lower left).

area for most vertebrate groups, and argues that such dominant forms have migrated centrifugally to other smaller and less favorable areas, including Europe, North and South America, and Australia.

Much of this classical biogeography assumed some permanence in the locations of continents; as a result interpretations of faunal similarities between them often rely on hypothetical mechanisms of transport from one continent to another, such as "rafting" of organisms across water gaps. Recent advances in geology have forced a partial revision of this view. Thus, there is now strong evidence (Dietz and Holden, 1970) that the continents were once joined in a large southern landmass (Pangaea) and have "drifted" apart (see also Chapter 2, pp. 39–40), with a gradual breakup that began in the early Mesozoic Era (about 200 million years ago). Geological evidence that the continents have drifted (and are now drifting) is accumulating rapidly (J. T. Wilson, 1973); much of classical biogeography will have to be reinterpreted in light of new findings. For example, certain very ancient groups of freshwater fishes, amphibians, and insects that had spread before the breakup of the continents now occur on several continents, whereas many other more recently evolved groups of plants and animals, such as mammals and birds, are restricted to a particular biogeographic region. These latter, more recent groups follow the regional divisions much more closely than the older groups (Kurtén, 1969).

Classical biogeography has produced several so-called biogeographic rules based upon recurring patterns of adaptation of organisms. Thus, homoiotherms living in cold climates are often larger than those from warmer regions; such a trend or *cline* can even be demonstrated within some wide-ranging species. This tendency, termed Bergmann's rule after its discoverer, has a probable causal basis in that large animals have less surface per unit of body volume than small ones (see Chapter 6, p. 202), resulting in more efficient retention of body heat. Many other biogeographic rules have also been proposed, all of which are basically descriptive. Thus, Allen's rule states that the appendages and/or extremities of homoiotherms are either longer or have a larger area in warmer climates; an example might be a jackrabbit which has much longer and larger ears than an arctic hare. The presumed functional significance is that large appendages with a larger relative surface area are better heat dissipaters than smaller ones. Another rule (Gloger's) asserts that animals from hot, dry areas tend to be paler than those from colder and wetter regions. Still another biogeographic rule is that fish from cold waters often have more vertebrae than those from warmer waters. The adaptive significance of many of these biogeographic trends remains obscure, although such geographically variable phenotypic traits are frequently developmentally flexible and respond more or less directly to temperature.

Island Biogeography

Ecosystems are usually very difficult to manipulate experimentally, and hence much of modern ecology has had to rely upon exploitation of "natural" experiments—situations in which one (or a few) factor(s) affecting a community differ between two (or more) ecosystems. For this reason, ecologists have long been especially interested in islands, which constitute some of the finest of natural ecological experiments. Different islands in an archipelago often contain different combinations of the mainland species, allowing an investigator to observe both ecological and evolutionary responses, such as niche shifts, of various component species to the presence or absence of other species. Islands can be exploited as natural ecological experiments in numerous other ways as well. Thus, because islands support fewer predatory species than comparable mainland habitats (see next section), they can be used to study the effects of predator exclusion (see also Chapter 5, pp. 168–169). Moreover, reduced species densities on islands, such as the land birds of Bermuda (Chapter 5, p. 153), allow partial analysis of the effects of interspecific competition upon the ecologies of those species which have populated an island.

Islands of a sort are widespread in the terrestrial landscape as well; a patch of forest separated from a larger stand of trees can be considered a "habitat island." Similarly, isolated lakes and mountain tops (Chapter 5, p. 153), represent "islands." To a nonflying insect, plants in the desert or trees within an open forest may also approximate islands in that they are separated from one another by relatively vast open spaces of a different and relatively inhospitable environment. Likewise, cattle droppings scattered about a field are islands to the animals that inhabit them (Mohr, 1943). Indeed, even a teaspoon of water or the body of an insect may be an island to a bacterium.

Species-Area Relationships

Larger islands generally support more species of plants and animals than smaller ones. In fact, when plotted on a double log scale, the number of species in a given taxon typically increases more or less linearly with island size (Figure 8.3). In most cases, a tenfold increase in area corresponds to an approximate doubling of the number of species. The slope of a linear regression line through such points is designated that taxon's z-value in the particular island system. It has been found empirically that z-values generally range from about 0.24 to about 0.33 in a variety of taxa on many different island systems (Table 8.1). The z-value is the exponent in the equation

$$S = CA^z \quad (1)$$

where S is the number of species, C is a constant which varies between taxa and from place to place, and A is the area of the island(s) concerned. Taking logarithms and rearranging, one obtains the following linear equation in which z is the slope:

$$\log S = \log C + z \cdot \log A \quad (2)$$

Based upon the assumption that species importances are determined along numerous niche dimensions and are therefore normally distributed when plotted semilogarithmically (Figure 7.11, p. 234), Preston (1962a, 1962b) calculated a theoretical "expected z-value" of 0.263, in reasonably close accord with empirical values in Table 8.1. Values larger than this result from topographic diversity and spatial replacement of species, or "islands within islands"; lower values arise with reduced replacement of species in space, as for instance on very homogeneous islands, continents, or subsamples of large islands (see below). Area, in itself, is probably not the primary factor affecting species density in most situations, but it presumably operates

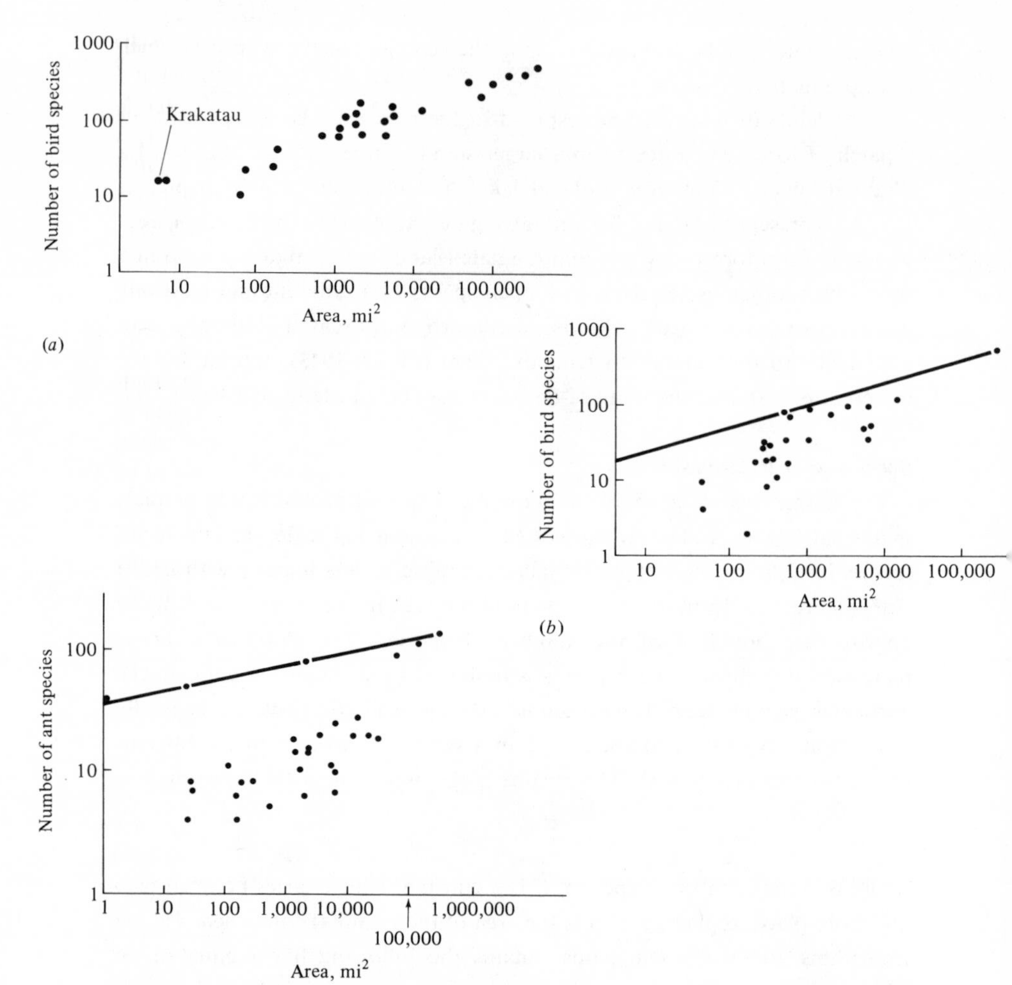

FIGURE 8.3 Various species-area relationships. (*a*) The numbers of species of land and freshwater birds on islands of the Sunda group in southeast Asia (Figure 8.2), with the Phillipines and New Guinea. Krakatau is plotted at the extreme left of the figure. (*b*) The numbers of species of land and freshwater birds on various, often remote, islands in the south Pacific, including the Moluccas, Melanesia, Micronesia, Polynesia, and Hawaii. The line is drawn through the two islands (Kei and New Guinea) nearest to source regions to demonstrate the degree of departure of species densities on the more remote islands. (*c*) The number of species of ponerine ants in the faunas of various Molluccan and Melanesian islands. The line represents the number of species with increasing area in subsamples of New Guinea, while points represent smaller islands. Note that the islands support fewer species than a comparable sized portion of New Guinea, but that the *rate* of increase of species with area is greater among the islands than it is within New Guinea. [*a*, *b* from MacArthur and Wilson (1967). *The Theory of Island Biogeography*. Copyright © 1967 by Princeton University Press. Reprinted by permission of Princeton University Press. *c* from Wilson (1961).]

TABLE 8.1 Estimated z-Values for Various Terrestrial Plants and Animals on Different Island Groups

Fauna or Flora	Island Group	z
Carabid beetles	West Indies	0.34
Ponerine ants	Melanesia	0.30
Amphibians and reptiles	West Indies	0.301
Breeding land and fresh-water birds	West Indies	0.237
Breeding land and fresh-water birds	East Indies	0.280
Breeding land and fresh-water birds	East-Central Pacific	0.303
Breeding land and fresh-water birds	Islands of Gulf of Guinea	0.489
Land vertebrates	Islands of Lake Michigan	0.239
Land plants	Galápagos Islands	0.325

Source: From MacArthur and Wilson (1967). *The Theory of Island Biogeography.* Reprinted by permission of Princeton University Press. Copyright © 1967 by Princeton University Press.

indirectly through increasing the variety of available habitats. Area can, however, directly affect species densities in some situations.

An area of mainland habitat comparable to and equal in size to an offshore island almost invariably supports more species, especially those at higher trophic levels, than does the island. In addition, the number of species in samples of a continental system also increases with the size (area) of the subsample, although not as rapidly as on islands (Figure 8.3). Typically, z-values in mainland situations range from about 0.12 to about 0.17. This difference arises because an island is a true "isolate" whereas a similar sized patch of mainland habitat is only a "sample"; rare species can occur in the mainland sample both due to migration from other areas and because areas immediately adjacent to the subsample also support other members of broad-ranging species. A mountain lion requiring a 20-square-mile territory would be unlikely to maintain a viable population on a small island of, say, less than 30 to 40 square miles, whereas these same cats are able to survive and replace themselves in a similar sized subsample of a larger landmass.

Equilibrium Theory

For many years, islands were considered to be in some sense "impoverished" with species both because of the obvious problems species have in colonizing them and because islands typically support fewer species than a comparable area of mainland habitat. Recently, however, the regularity of species–area patterns led MacArthur and Wilson (1963, 1967) to examine the possibility that islands might actually be supporting as many species as possible.

MacArthur and Wilson reasoned that the rate of immigration of *new* species to an island should decrease as the number of species on that island increases. The immigration rate must drop to zero as the species density of the island reaches the total number of species in the "species pool" available for colonization of the island at which no immigrant can be a new species. (The species pool corresponds to the total number of species in the source areas surrounding the particular island system.) Moreover, MacArthur and Wilson argue that the rate of extinction of species already present on the island should *increase* as the number of species on an island increases; this seems likely because, as more species invade an island, average population size must decrease and both the intensity of interspecific competition and the incidence of competitive exclusion should increase.

When the rate of immigation equals the rate of extinction (Figure 8.4), existing species go extinct at the same rate that new ones invade; thus species density reaches a dynamic equilibrium. Although species density stays constant, the continual turnover of species means that the species composition of the island can be changing.

MacArthur and Wilson's equilibrium theory is closely analogous to the model of the Verhulst-Pearl logistic equation for growth processes within a population (Chapter 4); thus, the number of individuals (density), N, is replaced by the number of species (species density), S, and the density dependent birth and death rates, b_N and d_N, are replaced by a falling immigration rate, λ, and a rising rate of extinction, μ, as the species density of an island increases. As a first approximation, we might assume that rates of

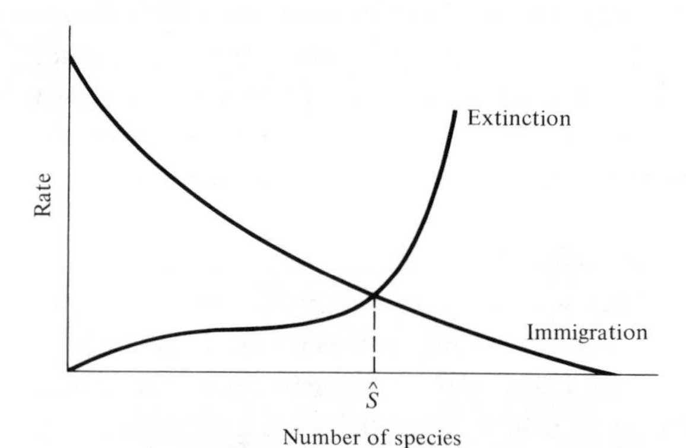

FIGURE 8.4 Illustration of the equilibrium theory for island species densities, with the immigration rate of new species falling and the rate of extinction of existing species rising as the total number of species on an island increases. At equilibrium, immigration just balances extinction and $\hat{S}$ different species exist on the island. The composition of the island's biota may change as some of the existing species go extinct and are replaced by other, different, species.

immigration (λ) and extinction (μ) vary linearly with species density according to the following equations:

$$\lambda_s = \lambda_0 - \alpha S \quad (3)$$

$$\mu_s = \mu_0 + \beta S \quad (4)$$

where λ_0 and μ_0 are, respectively, the rates of immigration and extinction with no species on the island, and α and β represent the rates of change in these rates as species density increases (Figure 8.5). (MacArthur and Wilson point out that this assumption of linearity is not as stringent as it might at first seem, since transformations of the ordinate may allow simultaneous straightening of immigration and extinction curves.) At equilibrium, or $\hat{S}$, the rate of immigration must exactly equal the rate of extinction, that is, λ_s must equal μ_s. Setting Equation (3) equal to (4),

$$\lambda_0 - \alpha\hat{S} = \mu_0 + \beta\hat{S} \quad (5)$$

and rearranging, one obtains the following expression for the number of species at equilibrium:

$$\hat{S} = \frac{\lambda_0 - \mu_0}{\alpha + \beta} \quad (6)$$

Equation (6) is, of course, identical in form to the expression for carrying capacity, K, in the logistic equation, which is $K = (b_0 - d_0)/(x + y)$ (see also Equation (25) in Chapter 4, p. 87).

As developed above, λ_s and μ_s represent total rates of immigration and extinction, and thus indicate little about the relative rates *per species* either already present on the island or available in the species pool (P).

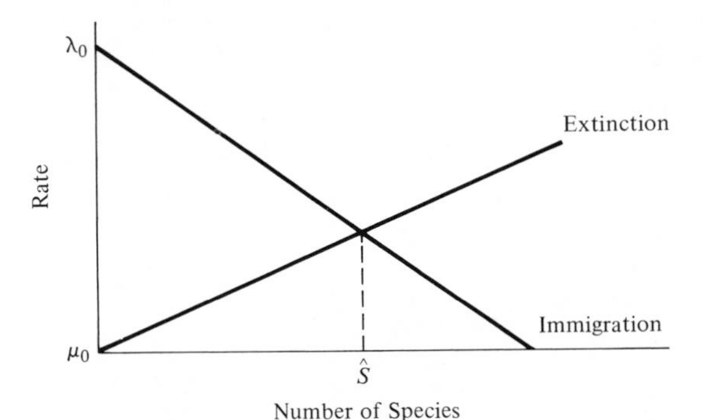

FIGURE 8.5 Diagram of immigration and extinction rates that change linearly with island species density. Equilibrium species density, $\hat{S}$, is a simple function of the slopes and intercepts of the two lines.

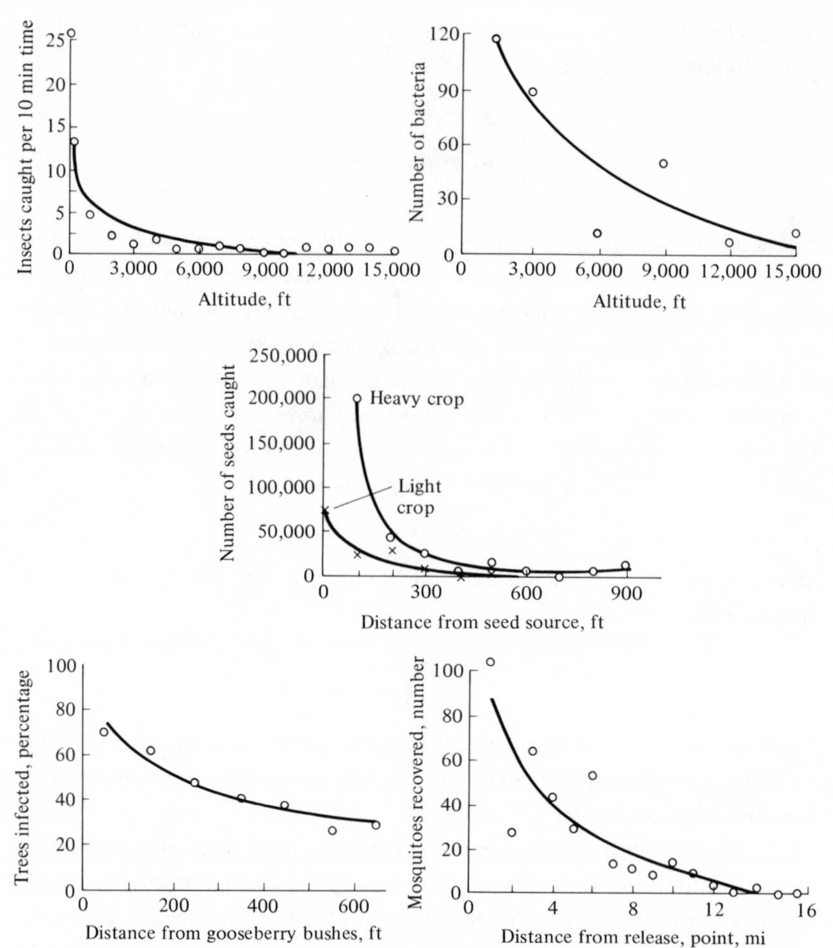

FIGURE 8.6 Some actual patterns of dispersal, both vertically and horizontally. The number of organisms decays rapidly at first and then more and more slowly with increasing distance. [From Odum (1959) after Wolfenbarger.]

The average rate of immigration per species, $\bar{\lambda}$, and the average rate of extinction per species, $\bar{\mu}$, can be obtained by dividing by, respectively, the number of species not yet on the island $(P-S)$ and the number already present on the island (S) as follows:

$$\bar{\lambda} = \frac{\lambda_S}{P-S} \quad \text{or} \quad \lambda_S = \bar{\lambda}(P-S) \tag{7}$$

$$\bar{\mu} = \frac{\mu_S}{S} \quad \text{or} \quad \mu_S = \bar{\mu}S \tag{8}$$

Again, at equilibrium, total extinction rate (μ_S) must equal the total rate of immigration (λ_S), that is $\lambda_S = \mu_S$, or in terms of the average rates per species (which are the rates with which an ecologist will usually be working):

$$\bar{\lambda}(P-\hat{S}) = \bar{\mu}\hat{S} \tag{9}$$

Solving for the equilibrium number of species, $\hat{S}$, gives

$$\hat{S} = \frac{\bar{\lambda}P}{\bar{\mu}+\bar{\lambda}} \tag{10}$$

Equation (10) demonstrates that $\hat{S}$ increases with increasing P and $\bar{\lambda}$ and decreases with increased $\bar{\mu}$. Notice also that $\bar{\lambda}$ is identical to α in Equation (3) and that μ is β in Equation (4).

Because dispersal falls off more or less exponentially with distance (Figure 8.6), MacArthur and Wilson reasoned that immigration rates should decrease with increasing distance from source areas (Figure 8.7). Further, they argued that rates of extinction should be largely unaffected by distance from source areas, per se, but instead should generally increase with decreasing island size because smaller islands support smaller, more tenuous, populations (Figure 8.8). [Because they present a smaller "target" for potential invaders, smaller islands might also have slightly lower immigration rates than other, equivalent, but larger islands. But this change should be minor compared with the expected decline due to the exponential decay in the number of immigrants with distance (Figure 8.6).] Moreover, simple islands with little topographic relief and relatively few different habitats should have generally higher extinction rates than more complex and more diverse islands with a greater variety of habitats, because the latter would provide a greater variety of immigrants with suitable opportunities for successful invasion and

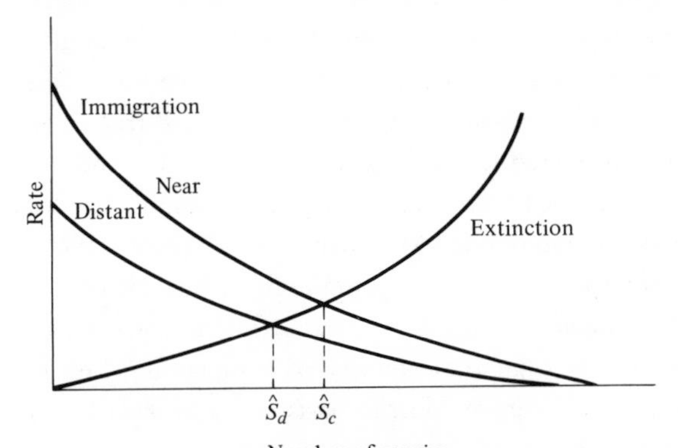

FIGURE 8.7 Immigration rates should decrease with increasing distance from source areas, so that distant islands should reach equilibrium with fewer species, $\hat{S}_d$, than close in islands, $\hat{S}_c$, all else being equal.

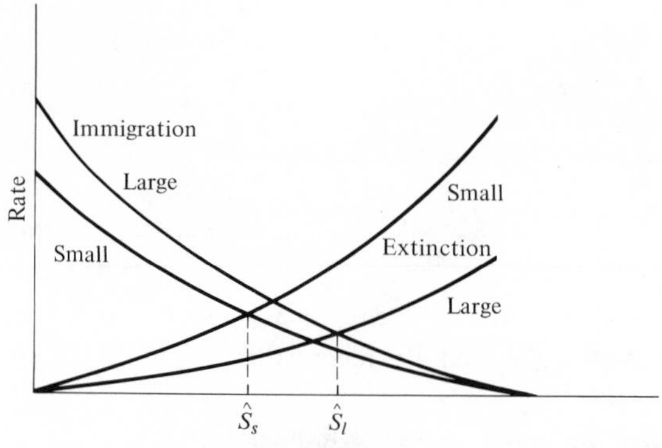

FIGURE 8.8 Extinction rates should be little affected by distances from source areas, but should often vary inversely with island size and/or complexity. Immigration rates may also be slightly higher on larger islands because they present a larger "target" for potential invaders. Thus, all else being equal, a small island should equilibrate with fewer species, $\hat{S}_s$, than a larger island, $\hat{S}_l$.

persistence on the island. Finally, clumped islands, such as archipelagos, should have higher rates of immigration than more scattered or isolated islands due to interchanges of plants and animals between islands.

Many of the predictions of equilibrium theory have now been supported by facts (see pp. 270–277). Turnover rates vary inversely with equilibrium species density, and, as indicated above, smaller islands support fewer species than larger ones.

The Compression Hypothesis

Organisms, when faced with more intense competition from other species, often, although not always, restrict their utilization of shared microhabitats and/or other resources (see Chapters 5 and 6). (These adjustments are those that take place in ecological time, during the lifetime of the organism concerned; longer term evolutionary adjustments were treated in Chapters 5 and 6 and are considered further under the next topic.) The fact of such niche contractions, coupled with theoretical considerations, has led to the so-called compression hypothesis (MacArthur and Wilson, 1967), which states that, as more species invade a community, place niches are compressed more than food niches (Figure 8.9). Any prey item worth eating should be acceptable no matter what the intensity of competition, but an animal must choose the places it forages on the basis of its *expectation* of yield, which will usually be markedly decreased in some patches of habitat by heightened competition (see also Chapter 6, pp. 209–210). Thus the compression hypothesis predicts that, during short-term, nonevolutionary

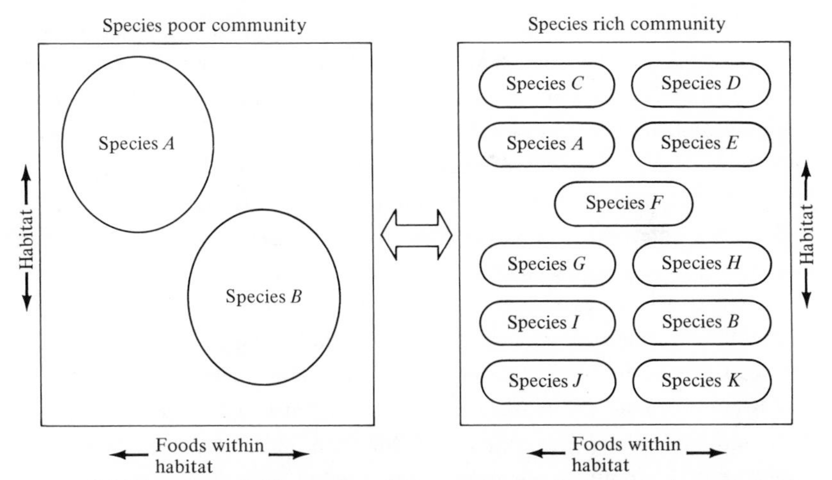

FIGURE 8.9 Diagrammatic illustration of the compression hypothesis. As more species invade a habitat, interspecific competition forces any single species to decrease the range of habitats it exploits; the range of foods eaten should often remain more or less constant, however. Conversely, if a species invades a habitat that is impoverished with species, reduced interspecific competition should often allow an expansion in the variety of habitats (or microhabitats) exploited. [After MacArthur and Wilson (1967). *The Theory of Island Biogeography.* Copyright © 1967 by Princeton University Press. Reprinted by permission of Princeton University Press.]

time, the habitats used should shrink with increased competition, whereas the range of foods eaten should remain relatively unaltered. However, should competitors reduce overall levels of available foods more or less equally among all the patches on a species' itinerary, food niche expansion will be favored (see pp. 108–109, 197, and 207–208).

The Morphological Variation–Niche Breadth Hypothesis

Van Valen (1965) noted that a species can be a generalist in two ways: (1) A population can contain a variety of different phenotypes, each using a smaller range of resources than the overall population [Roughgarden (1972) refers to this as the between-phenotype component of niche breadth], and/or (2) each individual in a population can itself be relatively flexible and generalized, with the resources utilized by any individual being similar to those exploited by the entire population [Roughgarden (1972) terms this the within-phenotype component of niche breadth]. Phenotypic variability within a population, by allowing different phenotypes to exploit different resources, should thus increase the overall range of resources exploited by the population. Moreover, by reducing niche overlap among the members of a population, between-phenotype niche expansion might be expected to reduce the average degree of interphenotypic competition.

Island species, freed from some interspecific competition, tend to exploit a wider range of habitats than mainland species, a phenomenon often referred to as "ecological release." Van Valen (1965) reasoned that reduced competition from other species should also favor an increased morphological variability because this would promote niche expansion. He postulated that island species should often be morphologically more variable than their mainland counterparts; moreover, Van Valen found evidence for just such an increased phenotypic variability in five out of six species of birds known to have broader niches on certain islands. Grant (1967), however, found less morphological variation (in length of wing, tail, tarsus, and bill) in some Mexican insular bird populations than in mainland ones. In a later study, Grant (1971) found no consistent trends in tarsal length variation between mainland and island populations of Mexican birds. He speculates that, under spatially uniform environmental conditions, selection favors little variation among individuals in feeding ecology and associated morphology by acting strongly against individuals that depart widely from the average phenotype, and that, under spatially varied (patchy) conditions, the opposite may be true.

Soulé and Stewart (1970) restate and somewhat reverse the hypothesis, as follows: generalized, broad-niched species should be phenotypically and morphologically more variable than more specialized and more narrow-niched species (they call this the niche-variation hypothesis). Soulé and Stewart, however, were unable to find any evidence that generalized African bird species such as crows are in fact morphologically more variable than more specialized species. Van Valen and Grant (1970) point out that the broad niche of these crows could well be primarily due to within-phenotype flexibility in resource utilization, and that one would not necessarily expect great morphological variability in such a situation.

Also of interest here is the fact that, although tropical species are often considered to be more specialized than temperate species, the bill dimensions of some tropical birds are at least as variable as those of some North Temperate species (Willson, 1969). Clearly much more information is needed before the relationships between morphological variation and niche breadth can be adequately assessed.

The Gene Flow–Variation Hypothesis

Among island populations of the lizard, *Uta stansburiana*, in the Gulf of California, both genetic and morphologic variability are greater on larger islands than on smaller ones (Soulé, 1971, and Figure 8.10). To explain these results, Soulé proposed an alternative to the niche-variation hypothesis,

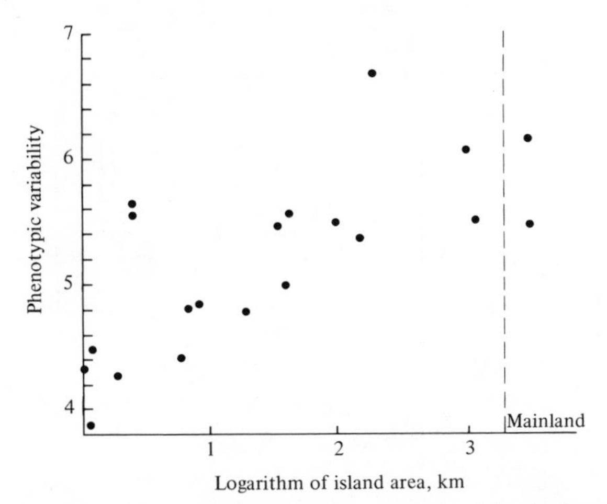

FIGURE 8.10 An index of overall variation in eight phenotypic characters of the lizard *Uta stansburiana* plotted against the logarithm of the area of 18 islands in the Gulf of California and 2 mainland areas. Variation increases with area. [After Soulé (1971).]

termed the gene flow-variation hypothesis, which states that gene flow between habitats with differing selective milieus generates genetic variability in spatially intermediate populations (see also Chapter 4, p. 124). An important distinction between the niche-variation hypothesis and the gene flow-variation hypothesis is that the former predicts that increased genotypic and phenotypic variation is adaptive, whereas the latter suggests that it is nonadaptive and merely an unavoidable byproduct of environmental heterogeneity and movement of organisms. Not enough is known at present to assess the importance of gene flow in maintaining variability of natural populations.

Islands as Ecological Experiments: Some Examples

Darwin's Finches

The Galápagos Islands, an archipelago of relatively small and remote, deep water, volcanic islands located about 600 miles west of the Ecuadoran coast (Figure 8.11), support a remarkable group of birds that nicely illustrate a number of evolutionary and ecological principles. Named Darwin's finches after the first evolutionist to appreciate and study them, these birds dominate the avifauna of the Galápagos. Only 26 species of land birds occurred in the archipelago naturally (i.e., before human introductions), and 13 of these are finches (the islands also support 4 species of mockingbirds, 2 flycatchers, 2 owls, a hawk, a dove, a cuckoo, a warbler, and a swallow).

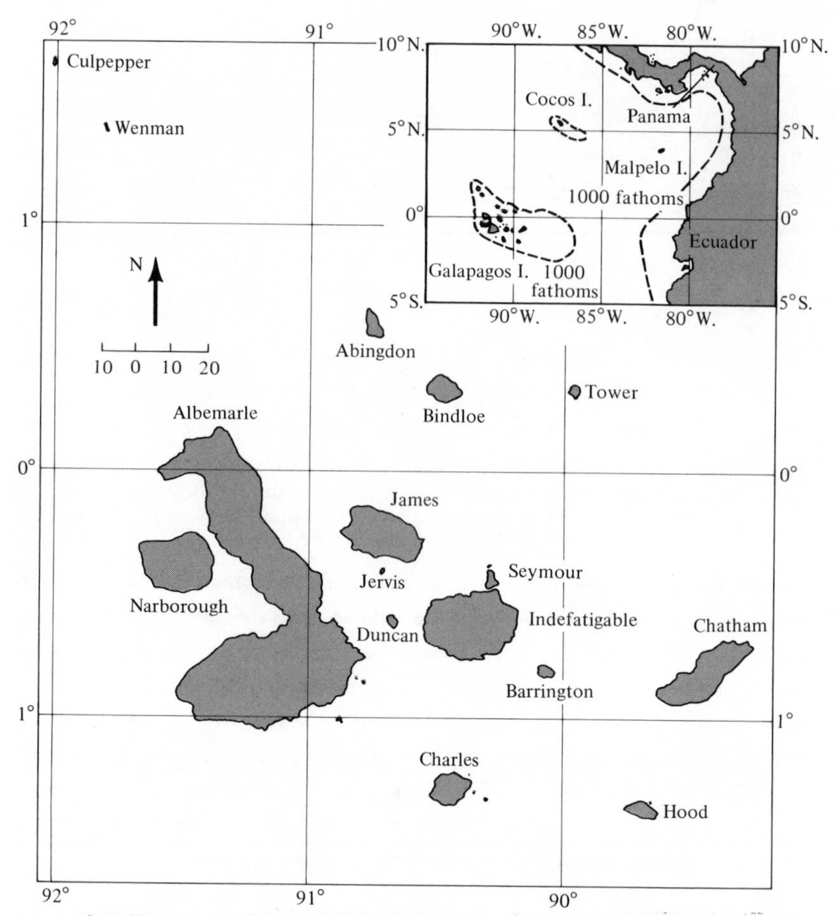

FIGURE 8.11 Two maps of the main islands in the Galápagos Archipelago. [Inset from Lack (1947). Larger map from Bowman (1961), originally published by the University of California Press. Reprinted by permission of The Regents of the University of California.]

These islands (16 major ones and a sprinkling of tiny islets) were formed from volcanic eruptions of the ocean floor about a million years ago; thus, originally, there were no organisms on the islands and their entire biota has been derived from mainland species. Because of their remoteness, relatively few different plant and animal stocks have been able to colonize the Galápagos. (The position of these islands on the equator, however, presumably makes them particularly vulnerable to invasions from rafts and floating islands carried out to sea in the equatorial current.)

These 13 species of finches are thought to have evolved from a single mainland finch ancestor that reached the islands long ago. [The reason for this belief is that these birds are similar enough to each other that they are classified as a distinct subfamily of finches, endemic to the Galápagos and Cocos Islands (below).] Archipelagos are ideal for geographic isolation and speciation, especially in land birds like finches that do not readily fly across wide stretches of water. In such effectively isolated populations, different selective pressures on different islands lead to divergent evolution and adaptations; moreover, occasional interchanges between islands result in competition which promotes niche diversification.

The necessity of geographic isolation and subsequent interisland colonization for the occurrence of speciation and adaptive radiation is nicely demonstrated by the Cocos Island finch, *Pinaroloxias inornata.* Cocos Island is a remote but solitary island several hundred miles north of the Galápagos and about the same distance from the mainland (Figure 8.11, inset). Although there are a fair variety of habitats on it, Cocos supports only one species of finch. There has been no opportunity for geographic isolation or reduced gene flow and the *Pinaroloxias* gene pool has never split. One would predict that this finch should, of necessity, be a generalist, probably with a high degree of phenotypic variability between individuals.

Adaptive radiation of these finches in the Galápagos has produced 3 different genera that differ in where they forage, how they forage and what they eat. The so-called ground finches (*Geospiza*) include 6 ground-foraging species with broad beaks which eat seeds of different sizes and types as well as the flowers of *Opuntia* cactus. The genus *Camarhynchus*, termed "tree finches" because they forage in trees, also contains 6 species with generally somewhat narrower beaks; one species is a vegetarian, while the other 5 eat different sized insects in different ways (one of these, *Camarhynchus pallidus*, the "woodpecker finch," uses sticks and cactus spines to probe cracks and crevices for insects much like a woodpecker uses its long pointed tongue). One very distinctive and monotypic genus, the so-called warbler finch, *Certhidea olivacea*, occurs on almost all the islets and islands in the archipelago and breeds throughout most habitats.

From 3 to 10 species of finches occur on any given island (Table 8.2) in various combinations. Beak lengths and depths are highly variable from island to island (Figure 8.12), presumably reflecting different environmental conditions among the islands, including interspecific competitive pressures. Indeed, Figure 8.12 nicely illustrates character displacement in beak depths; the tiny islets of Crossman and Daphne support only one member of a pair of very similar species, either *Geospiza fuliginosa* or *G. fortis*, respectively. On these two small islands, the beaks of both species are nearly the same size (about 8.5 to 11 mm deep), whereas on larger islands where the two

TABLE 8.2 Distributions of Darwin's Finches on Various Islands in the Galápagos Archipelago

Species	Abingdon	Albemarle	Barrington	Bindloe	Charles	Chatham	Culpepper	Duncan	Hood	Indefatigable	James	Jervis	Narborough	Seymour	Tower	Wenman
Geospiza magnirostris	×	×	×	×	×	—	×	×	—	×	×	×	×	×	×	×
Geospiza fortis	×	×	×	×	×	×	—	×	—	×	×	×	×	×	—	—
Geospiza fuliginosa	×	×	×	×	×	×	—	×	×	×	×	×	×	×	—	×
Geospiza difficilis	×	—	—	—	—	—	×	—	—	×	×	—	×	—	×	×
Geospiza scandens	×	×	×	×	×	×	—	×	—	×	×	×	—	×	—	—
Geospiza conirostris	—	—	—	—	—	—	×	—	×	—	—	—	—	—	×	—
Camarhynchus crassirostris	×	×	—	×	×	×	—	×	—	×	×	×	×	—	—	—
Camarhynchus psittacula	×	×	×	×	×	—	—	×	—	×	×	×	×	×	—	—
Camarhynchus pauper	—	—	—	—	×	—	—	—	—	—	—	—	—	—	—	—
Camarhynchus parvulus	×	×	×	—	×	×	—	×	—	×	×	×	×	×	—	×
Camarhynchus pallidus	—	×	—	—	—	×	—	×	—	×	×	×	—	×	—	—
Camarhynchus heliobates	—	×	—	—	—	—	—	—	—	—	—	—	×	—	—	—
Certhidea olivacea	×	×	×	×	×	×	×	×	×	×	×	×	×	×	×	—
Total number of species per island	9	10	7	7	9	7	4	9	3	10	10	9	9	8	4	5

Source: After Bowman (1961). Originally published by the University of California Press. Reprinted by permission of The Regents of the University of California.

species occur together in sympatry (Abingdon, Bindloe, James, Jervis, Albemarle, Indefatigable, Charles, and Chatham—upper part of Figure 8.12) beak depths are completely nonoverlapping with *fuliginosa* having a small beak (about 7 to 9 mm deep) and *fortis* a larger beak (about 10 to 14 or 16 mm deep). Beak dimensions, of course, determine in large part the size of the food items the birds eat.

Larger islands in the Galápagos Archipelago contain a greater variety of habitat types and, as a result, support more species of finches than do smaller islands. Moreover, the total number of finch species decreases with "average isolation," or the mean distance from other islands, while the number of endemic species increases with isolation (Hamilton and Rubinoff, 1963, 1967).

Krakatau

In 1883, the small volcanic island of Krakatau, located between Java and Sumatra (see Figure 8.2), erupted repeatedly over a three-month period. All of Krakatau and two adjacent islands were covered with red-hot lava, pumice, and ash to a depth of 100 feet or more. The islands were so hot that months afterward falling rain turned to steam upon contact. It is most unlikely that any organisms survived. Repopulation from adjacent Sumatra

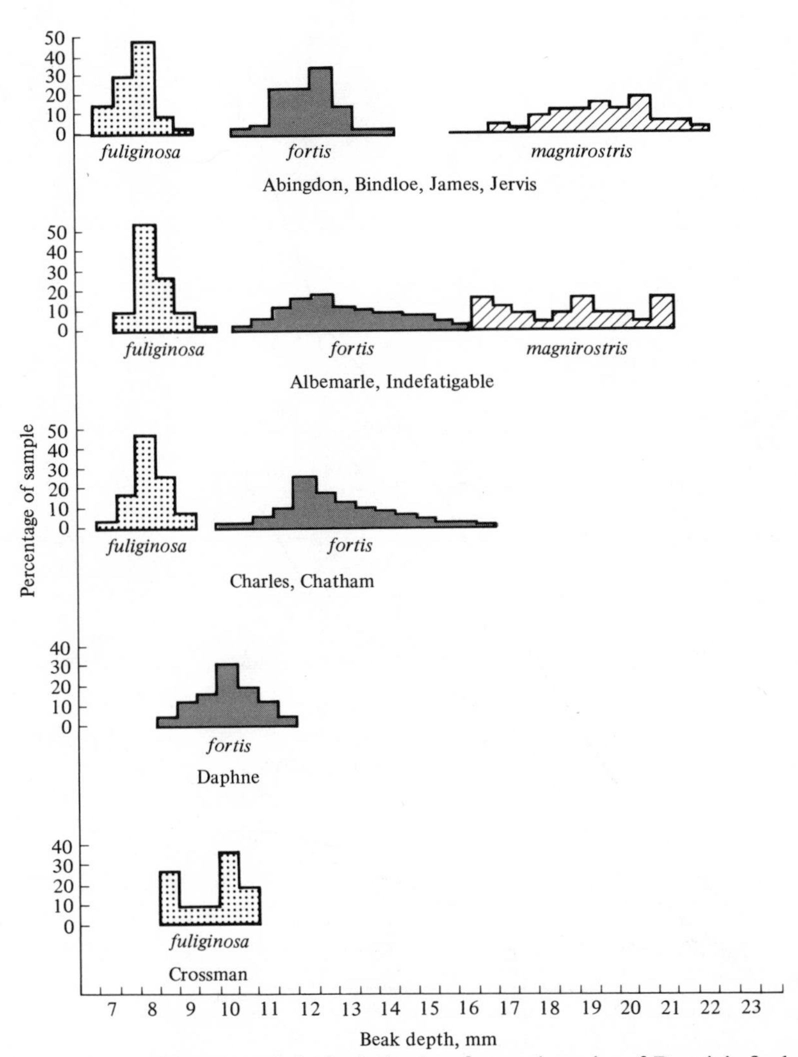

FIGURE 8.12 Histograms of the beak depths of several species of Darwin's finches, genus *Geospiza*, on different islands. In allopatry on the islets of Daphne and Crossman (bottom of figure), *G. fortis* and *G. fuliginosa* have beaks of very similar size, whereas in sympatry (upper 3 sets of histograms), their beak depths are entirely nonoverlapping. [From Lack (1947).]

(about 15 miles away) and Java proceeded rapidly and by 1921 the number of resident species of birds was comparable to that expected on a small island of eight square miles (the size of Krakatau after the eruptions) in the general region (see Figure 8.3). Moreover, the total number of bird species did not

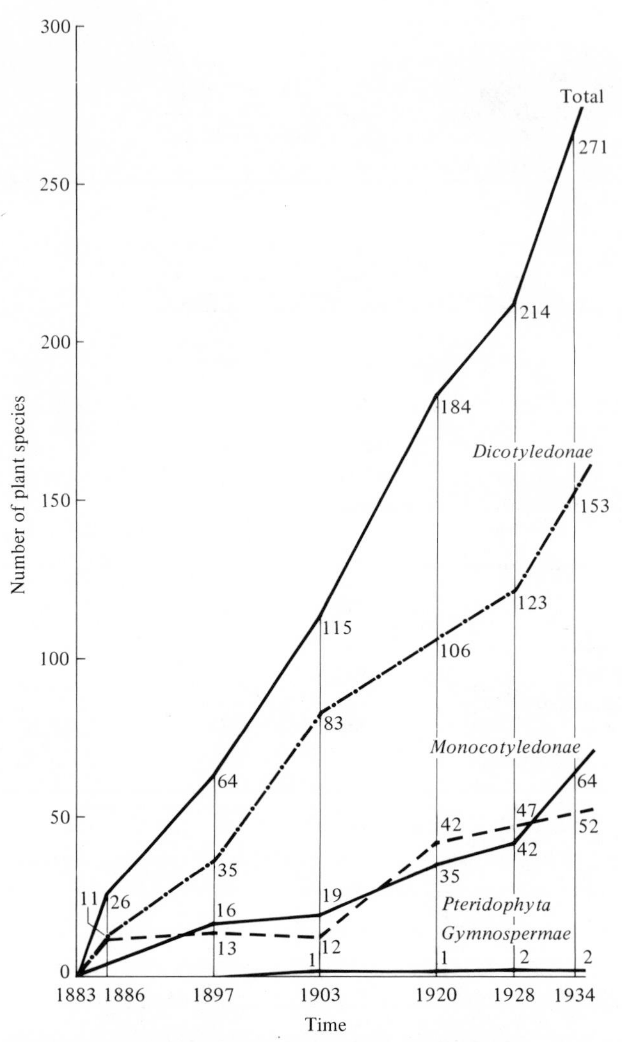

FIGURE 8.13 The numbers of plant species recorded on the three islands of the Krakatau group from 1883 through 1934. [From MacArthur and Wilson (1967). *The Theory of Island Biogeography*. Copyright © 1967 by Princeton University Press. Reprinted by permission of Princeton University Press.]

change much between 1921 and 1933, although the composition of the avifauna did (Table 8.3). This example suggests that mobile organisms like birds rapidly reach an equilibrium species density. Plant species, on the other hand, were still being added rapidly at last count in 1933 (Figure 8.13).

TABLE 8.3 Number of Species of Land and Freshwater Birds on Krakatau and Verlaten During Three Collection Periods, and the Number of Species "Lost" Between Intervals

	1908			1919–1921		
	Non-migrant	Migrant	Total	Non-migrant	Migrant	Total
Krakatau	13	0	13	27	4	31
Verlaten	1	0	1	27	2	29

	1932–1934			Number "lost"	
	Non-migrant	Migrant	Total	1908 to 1919–1921	1919–1921 to 1932–1934
Krakatau	27	3	30	2	5
Verlaten	29	5	34	0	2

Source: From MacArthur and Wilson (1967) after Dammerman.

Species Turnover Rates

At equilibrium, the total rate of immigration of species must equal the total extinction rate. However, because the species going extinct will undoubtedly often differ from those that successfully invade an island, the *composition* of an island's biota will be continually changing, even at equilibrium. Diamond (1969) studied bird species turnover on nine channel islands off the California coast, by comparing the composition of avian communities in 1968 with censuses published in 1917 (Table 8.4). Total bird species densities changed very little during the 51 years on seven islands (uncooperative owners prevented complete surveys on the other two islands, Santa Rosa and San Miguel, in 1917), showing that these islands have indeed reached an equilibrium; however, the species composition on the islands changed radically between 1917 and 1968. Estimates of turnover rate during this period of 51 years range from 17 to 62 percent of the average species pool of a given island (these estimates are conservative since some species undoubtedly went extinct and subsequently recolonized during the 51 years, while still others may have invaded and since gone extinct). Diamond found that turnover rate varied inversely with equilibrium species density (Figure 8.14).

The Taxon Cycle

Many island species are thought to progress through a series of evolutionary changes, termed a taxon cycle, that may eventually greatly increase their probability of going extinct (Wilson, 1961; MacArthur and

TABLE 8.4 Avifaunal Turnover on the Channel Islands off Southern California

Island	Area	Distance	1917 species	1968 species	Extinctions	Additions	Introductions	Immigrations	Turnover
	A	*B*	*C*	*D*	*E*	*F*	*G*	*H*	*I*
Los Coronados	1.0	8	11	11	4	4	0	4	36
San Nicolas	22	61	11	11	6	6	2	4	50
San Clemente	56	49	28	24	9	5	1	4	25
Santa Catalina	75	20	30	34	6	10	1	9	24
Santa Barbara	1.0	38	10	6	7	3	0	3	62
San Miguel	14	26	11	15	4	8	0	8	46
Santa Rosa	84	27	14	25	1	12	1	11	32
Santa Cruz	96	19	36	37	6	7	1	6	17
Anacapa	1.1	13	15	14	5	4	0	4	31

Source: From Diamond (1969).

Note: For each island, column *A* gives the area in square miles; *B*, the distance in miles from the nearest point on the mainland; *C*, the number of species of land and freshwater birds breeding in 1917; *D*, the number of breeding species in 1968; *E*, the number of species that were breeding in 1917 but not in 1968 and hence must have gone extinct in the interim; *F*, the number of species breeding in 1968 but not in 1917 ("additions"); *G*, the number of species present in 1968 that had been successfully introduced by man between 1917 and 1968 (all of these are game birds: California quail, Gambel's quail, pheasant, or chukar); *H*, the number of species present in 1968 but not in 1917 that had immigrated under their own power between 1917 and 1968, calculated as *F* minus *G*; and *I*, the turnover rate expressed in percent of the species pool per 51 years, calculated as $100\ (E+H)/(C+D-G)$.

Wilson, 1967; and Ricklefs and Cox, 1972). Under this hypothesis, early in the taxon cycle a species is widespread and occurs on many islands, is often in the process of invading new islands, and is only slightly, if at all, differentiated into distinct populations on the various islands. It is adapted to marginal, relatively unstable, habitats, such as riverbanks and forest clearings. Later in the cycle, populations of a species become progressively more and more differentiated on different islands, at first remaining widespread. At this stage, it penetrates the more stable habitats, such as old forests, where it must coexist with larger numbers of native species. Still later, after local extinction on some islands, a differentiated species becomes more restricted and its geographic range is fragmented. Finally, species at the end of the taxon cycle are found only on a single island (that is, they are *endemic* to one island). Occasionally, some species are able to shift back into marginal, species-poor habitats, thus restarting the cycle. Ricklefs and Cox (1972) argue that the taxon cycle for a particular species is driven by counteradaptations of the other members of an island biota against the species concerned. There

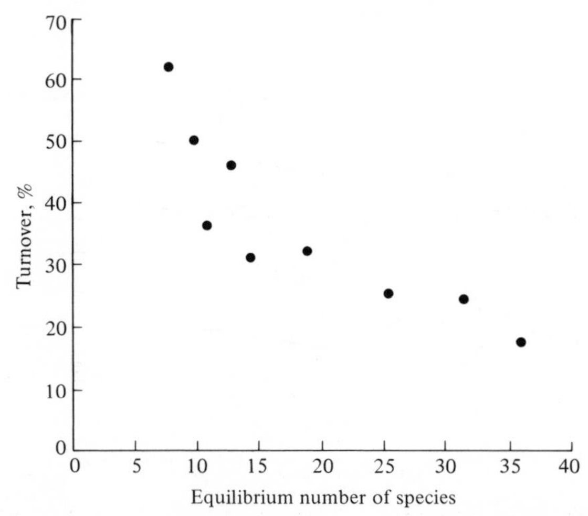

FIGURE 8.14 Estimated turnover rates of bird species plotted against the estimated equilibrium number of species of birds on the nine channel islands off southern California. [After Diamond (1969).]

is a shift from *r* selection to *K* selection as the taxon cycle progresses. Dispersal ability decreases as species depend less and less on the ability to colonize marginal habitats and more on their capacity to coexist with competing species in stable habitats. Such reduced dispersal and more pronounced local adaptation, in turn, favors speciation and endemism. Newly arrived colonists are relatively free of a counteradaptive load, allowing them to spread successfully throughout an island system in relatively unstable habitats. On small and remote islands, old populations of endemics (p. 154), such as the Cocos Island finch, may persist. Although the taxon cycle may not apply to all species, the concept underlying it presumably could be operative in mainland faunas as well as on islands. Little attempt has yet been made to interpret the ecology of mainland populations in terms of such counteradaptations (see, however, Chapter 4, p. 125; Chapter 5, pp. 169–175; and Chapter 7, pp. 249–250).

A Defaunation Experiment

An interesting ecological experiment was performed by Simberloff and Wilson (1970 and included references). After carefully censusing the entire arthropod faunas of several very tiny mangrove islets in the Florida keys, they had all arthropods exterminated by fumigation with methyl bromide, and then monitored the process of recolonization over a two-year period. Arthropods rapidly recolonized and within 200 days the numbers of species on the islands had stabilized (Figure 8.15). Although turnover rates remained

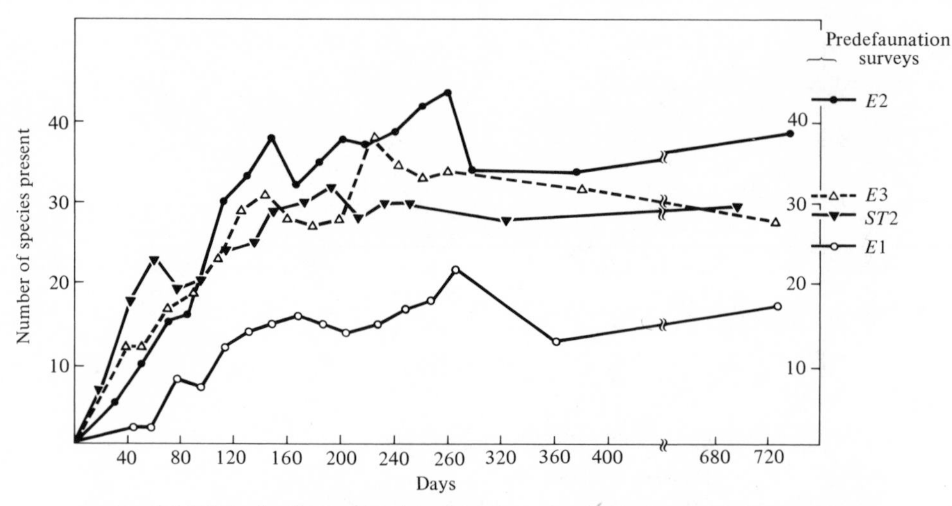

FIGURE 8.15 Colonization curves (number of arthropod species through time) of four tiny mangrove islets in the Florida keys following complete extermination of the arthropod fauna, but relatively little damage to the vegetation. The numbers of species on each islet first rose rapidly, but soon leveled off to species densities fairly close to those before defaunation (plotted on the right vertical axis). [After Simberloff and Wilson (1969). By permission of Duke University Press.]

quite high throughout the experimental period, the numbers of species on the islands remained relatively constant over a period of nearly two years, providing strong evidence that the islands have indeed reached an equilibrium. Two islands, *E*1 and *E*2, seem to have reached equilibrium at a slightly lower species density after defaunation. The apparent depression in the number of species at equilibrium may indicate that the species that reinvaded these islands interfere with one another more than the members of the original communities did; alternatively, this reduction in species density may be due to the fact that the new immigrants are not as well adapted to exploit the island's resources as were the original inhabitants. In any case, these results demonstrate that the actual composition of an island's fauna may itself partially determine the equilibrium number of species an island will support.

Wilson (1969) suggests that an island community may experience a sequence of several distinct sorts of equilibria through time (Figure 8.16). First, a "non-interactive" equilibrium in the number of species present may be reached even before the component populations come to equilibrium demographically and with one another. Then, because competitive and predatory intractions are intensified as populations saturate the island with individuals, a second "interactive" equilibrium is reached. Both these stages should occur relatively rapidly. Wilson envisions two other types of equi-

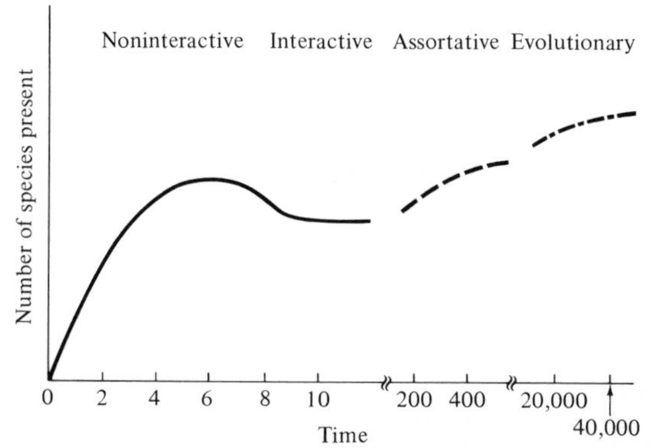

FIGURE 8.16 Hypothetical sequence of community equilibria on an island through time. The time scale is arbitrary and intended only to emphasize the vastly greater time spans required for the assortative and evolutionary equilibria. See text for explanation. [From Wilson (1969).]

libria, which would require considerably greater time spans to attain. As various species go extinct and others invade, the composition of an island's biota may gradually change until a certain set of species from the available species pool is reached that is composed of species with particularly low extinction rates; Wilson terms this the "assortative" equilibrium. Finally, given a much longer time span, the component species could actually evolve minimal extinction rates and an "evolutionary" equilibrium might be reached (Wilson, 1969).

Selected References

Introduction

Andrewartha and Birch (1954); Krebs (1972); MacArthur (1972); MacArthur and Wilson (1967); Udvardy (1969); Watts (1971).

Classical Biogeography

Dansereau (1957); Darlington (1957, 1959, 1965); Dietz and Holden (1970); Hesse, Allee, and Schmidt (1951); Kurtén (1969); MacArthur (1959); Newbigin (1936); Terborgh (1971); Udvardy (1969); Wallace (1876); J. T. Wilson (1971, 1973).

Island Biogeography

Carlquist (1965); MacArthur and Wilson (1963, 1967); Wilson (1969); Wilson and Bossert (1971).

Species-Area Relationships
Gleason (1922, 1929); Krebs (1972); MacArthur and Wilson (1967); Odum (1959, 1971); Preston (1948, 1960, 1962a, 1962b).

Equilibrium Theory
Brown (1971); MacArthur and Wilson (1963, 1967); Wilson (1969); Wilson and Bossert (1971).

The Compression Hypothesis
Crowell (1962); MacArthur (1972); MacArthur and Pianka (1966); MacArthur, Diamond, and Karr (1972).

The Morphological Variation–Niche Breadth Hypothesis
Grant (1967, 1971); Roughgarden (1972); Soulé and Stewart (1970); Van Valen (1965); Van Valen and Grant (1970); Willson (1969).

The Gene Flow–Variation Hypothesis
Ehrlich and Raven (1969); Ford (1964); Gilbert and Singer (1973); Soulé (1971); Wilson and Bossert (1971).

Islands as Ecological Experiments: Some Examples

Carlquist (1965); MacArthur (1972); MacArthur and Wilson (1967).

Darwin's Finches
Bowman (1961); Hamilton and Rubinoff (1963, 1967); Lack (1947).

Krakatau
Dammerman (1948); Docters van Leeuwen (1936); MacArthur and Wilson (1967).

Species Turnover Rates
Diamond (1969); Wilson (1969); Wilson and Bossert (1971).

The Taxon Cycle
MacArthur and Wilson (1967); Ricklefs and Cox (1972); Wilson (1961).

A Defaunation Experiment
Simberloff and Wilson (1970); Wilson (1969).

9 Man and His Environment

Some Ecological Principles Restated

Time and time again, we have noted that ecological systems, as well as their components, are normally near steady states and in dynamic equilibrium. At one extreme, the amount of heat energy entering earth's atmosphere must equal exactly that radiated back into space, or else the planet would either warm up or cool down. Other physical equilibria are widespread in, for example, the hydrologic cycle and the various other biogeochemical cycles. Within any particular community of organisms, production and respiration ultimately must balance (Figure 9.1); in successional stages, production at first exceeds respiration but eventually the two become equal when soil formation is finished and the climax, steady-state forest is reached. Even the extent and distribution of various nonclimax communities (which have not reached a steady state) presumably are in some sort of equilibrium, which is determined by the frequency of disturbances and destruction of other successional stages as well as the rate of successional change. Similarly, in most natural ecological communities, the rate of energy flow into each trophic level is exactly balanced by the rate of energy flow out of that level. On islands, rates of extinction of old species balance rates of immigration of new ones, with the total number of species on an island remaining relatively constant (even though the composition may change). Likewise, populations that go extinct locally are presumably replaced periodically by colonists from other populations, so that certain species may be viewed as a set of populations

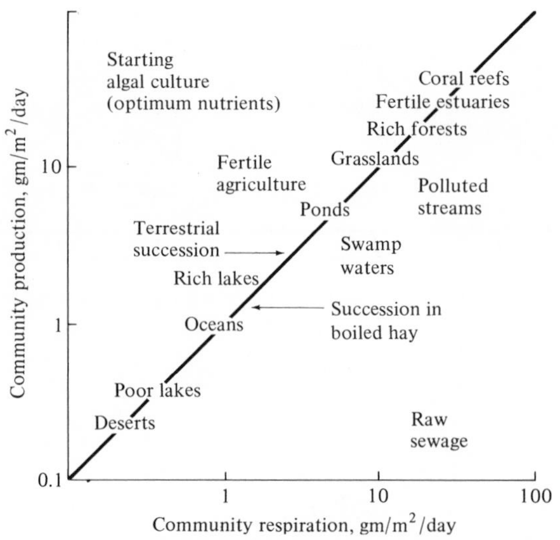

FIGURE 9.1 Total primary production of various communities plotted against the community's total respiration, in grams per square meter per day (proportional to cal/m²/day). Communities along the diagonal line are in equilibrium, with production equal to respiration. Production exceeds respiration (autotrophy) in those above the line, while respiration exceeds production (heterotrophy) in those below the line. [From Odum (1959) after H. T. Odum.]

with local extinctions more or less balanced by inoculations; otherwise such a species would either go extinct or its geographic range would expand indefinitely. Populations of organisms are also usually in some sort of balance, with births equal to deaths; over the long term the average actual rate of increase ultimately must average exactly zero, with the net reproductive rate equal to one, or else a population either overshoots its carrying capacity or declines to extinction. Even pairs of competing, commensal, prey–predator, and host–parasite populations must be in some sort of ecological and evolutionary balance in order to coexist with one another over any period of time. Similarly, any individual organism has a carefully regulated time and energy budget that must balance, with the total amount of energy gathered being equal to the summed amounts spent on various, often conflicting, organismic activities such as growth, maintenance, and reproduction.

Human Population Growth

The above points are obvious and incontestable, yet modern man has largely failed to appreciate their relevance to his own existence. The phenomenal growth of the human population in the last five centuries (Figure 9.2) is an incredible fact, and one that is rapidly becoming quite perilous. Since the middle ages there has been no decline in population, births have exceeded deaths, the intrinsic rate of increase has always been positive, and the human population has increased exponentially. At the present growth rate, the world population will double in about 35 years, and in some countries current rates will lead to a doubling in only 15 years. This means that only 35 years from now, during the lifetimes of many of us, we could produce enough new people to populate another earth-sized planet to the density of our own globe. However, these new humans would be right here on earth with us. At the time of this writing, there are approximately 4 billion people on earth; even if every couple now in existence were to limit themselves to no more than two children, *effective immediately*, the age distribution of humans contains so many young people that the world population would not stop growing for another 50 to 60 years. What's more, it would stabilize in size at about 9 or 10 billion people, more than twice the current population. This tremendous surge of humanity is a result of the agricultural, industrial, and medical revolutions.

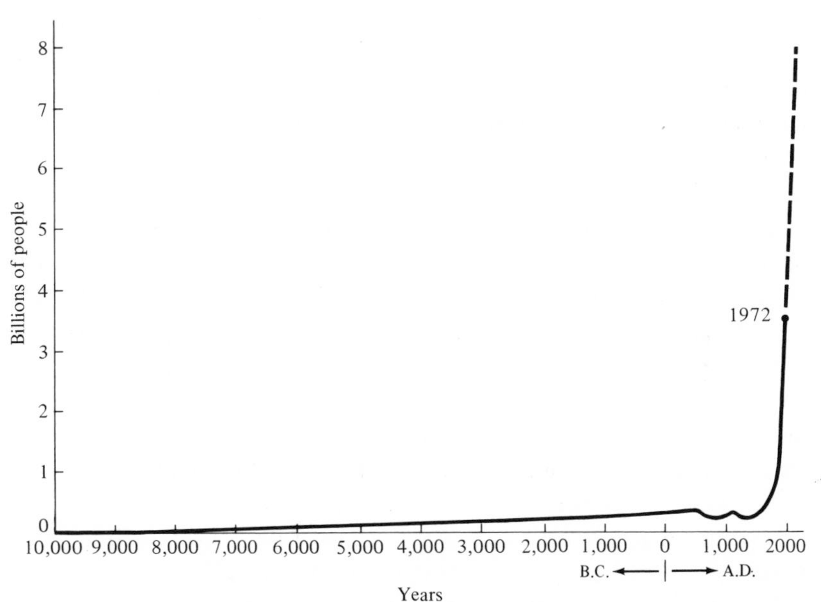

FIGURE 9.2 Estimated size of the human population during the interval from the Pleistocene ice age to present. The dashed line is projected assuming that every couple now in existence limits their reproduction to two offspring, *effective immediately*.

Many students of these matters say that it is already too late to prevent mass starvation and pestilence in many areas, such as China and India. Titles of recent books are indicative: *Standing Room Only*, *The Population Bomb*, and *Famine 1975*. Even now, it is estimated that, on the average, a human starves to death every 3 seconds somewhere in the world. Clearly we are close to or even above carrying capacity in some regions.

Americans still enjoy a fairly high standard of living by most people's criteria. However, gross inequalities both within and between nations can easily and often do generate domestic and international unrest. Five percent of the world's population now consume about 30 percent of earth's non-renewable resources. The underdeveloped countries and the underprivileged are beginning to demand their "share" of these resources and a decent standard of living. A major reason nations have gone to war is that dense populations need more space and other resources. Considering the vast ramifications of war, population pressures are an important root of many of man's problems.

Natural Selection and Man

Homo sapiens is a truly spectacular animal, with an amazing inventiveness and ability to learn and create. A human brain is a beautiful miniature computer with incredible potential. Indeed, technology has been so successful that many think man can accomplish nearly anything he desires. A great deal is already known about the organized reality around us and many of the laws of nature are understood; given time, we could potentially understand much more, including minute details of the intricate workings of our own brains and bodies. We already know much about the very stuff we are made of—matter itself. Man has come a long way both intellectually and culturally in an extremely short period of time, but he is also most impatient. He has subdued natural phenomena and killed or exterminated many organisms in the name of "progress," before either fully appreciating or understanding them. Rapidity of change is concomitant with man's population growth: we are changing our own environment and that of virtually all other organisms at an alarming rate, and one that is steadily accelerating.

The transition from small tribal groups of stone-age cave men in the Pleistocene, living a relatively simple hunting life, to the very complex industrialized world of today has taken place over a period of 10,000 years, or about 500 human generations. All of recorded history includes only 300 generations, and only a hundred generations have lived since the time of

Christ. A mere 25 generations have passed since 1500, the time of the onset of the present surge of human population growth. It appears that we have changed our environment faster than we have been able to adapt to it; we may, in many ways, be partial misfits in our own man-made environment. (Man has, of course, partially adapted to his ever-changing environment by making certain cultural adjustments.)

Man evolved slowly and gradually, with natural selection operating first to adapt him to his relatively simple hunting, and then agrarian, way of life. Certain human emotions and behavior that no doubt had real survival value in these past environments are now dangerously out of place in the context of our present, very different, environment. For instance, revenge must certainly have been adaptive at the level of small bands of men if they were in competition for limited supplies of, say, food, water, or shelter. (Through most of early human history, men cooperated among themselves in small bands which interacted and fought with one another.) Presumably the revenging cave man benefited because others thought twice before infringing on him again. Such actions thus protect one's own interests and they doubtless had clear-cut selective value in the cave, where they were probably programmed into our instinctive behavior over evolutionary time. Our tendency to seek revenge, however, makes little selective sense in high-speed automobiles on our highways or at the level of nuclear warfare. (What *will* we gain from a "second strike" capability?) Yet no one will dispute that such deep-seated revengeful human emotions and behavior are most definitely here.

Exactly analogous considerations hold in reproduction. It is tautological that natural selection has favored the human phenotypes with highest reproductive success. We have thus been programmed to enjoy sex and to want to have children. (Some would argue that such desires are environmentally controlled and culturally manipulatable, but they surely have a genetic component.) Men who leave more successful (that is, breeding) offspring pass proportionately more of their genes into the population gene pool of later generations than do those who leave either fewer or less fit progeny. Thus we are not surprised to find that most people like kids, especially their own, and that they often want to have many children.

Before the agricultural and medical revolutions, human death rates, especially among infants, were high enough to balance birth rates. But now, by killing our own predators and curing our diseases, we have reduced our death rate strikingly, while our birth rate has remained high as would be anticipated from principles of natural selection. As our environment deteriorates around us due to our burgeoning population, we thus find ourselves victims of natural selection. Man's great intelligence, which has given him both the power to avoid early death and an uncanny ability to inhabit new areas and to exploit new foods and other resources, will be his undoing. We use our brain to try to live outside of many ecological principles; such practices are of necessity temporary. Thus we defeat the pyramid of energy by killing top predators with traps, guns, and dogs, all ecologically unfair tactics in that no other animal has recourse to them, fortunately for us. We have few remaining natural predators, except other men and some parasitic and pathogenic organisms. We live almost anywhere on earth's surface and we eat a vast variety of foods. Man is a very versatile and opportunistic generalist. He exploits resource after resource, and then quickly changes to another as supplies of one are depleted or entirely used up. A prime example is the whaling industry: in the 1930's the largest whales, the blues, were hunted until stocks gave out, then whalers shifted to killing fin whales. After the fin whale population disappeared in the 1960's, the number of sei whales and sperm whales taken have steadily risen. It seems that these species will soon be overexploited as well.

No population's growth rate can continue to remain positive indefinitely in a finite world. Clearly, something is going to have to change. Provided

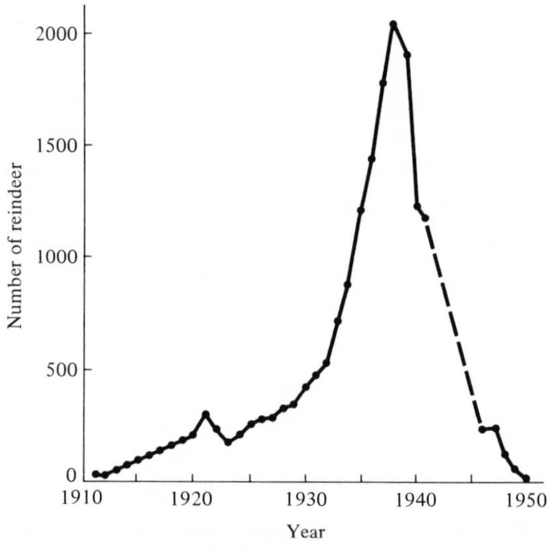

FIGURE 9.3 In 1911, 25 reindeer were introduced on Saint Paul Island in the Pribolofs off Alaska. The population grew rapidly and nearly exponentially until about 1938 when there were over 2000 animals in the 41 square mile island. The reindeer badly overgrazed their food supply (primarily lichens) and the population "crashed." Only 8 animals could be found in 1950. A very similar sequence of events occurred on Saint Matthew Island from 1944 through 1966. [After Krebs (1972) after Scheffer.]

that we do not first annihilate ourselves with nuclear weapons, our population may well eventually crash somewhat as did the reindeer on Saint Paul Island in Alaska (Figure 9.3). Of course, man *could* theoretically stabilize his population at sustainable densities, but it is unlikely that he will. In the event that we do not control our own population, one can only speculate as to what will finally limit the number of men on this globe. Possible limiting factors are numerous, including light, space, heat dissipation, water, pathogens, food, and/or nonrenewable resources. Human populations will probably eventually be limited by several of these factors acting simultaneously.

Selected Ramifications: The Rape of Planet Earth

In the next few pages, I briefly present a few selected examples of man's exploitation of nature, much of which is obvious and/or well known. Informed readers should simply skip to the next section. The most important single point is that the extent of man's consumption, modification, and destruction of resources is directly proportional to human population size. In an overdeveloped country, each citizen places a much greater strain on earth's resources than does each citizen of an underdeveloped country.

For years, the attitude of men has been that earth's resources, including other species, were put here for human use, to be exploited by men, as they most certainly have been. Often we do not even begin to know the full ramifications of what we are doing and yet we proceed at an alarming rate.

As one example, among many, consider the fossil fuels: coal, gas, and oil. It took some 50,000,000 years of primary production to form earth's invaluable coal and oil deposits, yet we will deplete them entirely in less than 1/100,000th of this time (only 500 years); each year we burn up fossil fuel which took *hundreds of thousands* of years to form. As we come closer and closer to completely exhausting these fuels, our rate of consumption skyrockets. Beyond the staggering fact of depletion of this huge and yet irreplaceable natural resource, ramifications of expending so much energy so rapidly have been profound. The smog problem is familiar to everyone. Releasing such vast quantities of carbon dioxide so rapidly has raised the CO_2 content of the upper atmosphere and increased the greenhouse effect (Chapter 2); perhaps this is why, from the late nineteenth century up until about 1940, earth's temperature began to rise, as much as 2°F in places. Fortunately in this case, by sheer accident, the warming trend was reversed in the 1940s, perhaps because enough particulate matter (carbon, soot, etc.) was put into the atmosphere to decrease the amount of incident solar radiation penetrating the atmosphere, which *so far* has more or less balanced the increased greenhouse effect. No one knows how much longer earth's precarious thermal balance will persist, but it is highly likely that our planet will soon be getting noticeably warmer. Man is tampering with the globe's temperature balance without even knowing or considering the long term consequences. No one knows exactly what effects such overall warming or cooling will have on local climates, but agriculture will almost certainly suffer. A warming of earth could also melt polar ice caps, raise sea levels, and innundate coastal cities.

Regular summer "brown outs" are becoming a familiar occurrence and the impending power shortage has received wide attention. Yet most people are really not very alarmed; they know that before long, certainly by the time we have depleted the supply of fossil fuels, we will be depending on nuclear energy which has virtually no limits. Man is so smart, he can convert matter into energy! (As an aside, I might note that we cannot *eat* atomic energy, although one could, I suppose, imagine using it to power tiers of lights in giant skyscrapers burning 24 hours a day to grow stacks of food crops. Heat dissipation would set a limit on even such a grandiose plan.) However, production of nuclear energy in many ways poses greater problems than burning of coal and oil, since disposing of radioactive wastes with a long half-life is exceedingly difficult. One way to dispose of undesirable pollutants is to inject them deep in the ground; lubrication of faults in Colorado by such means produced earthquakes in an area where they were unknown before. Radioactive wastes are usually encased in concrete and either buried or dropped into the depths of the oceans. Recent proposals to dispose of them in deep salt mines have been severely criticized. No matter what is done with it, much of this radioactive material gets back into circulation. Rainwater leaches out buried wastes and puts it back into our water supplies. Fish hundreds of miles out in the ocean are sometimes highly radioactive. Fallout of strontium and other isotopes from the atmosphere assures that all vegetables grown above ground and all milk, including human mother's milk, are radioactive. Hence, radioactive isotopes from nuclear testing and nuclear reactors already contaminate much of the atmosphere, most of our foods, and our own bodies. We do not yet know all the implications of or the effects of these substances on our own health, although we do know that radioactive materials can cause cancers. A recent proposal to shoot these wastes out into space might actually solve the radioactivity problem for the future, except for that already present in our environment and ourselves. But the ever present thermal problem can not be solved so easily.

Nuclear reactors heat up and must somehow be cooled, usually by a convenient nearby river. Initially these rivers were allowed to warm up, leading to so-called thermal pollution, that often either changed their fish faunas in undesirable ways or exterminated them entirely. Now, in this age of "ecological awareness," large cooling towers are built which cool the river water back down to a tolerable temperature. The reactor's excess heat, as well as some of the river's water, is given off to the atmosphere. Although much more heat can probably be put into earth's atmosphere, there is a *definite* upper limit on the rate at which heat can be dissipated into the atmosphere and from it into outer space. Thus, cooling towers have not solved the problem of thermal pollution, but have merely postponed it until some later time. Ultimately, the second law of thermodynamics sets the upper limit on how much "waste" heat can be dissipated and therefore sets an upper limit on human use of energy of all sorts.

One of the more tragic, and potentially irreparable, of recent human practices is the massive use of certain pesticides, particularly the chlorinated hydrocarbons such as DDT, DDE, and Dieldrin. Since the 1940s these chemicals have been produced in large amounts and used extensively in agriculture. They provide an inexpensive and very effective means of insect control and were therefore widely acclaimed as a major breakthrough in food production and pest control. Pesticides are sometimes necessary to control mass outbreaks of pests in the traditional agricultural crop of a single plant species, since these simple communities may usually be less stable than more diverse ones with more checks and balances between and among component species (Chapter 7). In a pure stand of its food species, a pest outbreak can rapidly become an epidemic with the pest population growing exponentially. However, some severe problems arise from the use of pesticides. They tend to kill off all insects indiscriminately, including beneficial predatory species and innocuous species like bees and butterflies. Indeed, predatory and parasitic insects are often more susceptible to pesticides than herbivorous pests, which have coevolved with chemical poisons because of the chemical defenses of their plant foods. A pest population that recovers from the effects of an insecticide is frequently freed from its competitors and predators, which allows its population to expand rapidly. The short generation times of most insects allows them to evolve rapidly. Under the very strong selection imposed upon them by lethal pesticides, pest populations have evolved highly resistant strains. Some such resistant strains may have actually evolved enzymes that break down pesticide molecules.

By far the biggest problem with chlorinated hydrocarbons, however, is that these very stable molecules do not disintegrate readily either on their own or by the action of physical factors; moreover, almost all organisms, including bacteria, have great difficulty in breaking them down. As a result, such pesticides persist and accumulate. These molecules now occur over the entire globe, including the Arctic and the Antarctic (there is even a "DDT belt" in the atmosphere). A terrifying fact is that chlorinated hydrocarbons are *known to interfere with photosynthesis in marine algae* (Wurster, 1968). Paul Ehrlich's essay "Ecocatastrophe" begins with the disruption of oceanic primary production due to chlorinated hydrocarbon interference with phytoplankton photosynthesis. The really frightening thing is that no one can say that Ehrlich's nightmare might not come true! Obviously, it should be illegal to make, let alone use, material with such devastating potential. Even though the United States has greatly reduced its own use of DDT and related chemicals, we still produce them in large amounts and ship them off to other countries. The DDT problem is an international one, for what one country does markedly affects the well-being of distant nations. In spite of all this, the World Health Organization recently recommended continued massive use of these poisons in underdeveloped and overpopulated countries —otherwise, more men would starve and die of malaria.

The solubility properties of the DDT family of molecules make them both highly specific to plants and animals and extremely easily transported. These chemicals are soluble in both water and lipids (fats), but, because they are differentially attracted to the latter, they are concentrated in the tissues of living organisms. Animals at higher trophic levels accumulate more pesticides than those at lower trophic levels because each time a prey item is consumed, most of the pesticide content of the prey is retained in the fatty tissues of the predator. Due to such amplification, concentrations as high as 120 parts per million have been recorded in the fats of some tertiary and quaternary carnivores. Humans are extremely tolerant to DDT and contain high levels of DDT residues; mother's milk, with its high fat content, is especially rich in these poisons. The effects of such retained pesticides are several: animals may simply die from an overdose, as frequently happens after a period of starvation, when fats are mobilized and the poisons released. Subtler effects of pesticide contamination of animals, including man, are little known; they could well be carcinogenic.

Birds are especially vulnerable to the DDT group of pesticides because these substances both mimic estrogen and depress the activity of the enzyme carbonic anhydrase, which plays a critical role in calcium deposition; the ultimate result is a decreased deposition of calcium and a thinning of their eggshells. As a result, there has been a steady decrease in eggshell thickness, and subsequent death of many embryos before hatching, in most predatory bird species since the introduction of DDT (Figure 9.4). Many such species,

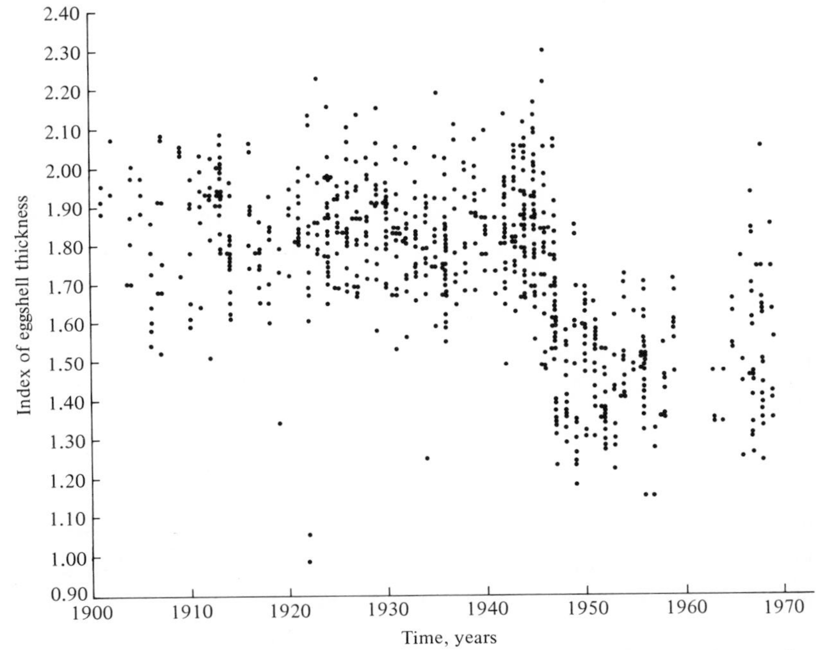

FIGURE 9.4 Plot showing the precipitous decrease in eggshell thickness of peregrine falcons in England after the first widespread use of DDT about 1945. These birds nearly became extinct in the early 1960's, but with reduced use of chlorinated hydrocarbons in Great Britain during this period, the thickness of their eggshells has increased and English peregrine populations seem to be recovering. [After Ratcliffe (1970).]

including the American bald eagle, the brown pelican, the osprey, the Bermuda petrel (a fish-eating sea bird endemic to Bermuda), and the peregrine falcon, have had their reproduction so curtailed that many populations are now in great danger of going extinct. The peregrine, the premier species of falconry, was once found from pole to pole and on all continents, but is now virtually extinct from the mainland U.S. Other populations may soon follow the American birds to extinction.

Man has exterminated hundreds of other species and many more are endangered. Species go extinct naturally during evolutionary time but the alarming thing is the great rapidity and the large number of extinctions due to human interference. What real value is another species? On a practical side, natural game in Africa produce a greater yield in biomass of meat than do domestic cattle and sheep on the same terrain; moreover the native fauna does less damage to the range. It is hardly surprising that species evolutionarily adapted to an area are more efficient at exploiting it. Secondly, the fact that diversity often leads to stability itself suggests that men may someday have use for a variety of species. Even now, we depend upon genetic variability to breed new and better farm animals and plants. Finally, it takes thousands or millions of years to evolve a species—each is unique and irreplaceable. For many of us, each species has a certain aesthetic value; I greatly enjoy watching a peregrine fly and I deeply regret that a time could come when no one will be able to see this magnificent bird again. To a naturalist and an ecologist, each species has its own particular adaptations that are worthy of study and that may reveal something yet unknown about the organized reality around us.

Another major repercussion of man's expanding population and industrialization concerns his water supply. There is a finite amount of water in the hydrologic cycle. The amount of water evaporated and transpired into the atmosphere sets the limit on how much precipitation can be given up by that same atmosphere. Water that falls on one place, as on the wet side of a mountain, is unavailable for precipitation in another, such as in the rainshadow on the leeward side of the same mountain (indeed, as such dry air falls and warms, it actually extracts water from ecosystems in its path). Man is, of course, very dependent upon the hydrologic cycle. We use huge amounts of water in industry, to grow crops and other foods, for sewage and waste removal, for drinking and bathing, for recreation, and as a power source.

For centuries men have used underground water. With the advent of water pumps, the use of ground water has skyrocketed, and, as a result the water table is falling in many places. Such nonequilibrium use of this water source obviously cannot continue for very long. If we want to continue to use underground water, it must be allowed to replenish itself. Indications are that a severe water shortage is impending. A few estimates of how much water is needed to produce some familiar commodities might underscore how much water each human needs: growing a pound of wheat requires 60 gallons, a pound of rice takes 250 gallons, a quart of milk 1000 gallons, a pound of meat from 2500 to 6000 gallons, and one average automobile about 100,000 gallons. Here again, of course, believers in the omnipotence of technology argue that distillation of sea water will save the day—accompanied, however, by the ever present increase in heat to be dissipated somewhere.

Recall again that a finite amount of water falls on earth's surface and that water used in one place must always come from somewhere else. Men have often gone to extremes to obtain water. Aqueducts and canals have been used for centuries and are still in vogue, with water being moved from an area with an "excess" by human standards to another area with a "deficit."

Such movement of water inevitably alters the hydrologic cycle and earth's weather patterns, however subtly. One neat way to obtain water has been advocated by the U.S. Forest Service in a remarkable pamphlet entitled "More Water by Cutting Trees." Removal of trees and other vegetation from hillsides of a drainage basin allows more water to run off into the creeks and rivers which drain the area, thus providing an increase in surface water for human consumption. However, the same amount of water gained in increased runoff is lost as actual evapotranspiration and therefore is not returned to the atmosphere at that locality. Hence somewhere else precipitation must be decreased, often to the consternation of another party. The city of Buenos Aires, Argentina, provides an example of such weather modification by human tampering with water supplies. Extensive cultivation and irrigation of a formerly arid grassland region immediately west of the city has, in recent years, increased the average annual precipitation by a full five inches. Typically, we modify climate first and then assess its ramifications and implications later.

Another realm of exploitative human activity is land destruction. As pointed out in Chapter 3, soil formation and primary succession take a very long time. Mountains of topsoil that took centuries to form are washed into the oceans annually due to careless man-made erosion. All too often, an area is first denuded of its vegetation by, say, overgrazing, and then much of its soil is lost to erosion. Centuries pass before recovery. Are the short-term gains during one individual's lifetime worth the long-term costs to future men?

As our population burgeons, natural communities are gradually replaced by overgrazed pastures, eroded fields, artificial lakes, golf courses, roads, parking lots, and housing developments. Some natural communities, such as the midwestern tall grass prairie, that once covered many thousands of square miles, have now virtually disappeared. Many other communities are being rapidly destroyed and it is now impossible to find an "undisturbed" natural community. All communities have been contaminated by man-made pesticide molecules and radioisotopes. Natural communities are on their way out. There simply is not room enough for them. This is tragic to ecology and to human knowledge, since so much remains to be both learned and appreciated about natural communities of plants and animals. Moreover, they could well be important to our own survival as well as the quality of life of future men.

Technology and the "Green Revolution"

Believers in the omnipotence of technology dispute much of the preceding and argue that it is technologically possible to support a much larger population of humans on earth. These observers see the plight of people in underdeveloped countries as a failure of technology to fulfill its potential. We have seen that technological "advances," such as DDT and gasoline engines, have proven to be extremely detrimental to our environment; they may ultimately actually *decrease* the maximal sustainable yield. Other aspirations for technological solutions are simply impossible. Thus, it is sometimes suggested that excess people could be shipped off to another planet. Even if Mars were inhabitable *and* we could ship people to that planet, at our present rate of growth we would populate Mars to the density of our own globe in only 35 years; it is energetically impossible to send any significant number of people off into space—they must remain on earth. Indeed, it would be impossible even to redistribute people on earth as needed, using all of our present and planned transportation facilities (Ehrlich and Ehrlich, 1972).

A recent technological breakthrough of considerable importance is the new high-yield grain crops, particularly the so-called miracle wheats and rices, that have started a so-called green revolution. Certainly these crops, coupled with more extensive fertilization, will increase immediate agricultural output and allow us to feed more people (in the long term, overfertilization often decreases maximal sustainable yields). It should be clear, however, that this new potential for food production will only buy us more time (perhaps 15 years) to come to grips with the necessity of population control. To the extent that such technological advances postpone widespread recognition that our population simply cannot continue to grow indefinitely on a finite planet, they are detrimental in the long run even though they are clearly beneficial over a shorter term. The larger we allow our population to become, the more difficult it will be to control it by decreasing birth rates and the more likely it becomes that population control will ultimately be accomplished by increased death rates. Not even technology can allow the human population to grow indefinitely.

Applied and Systems Ecology

We have noted repeatedly that man recognizes the implications, which are often vast and devastating, of his actions only after the fact. Indeed, the great complexity and intricacy of the interrelationships between and among resources, populations, environments, and communities makes it next to impossible to know the full ramifications of an act until after it has been performed. A young and vigorous branch of ecology, termed "systems ecology," deals with such interrelated components of ecological systems. Systems ecology relies heavily on computers, and systems ecologists build computer models of complex ecological systems that allow for various sorts of interactions between components, which components themselves often have many interacting subcomponents. Using actual data on how each component, at each level, affects others (in practice such data are *very* difficult to obtain), systems ecologists hope one day to model ecological systems accurately enough to predict their responses to any particular perturbation. Eventually, as dependable systems models are developed for various ecosystems, they will be merged into still larger systems that model human population, distribution, economy, etc. One day, we may have a giant systems model of the earth, which will predict many of the ramifications of any given ecological perturbation. It will be difficult to prevent such an omniscient machine from being misused for private interests. Of course, the systems approach is basically descriptive and deductive, rather than inductive, and, as such, it is limited in that the behavior of a system cannot be predicted with accuracy at states outside those used in the original data on responses between components.

Long-term versus Short-term Goals: Quality or Quantity?

A fundamental conflict of interests underlies the entire subject of man and his environment. Basically it is a question of short-term versus long-term goals. Men tend to think in a time scale proportional to their own life spans, and, as a result, they often take actions that are short sighted. For example, some organic chemists in industry now think that many of the complex molecules of petroleum products could be used much more wisely to make plastics and other products; yet we continue to burn up these valuable reserves merely moving matter back and forth. Future generations may well curse us for depleting the fossil fuels and for our general shortsightedness, which pervades so many of our activities. A very damaging short-term goal is pursuit of monetary profit because it often results in exploitation without regard to the future.

We might be wise to consider what would be required to maximize the *quality* of human lives, rather than the *quantity* of them. It should be clear from the preceding pages that we cannot have both quantity and quality. Let us briefly consider the rights a man would need to have to live a "good" life. Paul Ehrlich (1968) called some of these "inalienable rights" of man.

1. *The right to decent, uncrowded shelter.* This is already being denied many in underdeveloped countries and the underprivileged and poor people in overdeveloped countries. One reason for this denial is that populations are growing faster than adequate new housing can be constructed.

2. *The right to avoid regimentation.* Waiting in line, waiting for the traffic lights to change, being delayed in heavy traffic or caught in a traffic jam, are familiar events to all of us. As population grows, the incidence of such bothersome delays, and regimentation in general, must increase. A friend once jokingly remarked that we may soon have to stagger Christmas throughout the year so that all will be able to do their shopping.

3. *The right to enjoy natural beauty or to study nature* (should one wish to do so). The great pleasure of watching a peregrine, of course, comes to my mind. Obviously as more and more people are packed onto planet earth, the opportunity for any one of them to experience relatively undisturbed natural beauty diminishes.

4. *The right to eat well, and to eat pure foods* (that is, to eat without being poisoned with pesticides, etc.). Many Americans eat pretty well, but none are able to eat completely uncontaminated foods. The recent "health food" and organically grown food movements are evidence that many people are concerned about what they eat and that they are doing their best to minimize the poisons they ingest. Eating well, such as eating a steak, requires considerable space in which to raise the cattle (or their food). Persons in densely populated places, such as China, India, or Pakistan, are seldom able to afford the luxury of meat; of necessity these people are largely herbivorous. A time may soon come when even Americans will be unable to have steak unless they are extremely wealthy.

5. *The right to drink pure water.* Again this is being denied the majority of men. Water pollution is ultimately caused by overpopulation and overdevelopment coupled with inadequate controls. Our finite hydrologic cycle is stressed more and more with the addition of each new human being.

6. *The right to breathe clean air.* There are contaminants of one sort or another in the air everyone breathes. Air masses circulate around the globe approximately once each month, assuring that pollutants put into the air in one place are breathed by people living in another. Once again, air pollution is ultimately caused by people and the more people there are the worse it is likely to get.

7. *The right to replace oneself* (that is, for each couple to have *two* children which live to propagate themselves). Everyone should be allowed to have his own children should he so desire. But reproduction at more than the level of replacement on a finite earth dictates that, sometime in the future, others will not be able to enjoy this "right." Indeed, the *only* way that the right to reproduce can be passed on to future generations is by curtailing the amount of reproduction in the present population to replacement. Thus the seventh alienable right is self-perpetuating and guarantees that all men will have it. Moreover, by limiting reproduction and therefore population growth, enactment of this right (or its enforcement) goes a long ways toward assuring the other six inalienable rights. In fact, the list of rights could almost be reduced to a single right—if men and women were content to replace themselves, the other six rights could be achieved for all mankind.

An Ethic of Equilibrium

We are going to have to become more responsive to long-term consequences of our actions if man is to persist on earth and enjoy these seven inalienable rights. To guarantee these rights for future men, we will have to develop a new ethic, an ethic of equilibrium. Acceptance of this ethic is tantamount to accepting ecological principles as facts and as commandments. Such an ethic would dictate that we consider it *immoral* to disturb existing equilibria (at the very least they should not be perturbed beyond a state where equilibrium can be reached once again in a reasonable amount of time); moreover, we will have to do our best to restore equilibria that have been upset previously by men without our vision of equilibrium.

Innumerable changes will be necessary to meet the new ethic. Waste will have to be reduced to the barest possible minimum. Unnecessary consumption must cease. Recycling will have to be the rule rather than the exception. Built-in obsolescence must be eliminated. Litter must disappear. Either public transportation must replace private transport or the latter must be accomplished by manpower wherever possible. The use of pesticides will have to be greatly reduced or even abolished. Perhaps "organic" gardens with a mixture of species and biological control of pests will become commercial enterprises and replace giant crops of a single species with pesticide control of pests. Water will have to be respected and conserved and ground water must be allowed to stabilize. Innumerable other changes are necessary, but these serve to make the point.

Volition or Enforcement?

Even now, many Americans voluntarily recycle some of their own wastes, but recycling is far from complete. There is also a resurgence of bike riding in America, yet the majority of people still drive cars even for short distances. We ride elevators rather than walk up stairs. Organizations like Zero Population Growth (Z.P.G.) exist and some of the populace have actually decided to have only two children (a few even get sterilized after having them); however, many Americans have not limited their families to two. A few concerned persons grow their own private vegetable gardens without pesticides or inorganic fertilizers and some organic foods are sold commercially, but the vast majority continue to buy mass-produced produce that is grown with both pesticides and inorganic fertilizers. Large-scale commercial production of "organic" foods is at present unfeasible and expensive, and therefore unprofitable, except at exorbitant consumer prices. The question is: Can people be expected to adopt the new ethic of equilibrium of their own volition, or must it be enforced? Natural selection seems to dictate enforcement. The ethic of equilibrium requires that individuals restrain themselves from reproducing maximally, which essentially requires a "group" effort and would require group selection to evolve. Selection acting on individuals always favors increased reproductive success, provided that one's offspring are themselves reproductively successful. But the ethic of equilibrium and the seven inalienable rights cannot be achieved unless *individuals* act for the benefit of the group and future generations of men. Because those with the new ethic leave fewer offspring than those without it, the former are at a selective disadvantage. I strongly doubt that natural selection has programmed man (or any other species) to accept such a group-oriented reduction in individual fitness. Thus, widespread voluntary acceptance of the new ethic is unlikely; and, at this juncture, one can only speculate as to possibilities for culturally-induced or enforced control. Strict acceptance of the ethic, the sooner the better, is absolutely necessary; however, its enforcement does not seem to be imminent either in America or anywhere else. This fact, coupled with my opinion that men will not voluntarily restrict their own reproduction, portends a worsening of the already dire situation described in this chapter. I can only hope that I am wrong.

A Broader Perspective

Rather than become overly depressed by this grim prognosis for the future of mankind, we might instead briefly reconsider in a broader perspective that organized reality with which this book began. Clearly the real world exists independently of our inaccurate and incomplete perception of it. The following quotation from H. Smith (1952) nicely describes Man's insignificance to its existence:

> Man did not have forever to harness the forces of the sun and stars. The sun was an elderly light, long past the turbulent heat of youth, and would someday join the senile class of once-luminiferous bodies. In some incredibly remote time a chance collision might blow it up again into incandescent gas and start a new local cosmic cycle, but of man there would be no trace. In Balfour's terms, he "will go down into the pit, and all his thoughts will perish. The uneasy consciousness, which in this obscure corner has for a brief space broken the contented silence of the universe, will be at rest. Matter will know itself no longer. 'Imperishable monuments' and 'immortal deeds', death itself, and love stronger than death, will be as though they had never been. Nor will anything that *is* be better or be worse for all that labour, genius, devotion and suffering of man have striven through countless generations to effect."

The principles of natural selection developed in this book, however, will persist as long as assemblages of molecules replicate themselves anywhere in the universe.

Selected References

Some Ecological Principles Restated

Collier *et al.* (1973); Krebs (1972); Odum (1959, 1963, 1971); Ricklefs (1973); Watt (1973).

Human Population Growth

Appleman (1965); Borgstrom (1969); Day and Day (1965); Ehrlich (1968); Ehrlich and Ehrlich (1970, 1972); Heer (1968); Hopcraft (1968); Keyfitz and Flieger (1971); Krebs (1972); Meadows *et al.* (1972); Paddock and Paddock (1967); Sax (1955); United Nations Demographic Year Book (series by years).

Natural Selection and Man

Ardrey (1966); Bajema (1971); Borgstrom (1967, 1969); Lee and De Vore (1972); Morris (1967).

Selected Ramifications: The Rape of Planet Earth

Barkley and Seckler (1972); Benarde (1970); Beverton and Holt (1957); Blake (1964); Bresler (1968); Campbell and Wade (1972); Carson (1962); Day, Fost, and Rose (1971); Detwyler (1971); Ehrlich and Ehrlich (1972); Esposito (1970); Falk (1971); Gates (1972); Hickey (1969); Hickey and Anderson (1968); Hinrichs (1971); Istock (1971); Jarrett (1969); C. E. Johnson (1970); Marine (1969); Marx (1967); Murdoch (1971); Osborn (1948); Ratcliffe (1970); Rudd (1964); Shepard and McKinley (1969); Turk, Turk, and Wittes (1972); Wagner (1971); Wurster (1968, 1969); Wurster and Wingate (1968).

Technology and the "Green Revolution"

Brown (1970); Cloud (1969); Ehrlich and Ehrlich (1970, 1972); Gorden and Gorden (1972); Huffaker (1971); Krebs (1972); Mishan (1970); E. Odum (1971); H. Odum (1971); Spicer (1952); Watt (1973).

Applied and Systems Ecology

Bertalanffy (1969); Bormann and Likens (1967); Caswell *et. al.* (1972); Chorley and Kennedy (1971); Clark *et al.* (1967); Dale (1970); Foin (1972); Forrester (1971); Holling (1959a, 1959b, 1963, 1964, 1965, 1966); Hubbell (1971, 1973a, 1973b); Huffaker (1971); H. Odum (1971); Patten (1971, 1972); Reichle (1970); Van Dyne (1966); Waterman (1968); Watt (1966, 1968, 1973).

Long-term versus Short-term Goals: Quality or Quantity?

Dasmann (1970, 1972); Ehrlich (1968); Ehrlich and Harriman (1971); Henshaw (1971); Holdren and Ehrlich (1971); H. D. Johnson (1970); Love and Love (1970); Meadows *et al.* (1972); Watt (1973).

An Ethic of Equilibrium

Cailliet, Setzer, and Love (1971); Disch (1970); Ehrlich (1968); Ehrlich and

Harriman (1971); Gorden and Gorden (1972); Hardin (1969, 1970); Harrison (1971); Potter (1971).

Volition or Enforcement?

Hardin (1968, 1969, 1970, 1973).

A Broader Perspective

H. Smith (1952).

Bibliography

Allee, W.C. 1951. *Cooperation among animals with human implications.* Schuman, New York (Revised Edition of *Social Life of Animals,* Norton, New York, 1938). 233 pp.

Allee, W.C., A.E. Emerson, O. Park, T. Park, and K.P. Schmidt. 1949. *Principles of animal ecology.* Saunders, Philadelphia. 837 pp.

Anderson, W.W. 1971. Genetic equilibrium and population growth under density-regulated selection. *Amer. Natur.* 105:489–498.

Andrewartha, H.G. 1961. *Introduction to the study of animal populations.* Methuen, London. 281 pp.

Andrewartha, H.G. 1963. Density dependence in the Australian thrips. *Ecology* 44:218–220.

Andrewartha, H.G., and L.C. Birch. 1953. The Lotka-Volterra theory of interspecific competition. *Aust. J. Zool.* 1:174–177.

Andrewartha, H.G., and L.C. Birch. 1954. *The distribution and abundance of animals.* University of Chicago Press, Chicago. 782 pp.

Appleman, P. 1965. *The silent explosion.* Beacon, Boston.

Ardrey, R. 1966. *The territorial imperative.* Atheneum, New York. 390 pp.

Arnold, S.J. 1972. Species densities of predators and their prey. *Amer. Natur.* 106:220–236.

Ashmole, N.P. 1963. The regulation of numbers of tropical oceanic birds. *Ibis* 103*b*:458–473.

Ayala, F.J. 1968. Genotype, environment, and population numbers. *Science* 162:1453–1459.

Baker, H.G. 1970. Evolution in the tropics. *Biotropica* 2:101–111.

Baker, J.R. 1938. The evolution of breeding systems. In *Evolution*, essays presented to E.S. Goodrich. Oxford Univ. Press, London.

Bajema, C.J. (Ed.). 1971. *Natural selection in human populations, the measurement of ongoing genetic evolution in contemporary societies.* Wiley, New York. 406 pp.

Barkley, P.W., and D.W. Seckler. 1972. *Economic growth and environmental decay: the solution becomes the problem.* Harcourt Brace Jovanovich, New York. 194 pp.

Bartlett, M.S. 1960. *Stochastic population models in ecology and epidemiology.* Methuen, London. 90 pp.

Beard, J.S. 1955. The classification of tropical American vegetation types. *Ecology* 36:89–100.

Beauchamp, R.S.A., and P. Ullyott. 1932. Competitive relationships between certain species of fresh-water triclads. *J. Ecol.* 20:200–208.

Benarde, M.A. 1970. *Our precarious habitat.* Norton, New York. 362 pp.

Benson, S.B. 1933. Concealing coloration among some desert rodents of the southwestern United States. *Univ. Calif. Publ. Zool.* 40:1–70.

Benson, W.W. 1972. Natural selection for Müllerian mimicry in *Heliconius erato* in Costa Rica. *Science* 176:936–939.

Bernal, J.D. 1967. *The origin of life.* World, Cleveland. 345 pp.

Bertalanffy, L. 1957. Quantitative laws in metabolism and growth. *Quart. Rev. Biol.* 32:217–231.

Bertalanffy, L. (Ed.). 1969. *General systems theory: foundations, development, applications.* Braziller, New York. 290 pp.

Beverton, R.J.H., and S.J. Holt. 1957. On the dynamics of exploited fish populations. *Great Brit. Min. Agr. Fish, Food, Fish. Invest. Ser. 2*, vol. 19:1–533.

Billings, W.D. 1964. *Plants and the ecosystem.* Wadsworth, Belmont, Calif.

Birch, L.C. 1948. The intrinsic rate of natural increase of an insect population. *J. Anim. Ecol.* 16:15–26.

Birch, L.C. 1953. Experimental background to the study of the distribution and abundance of insects. III. The relations between innate capacity for increase and survival of different species of beetles living together on the same food. *Evolution* 7:136–144.

Birch, L.C. 1957. The meanings of competition. *Amer. Natur.* 91:5–18.

Birch, L.C., and P.R. Ehrlich. 1967. Evolutionary history and population biology. *Nature* 214:349–352.

Black, C.A. 1968. *Soil-plant relationships* (2nd ed.). Wiley, New York. 792 pp.

Blair, T.A., and R.C. Fite. 1965. *Weather elements.* Prentice-Hall, Englewood Cliffs, N. J.

Blair, W.F. 1960. *The rusty lizard. A population study.* Univ. Texas Press, Austin. 185 pp.

Blake, P. 1964. *God's own junkyard, the planned deterioration of America's landscape.* Holt, Rinehart and Winston, New York.

Blum, H.F. 1968. *Time's arrow and evolution.* Princeton Univ. Press, Princeton, N.J. 232 pp.

Blumenstock, D.I., and C.W. Thornthwaite. 1941. Climate and the world pattern. pp. 98–127. In *Climate and man.* U. S. Department of Agriculture Yearbook. Washington, D.C. 1248 pp.

Bogue, D.J. 1969. *Principles of demography.* Wiley, New York.

Bonner, J.T. 1965. *Size and cycle: an essay on the structure of biology.* Princeton Univ. Press, Princeton, N.J. 219 pp.

Borgstrom, G. 1967. *The hungry planet.* Collier, New York.

Borgstrom, G. 1969. *Too many, a story of earth's biological limitations.* Macmillan, New York.

Bormann, F.H., and G.E. Likens. 1967. Nutrient cycling. *Science* 155:424–429.

Bowman, R.L. 1961. Morphological differentiation and adaptation in the Galápagos finches. *Univ. Calif. Publ. Zoology.* Vol. 58. Univ. of Calif. Press, Berkeley. 326 pp.

Braun-Blanquet, J. 1932. *Plant sociology: the study of plant communities* (translated and edited by G.D. Fuller and H.C. Conard). McGraw-Hill, New York. 439 pp.

Bresler, J.B. 1968. *Environments of man.* Addison-Wesley, Reading, Mass. 289 pp.

Brody, S. 1945. *Bioenergetics and growth.* Van Nostrand Reinhold, New York. 1023 pp.

Brooks, G.R., Jr. 1967. Population ecology of the ground skink, *Lygosoma laterale* (Say). *Ecol. Monogr.* 37:71–87.

Brower, L.P. 1969. Ecological chemistry. *Sci. Amer.* 220:22–29.

Brower, L.P., and J. Brower. 1964. Birds, butterflies, and plant poisons: a study in ecological chemistry. *Zoologica* 49:137–159.

Brown, J.H. 1971. Mammals on mountaintops: nonequilibrium insular biogeography. *Amer. Natur.* 105:467–478.

Brown, J.L. 1964. The evolution of diversity in avian territorial systems. *Wilson Bull.* 76:160–169.

Brown, J.L. 1966. Types of group selection. *Nature* 211:870.

Brown, J.L. 1969. Territorial behavior and population regulation in birds. *Wilson Bull.* 81:293–329.

Brown, J.L., and G.H. Orians. 1970. Spacing patterns in mobile animals. *Ann. Rev. Ecol. Syst.* 1:239–262.

Brown, L.R. 1970. *Seeds of change. The green revolution and development in the 1970's.* Praeger, New York.

Brown, W.L., and E.O. Wilson. 1956. Character displacement. *Syst. Zool.* 5:49–64.

Burges, A., and F. Raw (Eds.). 1967. *Soil biology.* Academic Press, New York. 532 pp.

Byers, H. G. 1954. The atmosphere up to 30 kilometers. In G.P. Kuiper (Ed.), *The earth as a planet.* Univ. Chicago Press, Chicago.

Cailliet, G., P. Setzer, and M. Love. 1971. *Everyman's guide to ecological living.* Macmillan, New York. 119 pp.

Cain, S.A. 1950. Life-forms and phytoclimate. *Bot. Rev.* 16:1–32.

Calvin, M. 1969. *Chemical evolution.* Oxford Univ. Press, New York. 278 pp.

Campbell, R. R., and J. L. Wade. 1972. *Society and environment: the coming collision.* Allyn and Bacon, Boston. 375 pp.

Carlquist, S. 1965. *Island life: a natural history of the islands of the world.* Natural History Press, Garden City, N.Y.

Carpenter, C. R. 1958. Territoriality: a review of concepts and problems. pp. 224–250. In A. Roe and G.G. Simpson (Eds.), *Behavior and evolution.* Yale Univ. Press, New Haven. 557 pp.

Carson, R. 1962. *Silent spring.* Houghton-Mifflin, Boston.

Caswell, H., H.E. Koenig, J.A. Resh, and Q.E. Ross. 1972. An introduction to systems science for ecologists. pp. 4–78. In B. Patten (Ed.), *Systems analysis and simulation in ecology.* Vol. II. Academic Press, New York. 592 pp.

Caswell, H., F. Reed, S.N. Stephenson, and P.A. Werner. 1973. Photosynthetic pathways and selective herbivory: a hypothesis. *Amer. Natur.* 107:465–480.

Caughley, G. 1966. Mortality patterns in mammals. *Ecology* 47:906–918.

Chambers, K. L. (Ed.). 1970. *Biochemical coevolution.* 29th Biology Colloquium, Oregon State Univ. Press, Eugene. 117 pp.

Charlesworth, B. 1971. Selection in density-regulated populations. *Ecology* 52:469–474.

Chitty, D. 1960. Population processes in the vole and their relevance to general theory. *Canad. J. Zool.* 38:99–113.

Chitty, D. 1967a. The natural selection of self-regulatory behavior in animal populations. *Proc. Ecol. Soc. Australia* 2:51–78.

Chitty, D. 1967b. What regulates bird populations? *Ecology* 48:698–701.

Chorley, R. J., and B. A. Kennedy. 1971. *Physical geography: a systems approach.* Prentice-Hall, London. 370 pp.

Christian, J. J., and D. E. Davis. 1964. Endocrines, behavior, and population. *Science* 146:1550–1560.

Clapham, W. B. 1973. *Natural ecosystems.* Macmillan, New York. 248 pp.

Clark, L. R., P. W. Geier, R. D. Hughes, and R. F. Morris. 1967. *The ecology of insect populations in theory and practice.* Methuen, London. 232 pp.

Clarke, B. C. 1972. Density-dependent selection. *Amer. Natur.* 106:1–13.

Clarke, G. L. 1954. *Elements of ecology.* Wiley, New York. 560 pp.

Clements, F. E. 1920. *Plant succession: an analysis of the development of vegetation.* Publication No. 290. Carnegie Institute, Washington, D.C. 388 pp.

Clements, F.E. 1949. *Dynamics of vegetation.* Hafner, New York. 296 pp.

Cloud, P.E., Jr. (Ed.). 1969. *Resources and man.* Freeman, San Francisco.

Cody, M.L. 1966. A general theory of clutch size. *Evolution* 20:174–184.

Cody, M.L. 1968. On the methods of resource division in grassland bird communities. *Amer. Natur.* 102:107–147.

Cody, M.L. 1970. Chilean bird distribution. *Ecology* 51:455–463.

Cody, M.L. 1971. Ecological aspects of reproduction. Chapter 10 (pp. 461–512). In D.S. Farner and J.R. King (Eds.), *Avian Biology.* Vol. I. Academic Press, New York. 586 pp.

Cody, M.L. 1973. *Competition and community structure.* Princeton Univ. Press, Princeton, N. J.

Cole, L.C. 1951. Population cycles and random oscillations. *J. Wildl. Manage.* 15:233–251.

Cole, L.C. 1954a. Some features of random cycles. *J. Wildl. Manage.* 18:2–24.

Cole, L.C. 1954b. The population consequences of life history phenomena. *Quart. Rev. Biol.* 29:103–137.

Cole, L.C. 1958. Sketches of general and comparative demography. *Cold Spring Harbor Symp. Quant. Biol.* 22:1–15.

Cole, L.C. 1960. Competitive exclusion. *Science* 132:348–349.

Cole, L.C. 1965. Dymanics of animal population growth. pp. 221–241. In M.C. Sheps and J.C. Ridley (Eds.), *Public health and population change.* Univ. Pittsburgh Press, Pittsburgh.

Colinvaux, P.A. 1973. *Introduction to ecology.* Wiley, New York. 621 pp.

Collier, B., G.W. Cox, A.W. Johnson, and P.C. Miller. 1973. *Dynamic ecology.* Prentice-Hall, Englewood Cliffs, N.J. 563 pp.

Colwell, R.K., and D.J. Futuyma. 1971. On the measurement of niche breadth and overlap. *Ecology* 52:567–576.

Connell, J.H. 1961a. The effects of competition, predation by *Thais lapillus*, and other factors on natural populations of the barnacle, *Balanus balanoides. Ecol. Monogr.* 31:61–104.

Connell, J.H. 1961b. The influence of interspecific competition and other factors on the distribution of the barnacle, *Chthamalus stellatus. Ecology* 42:710–723.

Connell, J.H. 1970. A predator–prey system in the marine intertidal region. I. *Balanus glandula* and several predatory species of *Thais. Ecol. Monogr.* 40:49–78.

Connell, J.H., and E. Orias. 1964. The ecological regulation of species diversity. *Amer. Natur.* 98:399–414.

Connell, J.H., D.B. Mertz, and W.W. Murdoch. 1970. *Readings in ecology and ecological genetics.* Harper & Row, New York. 397 pp.

Cott, H.B. 1940. *Adaptive coloration in animals.* Oxford Univ. Press, London. 508 pp.

Cowles, R.P., and C.E. Brambel. 1936. A study of the environmental conditions in a bog pond with special reference to the diurnal vertical distribution of *Gonyostomum semen. Biol. Bull. Mar. Biol. Lab. Woods Hole* 71:286–298.

Crocker, R.L. 1952. Soil genesis and the pedogenic factors. *Quart. Rev. Biol.* 27:139–168.

Crocker, R.L., and J. Major. 1955. Soil development in relation to vegetation and surface age at Glacier Bay, Alaska. *J. Ecol.* 43:427–448.

Crombie, A.C. 1947. Interspecific competition. *J. Anim. Ecol.* 16:44–73.

Crook, J.H. 1962. The adaptive significance of pair formation types in weaver birds. *Symp. Zool. Soc. London* 8:57–70.

Crook, J.H. 1963. Monogamy, polygamy and food supply. *Discovery* (Jan.): 35–41.

Crook, J.H. 1964. The evolution of social organization and visual communication in the weaver birds (Ploceinae). *Behaviour* 10:1–178.

Crook, J.H. 1965. The adaptive significance of avian social organization. pp. 181–218. In P.E. Ellis (Ed.), *Social Organization of animal communities.* Symp. Zool. Soc. London, Vol. 14. Zoological Society of London.

Crook, J.H. 1972. Sexual selection, dimorphism, and social organization in the primates. pp. 231–281. In B.G. Campbell (Ed.), *Sexual selection and the descent of man (1871–1971).* Aldine-Atherton, Chicago.

Crowell, K.L. 1962. Reduced interspecific competition among the birds of Bermuda. *Ecology* 43:75–88.

Dale, M.B. 1970. Systems analysis and ecology. *Ecology* 51:2–16.

Dammerman, K.W. 1948. The fauna of Krakatau 1883–1933. *Verhandel. Kon-Inkl. Ned. Akad. Wetenschap. Afdel. Natuurk.* 44:1–594.

Dansereau, P. 1957. *Biogeography: an ecological perspective.* Ronald, New York. 394 pp.

Darlington, C.D., and K. Mather. 1949. *The elements of genetics.* Allen and Unwin, London. 446 pp.

Darlington, P.J. 1957. *Zoogeography: the geographical distribution of animals.* Wiley, New York. 675 pp.

Darlington, P.J. 1959. Area, climate, and evolution. *Evolution* 13:488–510.

Darlington, P.J. 1965. *Biogeography of the southern end of the world.* Harvard Univ. Press, Cambridge. 236 pp.

Darlington, P.J. 1971. Nonmathematical models for evolution of altruism, and for group selection. *Proc. Nat. Acad. Sci.* 69:293–297.

Darwin, C. 1859. *The origin of species by means of natural selection* (numerous editions). Murray, London.

Darwin, C. 1871. *The descent of man, and selection in relation to sex* (numerous editions). Murray, London.

Dasmann, R.F. 1970. *A different kind of country.* Collier-Macmillan, London. 276 pp.

Dasmann, R.F. 1972. *Environmental conservation* (3rd ed.). Wiley, New York. 473 pp.

Daubenmire, R.F. 1947. *Plants and environment.* Wiley, New York. 424 pp.

Daubenmire, R.F. 1956. Climate as a determinant of vegetation distribution in eastern Washington and northern Idaho. *Ecol. Monogr.* 26:131–154.

Daubenmire, R.F. 1968. *Plant communities.* Harper & Row, New York. 300 pp.

Davidson, J., and H.G. Andrewartha. 1948. Annual trends in a natural population of *Thrips imaginis* (Thysanoptera). *J. Anim. Ecol.* 17:193–222.

Dawson, P.S., and C.E. King (Eds.). 1971. *Readings in population biology.* Prentice- Hall, Englewood Cliffs, N.J.

Day, J.A., F.F. Fost, and P. Rose. 1971. *Dimensions of the environmental crisis.* Wiley, New York. 212 pp.

Day, L.H., and A.T. Day. 1965. *Too many Americans.* Delta, New York.

Dayton, P.K. 1971. Competition, disturbance, and community organization: the provision and subsequent utilization of space in a rocky intertidal community. *Ecol. Monogr.* 41:351–389.

DeBach, P. 1966. The competitive displacement and coexistence principles. *Ann. Rev. Entomol.* 11:183–212.

Deevey, E.S., Jr. 1947. Life tables for natural populations of animals. *Quart. Rev. Biol.* 22:283–314.

Detwyler, T.R. (Ed.). 1971. *Man's impact on environment.* McGraw-Hill, New York. 731 pp.

Diamond, J.M. 1969. Avifaunal equilibria and species turnover rates on the channel islands of California. *Proc. Nat. Acad. Sci.* 64:57–63.

Dice, L.R. 1952. *Natural communities.* Univ. Michigan Press, Ann Arbor. 547 pp.

Dietz, R.S., and J.C. Holden. 1970. The breakup of pangaea. *Sci. Amer.* (Oct.):30–41.

Disch, R. (Ed.). 1970. *The ecological conscience: values for survival.* Prentice-Hall, Englewood Cliffs, N.J. 206 pp.

Dobzhansky, T. 1950. Evolution in the tropics. *Amer. Sci.* 38:208–221.

Docters van Leeuwen, W.M. 1936. Krakatau, 1833 to 1933. *Ann. Jard. Botan. Buitenzorg* 56–57:1–506.

Doeksen, J., and J. van der Drift. 1963. *Soil organisms.* North-Holland, Amsterdam. 453 pp.

Drake, E.T. (Ed.). 1968. *Evolution and environment.* Yale Univ. Press, New Haven, Conn. 478 pp.

Dressler, R.L. 1968. Pollination by euglossine bees. *Evolution* 22:202–210.

Dunbar, M.J. 1960. The evolution of stability in marine environments: natural selection at the level of the ecosystem. *Amer. Natur.* 94:129–136.

Dunbar, M.J. 1968. *Ecological development in polar regions.* Prentice-Hall. Englewood Cliffs, N.J. 119 pp.

Ehrlich, P. R. 1968. *The population bomb.* Ballantine, New York. 223 pp.

Ehrlich, P. R., and L. C. Birch. 1967. "The balance of nature" and "population control." *Amer. Natur.* 101:97–107.

Ehrlich, P.R., and A.H. Ehrlich. 1970. *Population, resources, environment: issues in human ecology.* Freeman, San Francisco. 383 pp.

Ehrlich, P. R., and A. H. Ehrlich. 1972. *Population, resources, environment: issues in human ecology* (2nd ed.). Freeman, San Francisco. 509 pp.

Ehrlich, P. R., and R. L. Harriman. 1971. *How to be a survivor: a plan to save spaceship earth.* Ballantine, New York. 208 pp.

Ehrlich, P. R., and R. W. Holm. 1963. *The process of evolution.* McGraw-Hill, New York. 347 pp.

Ehrlich, P. R., and P. H. Raven. 1964. Butterflies and plants: a study in coevolution. *Evolution* 18:586–608.

Ehrlich, P. R., and P. H. Raven. 1969. Differentiation of populations. *Science* 165:1228–1231.

Elton, C. S. 1927. *Animal ecology.* Sidgwick and Jackson, London. 209 pp.

Elton, C. S. 1942. *Voles, mice and lemmings: problems in population dynamics.* Oxford Univ. Press, London. 496 pp.

Elton, C. S. 1946. Competition and the structure of ecological communities. *J. Anim. Ecol.* 15:54–68.

Elton, C. S. 1949. Population interspersion: an essay on animal community patterns. *J. Ecol.* 37:1–23.

Elton, C. S. 1958. *The ecology of invasions by animals and plants.* Methuen, London. 181 pp.

Elton, C. S. 1966. *The pattern of animal communities.* Methuen, London.

Emlen, J. M. 1966. The role of time and energy in food preference. *Amer. Natur.* 100:611–617.

Emlen, J.M. 1968a. Optimal choice in animals. *Amer. Natur.* 102:385–390.

Emlen, J.M. 1968b. A note on natural selection and the sex ratio. *Amer. Natur.* 102:94–95.

Emlen, J. M. 1970. Age specificity and ecological theory. *Ecology* 51:588–601.

Emlen, J. M. 1973. *Ecology: an evolutionary approach.* Addison-Wesley, Reading, Mass. 493 pp.

Engelmann, M. D. 1966. Energetics, terrestrial field studies, and animal productivity. *Adv. Ecol. Res.* 3:73–115.

Errington, P. L. 1946. Predation and vertebrate populations. *Quart. Rev. Biol.* 21:144–177.

Errington, P. L. 1956. Factors limiting higher vertebrate populations. *Science* 124:304–307.

Errington, P. L. 1963. *Muskrat populations.* Iowa State Univ. Press, Ames. 665 pp.

Esposito, J. C. 1970. *Vanishing air.* Grossman, New York.

Evans, F.C., and F.E. Smith. 1952. The intrinsic rate of natural increase for the human louse, *Pediculus humanus* L. *Amer. Natur.* 86:299–310.

Eyre, S.R. 1963. *Vegetation and soils: a world picture.* Aldine, Chicago. 324 pp.

Faegri, K., and L. van der Pijl. 1971. *The principles of pollination ecology.* Pergamon Press, London. 248 pp.

Falk, R. 1971. *This endangered planet.* Random House, New York.

Falls, J.B. 1969. Functions of territorial songs in the white-throated sparrow. pp. 207–232. In R.A. Hinde (Ed.), *Bird vocalizations.* Cambridge Univ. Press, Cambridge, England. 394 pp.

Feeny, P.P. 1968. Effects of oak leaf tannins on larval growth of the winter moth *Operophtera brumata. J. Insect Physiol.* 14:805–817.

Feeny, P.P. 1970. Seasonal changes in oak leaf tannins and nutrients as a cause of spring feeding by winter moth caterpillars. *Ecology* 51:565–581.

Finch, V.C., and G.T. Trewartha. 1949. *Physical elements of geography.* McGraw-Hill, New York.

Fischer, A.G. 1960. Latitudinal variations in organic diversity. *Evolution* 14: 64–81.

Fisher, J., and R.T. Peterson. 1964. *The world of birds.* Doubleday, New York.

Fisher, R.A. 1930. *The genetical theory of natural selection.* Clarendon Press, Oxford. 272 pp.

Fisher, R.A. 1958a. *The genetical theory of natural selection* (2nd ed.). Dover, New York. 291 pp.

Fisher, R.A. 1958b. Polymorphism and natural selection. *J. Anim. Ecol.* 46: 289–293.

Fisher, R.A., A.S. Corbet, and C.B. Williams. 1943. The relation between the number of species and the number of individuals in a random sample of an animal population. *J. Anim. Ecol.* 12:42–58.

Flohn, H. 1969. *Climate and weather.* World Univ. Library, McGraw-Hill, New York.

Foin, T.C. 1972. Systems ecology and the future of human society. pp. 475-531. In B. Patten (Ed.), *Systems analysis and simulation in ecology.* Vol. II. Academic Press, New York. 592 pp.

Fons, W.L. 1940. Influence of forest cover on wind velocity. *J. Forestry* 38: 481–486.

Force, D.C. 1972. *r*- and *K*-strategists in endemic host-parasitoid communities. *Bull. Entomol. Soc. Amer.* 18:135–137.

Ford, E.B. 1931. *Mendelism and evolution.* Methuen, London. 122 pp.

Ford, E.B. 1964. *Ecological genetics.* Methuen, London. 335 pp.

Ford, R.F., and W.E. Hazen. 1972. *Readings in aquatic ecology.* Saunders, Philadelphia. 379 pp.

Forrester, J.W. 1971. Alternatives to catastrophe-understanding the counterintuitive behavior of social systems. *Technology Rev.* 73:52–68.

Fox, S.W., and K. Dose. 1972. *Molecular evolution and the origin of life.* Freeman, San Francisco. 359 pp.

Frank, P.W. 1968. Life histories and community stability. *Ecology* 49:355–357.

Fretwell, S.D. 1972. *Populations in a seasonal environment.* Princeton Univ. Press, Princeton, N.J. 217 pp.

Frey, D.G. 1963. *Limnology in North America.* Univ. of Wisconsin Press, Madison. 734 pp.

Fried, M., and H. Broeshart. 1967. *The soil-plant system in relation to inorganic nutrition.* Academic Press, New York. 358 pp.

Futuyma, D.J. 1973. Community structure and stability in constant environments. *Amer. Natur.* 107:443–446.

Gadgil, M., and W.H. Bossert. 1970. Life historical consequences of natural selection. *Amer. Natur.* 104:1–24.

Gadgil, M., and O.T. Solbrig. 1972. The concept of *r* and *K* selection: evidence from wild flowers and some theoretical considerations. *Amer. Natur.* 106:14–31.

Gallopin, G.C. 1972. Structural properties of food webs. pp. 241–282. In B. Patten (Ed.), *Systems analysis and simulation in ecology.* Vol. II. Academic Press, New York. 592 pp.

Gates, D.M. 1962. *Energy exchange in the biosphere.* Harper & Row, New York. 151 pp.

Gates, D.M. 1965. Energy, plants and ecology. *Ecology* 46:1–13.

Gates, D.M. 1972. *Man and his environment: climate.* Harper & Row, New York. 175 pp.

Gause, G.F. 1934. *The struggle for existence.* Hafner, New York (reprinted 1964). 163 pp.

Gause, G.F. 1935. Experimental demonstration of Volterra's periodic oscillations in the numbers of animals. *J. Exp. Biol.* 12:44–48.

Geiger, R. 1966. *The climate near the ground.* Harvard Univ. Press, Cambridge, Mass. 611 pp.

Gibb, J.A. 1956. Food, feeding habits, and territory of the Rock Pipit, *Anthus spinoletta. Ibis* 98:506–530.

Gibb, J.A. 1960. Populations of tits and goldcrests and their food supply in pine plantations. *Ibis* 102:163–208.

Gilbert, L.E. 1971. Butterfly-plant coevolution: has *Passiflora adenopoda* won the selectional race with Heliconiine butterflies? *Science* 172:585–586.

Gilbert, L.E. 1972. Pollen feeding and reproductive biology of *Heliconius* butterflies. *Proc. Nat. Acad. Sci.* 69:1403–1407.

Gilbert, L.E., and M.C. Singer. 1973. Dispersal and gene flow in a butterfly species. *Amer. Natur.* 107:58–72.

Gill, D.E. 1972. Intrinsic rates of increase, saturation densities, and competitive ability. I. An experiment with *Paramecium. Amer. Natur.* 106:461–471.

Gisborne, H.T. 1941. How the wind blows in the forest of northern Idaho. *Northern Rocky Mountain Forest Range Experimental Station.*

Gleason, H.A. 1922. On the relation between species and area. *Ecology* 3: 158–162.

Gleason, H.A. 1929. The significance of Raunkiaer's law of frequency. *Ecology* 10:406–408.

Gleason, H.A., and A. Cronquist. 1964. *The natural geography of plants.* Columbia Univ. Press, New York. 420 pp.

Golley, F.B. 1960. Energy dynamics of a food chain of an old-field community. *Ecol. Monogr.* 30:187–206.

Goodman, L.A. 1971. On the sensitivity of the intrinsic growth rate to changes in the age-specific birth and death rates. *Theoret. Pop. Biol.* 2:339–354.

Gorden, M., and M. Gorden. 1972. *Environmental management: science and politics.* Allyn and Bacon, Boston. 548 pp.

Gordon, H.T. 1961. Nutritional factors in insect resistance to chemicals. *Ann. Rev. Entomol.* 6:27–54.

Grant, P.R. 1967. Bill length variability in birds of the Tres Marías Islands, Mexico. *Canad. J. Zool.* 45:805–815.

Grant, P.R. 1971. Variation in the tarsus length of birds in island and mainland regions. *Evolution* 25:599–614.

Green, R.H. 1969. Population dynamics and environmental variability. *Amer. Zool.* 9:393–398.

Green, R.H. 1971. A multivariate statistical approach to the Hutchinsonian niche: bivalve mollusks of central Canada. *Ecology* 52:543–556.

Greig-Smith, P. 1964. *Quantitative plant ecology* (2nd ed.). Butterworth, London. 256 pp.

Grice, G.D., and A.D. Hart. 1962. The abundance, seasonal occurrence and distribution of the epizooplankton between New York and Bermuda. *Ecol. Monogr.* 32:287–307.

Grinnell, J. 1917. The niche relationships of the California thrasher. *Auk* 21: 364–382.

Grinnell, J. 1924. Geography and evolution. *Ecology* 5:225–229.

Grinnell, J. 1928. The presence and absence of animals. *Univ. Calif. Chronicle* 30:429–450. (Reprinted in *Joseph Grinnell's Philosophy of nature*, Univ. California Press, Berkeley, 1943. pp. 187–208.)

Grodzinski, W., and A. Gorecki. 1967. Daily energy budgets of small rodents pp. 295–314. In K. Petrusewicz (Ed.), *Secondary productivity of terrestrial ecosystems*, Vol. I. Warsaw.

Gunter, G. 1941. Death of fishes due to cold on the Texas coast, January, 1940. *Ecology* 22:203–208.

Haartman, L.V. 1969. Nest-site and evolution of polygamy in European passerine birds. *Ornis Fenn.* 46:1–12.

Hairston, N.G. 1951. Interspecies competition and its probable influence upon the vertical distribution of Appalachian salamanders of the genus *Plethodon. Ecology* 32:266–274.

Hairston, N.G., and G.W. Byers. 1954. The soil arthropods of a field in southern Michigan: a study in community ecology. *Contrib. Lab. Vert. Biol., Univ. of Michigan* 64:1–37.

Hairston, N.G., F.E. Smith, and L.B. Slobodkin. 1960. Community structure, population control, and competition. *Amer. Natur.* 94:421–425.

Hairston, N.G., J.D. Allan, R.K. Colwell, D.J. Futuyma, J. Howell, M.D. Lubin, J. Mathias, and J.H. Vandermeer. 1968. The relationship between species diversity and stability: an experimental approach with protozoa and bacteria. *Ecology* 49:1091–1101.

Haldane, J.B.S. 1932. *The causes of evolution* (reprinted 1966). Cornell Univ. Press, Ithaca, N.Y. 235 pp.

Hamilton, T.H. 1961. On the functions and causes of sexual dimorphism in breeding plumage of North American species of warblers and orioles. *Amer. Natur.* 45:121–123.

Hamilton, T.H., and I. Rubinoff. 1963. Isolation, endemism, and multiplication of species in the Darwin finches. *Evolution* 17:388–403.

Hamilton, T.H., and I. Rubinoff. 1967. On predicting insular variation in endemism and sympatry for the Darwin finches in the Galapagos archipelago. *Amer. Natur.* 101:161–172.

Hamilton, W.D. 1964. The genetical evolution of social behavior (two parts). *J. Theoret. Biol.* 7:1–52.

Hamilton, W.D. 1966. The moulding of senescence by natural selection. *J. Theoret. Biol.* 12:12–45.

Hamilton, W.D. 1967. Extraordinary sex ratios. *Science* 156:477–488.

Hamilton, W.D. 1970. Selfish and spiteful behaviour in an evolutionary model. *Nature* 228:1218–1220.

Hamilton, W.D. 1972. Altruism and related phenomena, mainly in insects. *Ann. Rev. Ecol. Syst.* 3:193–232.

Hardin, G. 1960. The competitive exclusion principle. *Science* 131:1292–1297.

Hardin, G. 1968. The tragedy of the commons. *Science* 162:1243–1248.

Hardin, G. (Ed.). 1969. *Population, evolution, and birth control.* Freeman, San Francisco. 386 pp.

Hardin, G. 1970. *Birth control.* Pegasus, New York.

Hardin, G. 1973. *Stalking the wild taboo.* Kaufmann, Los Altos, Calif. 240 pp.

Harper, J.L. 1961a. Approaches to the study of plant competition. *Soc. Exp. Biol. Symp.* 15:1–39.

Harper, J.L. 1961b. The evolution and ecology of closely related species living in the same area. *Evolution* 15:209–227.

Harper, J.L. 1967. A Darwinian approach to plant ecology. *J. Ecol.* 55: 247–270.

Harper, J.L. 1969. The role of predation in vegetational diversity. *Brookhaven Symp. Biol.* 22:48–62.

Harper, J.L., and J. Ogden. 1970. The reproductive strategy of higher plants. I. The concept of strategy with special reference to *Senecio vulgaris* L. *J. Ecol.* 58:681–698.

Harper, J.L., P.H. Lovell, and K.G. Moore. 1970. The shapes and sizes of seeds. *Ann. Rev. Ecol. Syst.* 1:327–356.

Harrison, G. 1971. *Earthkeeping*. Houghton-Mifflin, Boston.

Haurwitz, B., and J.M. Austin. 1944. *Climatology*. McGraw-Hill, New York. 410 pp.

Hazen, W.E. 1964. *Readings in population and community ecology* (1st ed). Saunders, Philadelphia. 388 pp.

Hazen, W.E. 1970. *Readings in population and community ecology* (2nd ed). Saunders, Philadelphia. 421 pp.

Heatwole, H. 1965. Some aspects of the association of cattle egrets with cattle. *Anim. Behaviour* 13:79–83.

Heer, D.M. 1968. *Society and population*. Prentice-Hall, Englewood Cliffs, N.J.

Heinrich, B., and P.H. Raven. 1972. Energetics and pollination ecology. *Science* 176:597–602.

Henshaw, P.S. 1971. *This side of yesterday, extinction or utopia*. Wiley, New York. 186 pp.

Hensley, M.M., and J.B. Cope. 1951. Further data on removal and repopulation of the breeding birds in a spruce-fir forest community. *Auk* 68: 483–493.

Hespenhide, H. 1971. Food preference and the extent of overlap in some insectivorous birds, with special reference to Tyrannidae. *Ibis* 113:59–72.

Hesse, R., W.C. Allee, and K.P. Schmidt. 1951. *Ecological animal geography* (2nd ed.). Wiley, New York. 715 pp.

Hickey, J.J. (Ed.). 1969. *Peregrine falcon populations, their biology and decline*. Univ. of Wisconsin Press, Madison. 596 pp.

Hickey, J.J., and D.W. Anderson. 1968. Chlorinated hydrocarbons and egg shell changes in raptorial and fish-eating birds. *Science* 162:271–272.

Hinrichs, N. (Ed.). 1971. *Population, environment and people*. McGraw-Hill, New York. 227 pp.

Holdren, J.P., and P.R. Ehrlich (Eds.). 1971. *Global ecology*. Harcourt Brace Jovanovich, New York. 295 pp.

Holdridge, L.R. 1947. Determination of world plant formations from simple climatic data. *Science* 105:367–368.

Holdridge, L.R. 1959. Simple method for determining potential evapotranspiration from temperature data. *Science* 130:572.

Holdridge, L.R. 1967. *Life zone ecology*. Tropical Science Center, San Jose, Costa Rica. 124 pp.

Holling, C.S. 1959a. The components of predation as revealed by a study of small-mammal predation of the European pine sawfly. *Canad. Entomol.* 91:293–320.

Holling, C.S. 1959b. Some characteristics of simple types of predation and parasitism. *Canad. Entomol.* 91:385–398.

Holling, C.S. 1961. Principles of insect predation. *Ann. Rev. Entomol.* 6:163–182.

Holling, C.S. 1963. An experimental component analysis of population processes. *Mem. Entomol. Soc. Canada* 32:22–32.

Holling, C.S. 1964. The analysis of complex population processes. *Canad. Entomol.* 96:335–347.

Holling, C.S. 1965. The functional response of predators to prey density and its role in mimicry and population regulation. *Mem. Entomol. Soc. Canada* 45:1–60.

Holling, C.S. 1966. The functional response of invertebrate predators to prey density. *Mem. Entomol. Soc. Canada* 48:1–87.

Hopcraft, A. 1968. *Born to hunger*. Houghton-Mifflin, Boston.

Horn, H.S. 1966. Measurement of overlap in comparative ecological studies. *Amer. Natur.* 100:419–424.

Horn, H.S. 1968a. Regulation of animal numbers: a model counter-example. *Ecology* 49:776–778.

Horn, H.S. 1968b. The adaptive significance of colonial nesting in the Brewer's blackbird (*Euphagus cyanocephalus*). *Ecology* 49:682–694.

Horn, H.S. 1971. *The adaptive geometry of trees*. Princeton Univ. Press, Princeton, N.J. 144 pp.

Howard, H.E. 1920. *Territory in bird life*. Murray, London. 308 pp. (reprinted in 1964 by Atheneum, New York. 293 pp.).

Hubbell, S.P. 1971. Of sowbugs and systems: the ecological bioenergetics of a terrestrial isopod. pp. 269–324. In B. Patten (Ed.), *Systems analysis and simulation in ecology*. Vol I. Academic Press, New York. 607 pp.

Hubbell, S.P. 1973a. Populations and simple food webs as energy filters. I. One-species systems. *Amer. Natur.* 107:94–121.

Hubbell, S.P. 1973b. Populations and simple food webs as energy filters. II. Two-species systems. *Amer. Natur.* 107:122–151.

Huffaker, C.B. 1958. Experimental studies on predation: dispersion factors and predator-prey oscillations. *Hilgardia* 27:343–383.

Huffaker, C.B. 1971. *Biological control*. Plenum, New York. 511 pp.

Hurd, L.E., M.V. Mellinger, L.L. Wolf, and S.J. McNaughton. 1971. Stability and diversity at three trophic levels in terrestrial successional ecosystems. *Science* 173:1134–1136.

Hutchinson, G.E. 1951. Copepodology for the ornithologist. *Ecology* 32: 571–577.

Hutchinson, G.E. 1953. The concept of pattern in ecology. *Proc. Nat. Acad. Sci.* 105:1–12.

Hutchinson, G.E. 1957a. Concluding remarks. *Cold Spring Harbor Symp. Quant. Biol.* 22:415–427.

Hutchinson, G.E. 1957b. *A treatise on limnology*. Vol. I. *Geography, physics, and chemistry*. Wiley, New York. 1015 pp.

Hutchinson, G.E. 1959. Homage to Santa Rosalia, or why are there so many kinds of animals? *Amer. Natur.* 93:145–159.

Hutchinson, G.E. 1961. The paradox of the plankton. *Amer. Natur.* 95:137–145.

Hutchinson, G.E. 1965. *The ecological theater and the evolutionary play.* Yale Univ. Press, New Haven, Conn. 139 pp.

Hutchinson, G.E. 1967. *A treatise on limnology.* Vol. II. *Introduction to lake biology and the limnoplankton.* Wiley, New York. 1115 pp.

Hutchinson, G.E., and R.H. MacArthur. 1959. A theoretical ecological model of size distributions among species of animals. *Amer. Natur.* 93:117–125.

Istock, C.A. 1967. The evolution of complex life cycle phenomena: an ecological perspective. *Evolution* 21:592–605.

Istock, C.A. 1971. Modern environmental deterioration as a natural process. *Int. J. Environ. Studies* 1:151–155.

Janzen, D.H. 1966. Coevolution of mutualism between ants and acacias in Central America. *Evolution* 20:249–275.

Janzen, D.H. 1967. Fire, vegetation structure, and the ant-acacia interaction in Central America. *Ecology* 48:26–35.

Janzen, D.H. 1970. Herbivores and the number of tree species in tropical forests. *Amer. Natur.* 104:501–528.

Janzen, D.H. 1971a. Euglossine bees as long-distance pollinators of tropical plants. *Science* 171:203–205.

Janzen, D.H. 1971b. Seed predation by animals. *Ann. Rev. Ecol. Syst.* 2:465–492.

Jarrett, H. (Ed.). 1969. *Environmental quality in a growing economy.* Johns Hopkins Press, Baltimore.

Jelgersma, S. 1966. Sea-level changes during the last 10,000 years. pp. 54–71. In J.S. Sawyer (Ed.), *World climate from* 8,000 *to* 0 B.C. *Proc. Int. Symp. on World Climate* 8,000 *to* 0 B.C. Imperial College, London 1966. Royal Meteorological Society, London.

Jenny, H. 1941. *Factors of soil formation.* McGraw-Hill, New York. 281 pp.

Joffe, J.S. 1949. *Pedology.* Pedology, New Brunswick, N.J. 662 pp.

Johnson, C.E. (Ed.). 1970. *Eco-crisis.* Wiley, New York. 182 pp.

Johnson, H.D. (Ed.). 1970. *No deposit-no return.* Addison-Wesley, Reading, Mass. 351 pp.

Johnson, M.P., and S.A. Cook. 1968. "Clutch size" in buttercups. *Amer. Natur.* 102: 405–411.

Johnson, M.P., L.G. Mason, and P.H. Raven. 1968. Ecological parameters and plant species diversity. *Amer. Natur.* 102:297–306.

Johnston, R.F. 1954. Variation in breeding season and clutch size in song sparrows of the Pacific coast. *Condor* 56:268–273.

Jones, D.A. 1962. Selective eating of the acyanogenic form of the plant *Lotus corniculatus* L. by various animals. *Nature* 193:1109–1110.

Jones, D. A. 1966. On the polymorphism of cyanogenesis in *Lotus corniculatus*. Selection by animals. *Canad. J. Genet. Cytol.* 8:556–567.

Jukes, T. H. 1966. *Molecules and evolution*. Columbia Univ. Press, New York. 285 pp.

Keith, L. B. 1963. *Wildlife's ten-year cycle*. Univ. of Wisconsin Press, Madison. 201 pp.

Kendeigh, S. C. 1961. *Animal Ecology*. Prentice-Hall, Englewood Cliffs, N.J. 468 pp.

Kershaw, K. A. 1964. *Quantitative and dynamic ecology*. Arnold, London. 183 pp.

Kettlewell, H. B. D. 1956. Further selection experiments on industrial melanism in the Lepidoptera. *Heredity* 10:287–301.

Kettlewell, H.B.D. 1958. Industrial melanism in the Lepidoptera and its contribution to our knowledge of evolution. *Proc. 10th Int. Congr. Entomol.* 2:831–841.

Keyfitz, N., and W. Flieger. 1971. *Populations: facts and methods of demography*. Freeman, San Francisco.

King, C.E. 1971. Resource specialization and equilibrium population size in patchy environments. *Proc. Nat. Acad. Sci.* 68:2634–2637.

King, C. E., and W. W. Anderson. 1971. Age-specific selection. II. The interaction between *r* and *K* during population growth. *Amer. Natur.* 105: 137–156.

Kircher, H. W., and W. B. Heed. 1970. Phytochemistry and host plant specificity in *Drosophila*. pp. 191–209. In C. Steelink and V. C. Runeckles (Eds.), *Recent advances in phytochemistry*, Vol. 3. Appleton-Century-Crofts, New York.

Kircher, H. W., W. B. Heed, J. S. Russell, and J. Grove. 1967. Senita cactus alkaloids: their significance to Sonoran desert *Drosophila* ecology. *J. Insect Physiol.* 13:1869–1874.

Klomp, H. 1970. The determination of clutch-size in birds. *Ardea* 58:1–124.

Klopfer, P. H. 1962. *Behavioral aspects of ecology*. Prentice-Hall, Englewood Cliffs, N.J. 171 pp.

Klopfer, P. H., and R. H. MacArthur. 1960. Niche size and faunal diversity. *Amer. Natur.* 94:293–300.

Klopfer, P. H., and R. H. MacArthur. 1961. On the causes of tropical species diversity: niche overlap. *Amer. Natur.* 95:223–226.

Knight, C. B. 1965. *Basic concepts of ecology*. Macmillan, New York. 468 pp.

Kohn, A. J. 1959. The ecology of *Conus* in Hawaii. *Ecol. Monogr.* 29:47–90.

Kohn, A. J. 1968. Microhabitats, abundance and food of *Conus* on atoll reefs in the Maldive and Chagos Islands. *Ecology* 49: 1046–1062.

Kolman, W. A. 1960. The mechanism of natural selection for the sex ratio. *Amer. Natur.* 94:373–377.

Kormondy, E. J. 1969. *Concepts of ecology*. Prentice-Hall, Englewood Cliffs, N.J. 209 pp.

Kozlovsky, D.G. 1968. A critical evaluation of the trophic level concept. I. Ecological efficiencies. *Ecology* 49:48–60.

Krebs, C.J. 1964. The lemming cycle at Baker Lake, Northwest Territories, during 1959–62. *Arctic Inst. of North America Tech. Paper No. 15.* 104 pp.

Krebs, C.J. 1966. Demographic changes in fluctuating population *of Microtus californicus. Ecol. Monogr.* 36:239–273.

Krebs, C.J. 1970. *Microtus* population biology: behavioral changes associated with the population cycle in *M. ochrogaster* and *M. pennsylvanicus. Ecology* 51:34–52.

Krebs, C.J. 1972. *Ecology: the experimental analysis of distribution and abundance.* Harper & Row, New York. 694 pp.

Krebs, C.J., and K.T. DeLong. 1965. A *Microtus* population with supplemental food. *J. Mammal.* 46:566–573.

Krebs, C.J., B.L. Keller, and J.H. Myers. 1971. *Microtus* population densities and soil nutrients in southern Indiana grasslands. *Ecology* 52:660–663.

Krebs, C.J., B.L. Keller, and R.H. Tamarin. 1969. *Microtus* population biology: I. Demographic changes in fluctuating populations of *M. ochrogaster* and *M. pennsylvanicus* in southern Indiana, 1965–1967. *Ecology* 50:587–607.

Kurtén, B. 1969. Continental drift and evolution. *Sci. Amer.* (March):54–64.

Lack, D. 1945. The ecology of closely related species with special reference to cormorant (*Phalacrocorax carbo*) and shag (*P. aristotelis*). *J. Anim. Ecol.* 14:12–16.

Lack, D. 1947. *Darwin's finches.* Cambridge Univ. Press, Cambridge, Eng. land. 204 pp. (Reprinted in 1961 by Harper & Row, New York. 204 pp.)

Lack, D. 1954. *The natural regulation of animal numbers.* Oxford Univ. Press, New York. 343 pp.

Lack, D. 1966. *Population studies of birds.* Oxford Univ. Press, New York. 341 pp.

Lack, D. 1968. *Ecological adaptations for breeding in birds.* Methuen, London. 409 pp.

Lack, D. 1971. *Ecological isolation in birds.* Blackwell, Oxford. 404 pp.

Lee, R.B., and I. De Vore (Eds.). 1972. *Man the hunter.* Aldine, Chicago. 415 pp.

Leigh, E.G., Jr. 1965. On the relation between the productivity, biomass, diversity, and stability of a community. *Proc. Nat. Acad. Sci.* 53:777–783.

Lerner, I.M., and F.K. Ho. 1961. Genotype and competitive ability of *Tribolium* species. *Amer. Natur.* 95:329–343.

Leslie, P.H., and T. Park. 1949. The intrinsic rate of natural increase of *Tribolium castaneum* Herbst. *Ecology* 30:469–477.

Levins, R. 1966. The strategy of model building in population biology. *Amer. Sci.* 54:421–431.

Levins, R. 1968. *Evolution in changing environments.* Princeton Univ. Press, Princeton, N.J. 120 pp.

Lewontin, R.C. 1965. Selection for colonizing ability. pp. 77–94. In H.G. Baker and G.L. Stebbins (Eds.), *The genetics of colonizing species.* Academic Press, New York. 588 pp.

Lewontin, R.C. 1969. The meaning of stability. *Brookhaven Symp. Biol.* 22: 13–24.

Lewontin, R.C. 1970. The units of selection. *Ann. Rev. Ecol. Syst.* 1:1–18.

Liebig, J. 1840. *Chemistry in its application to agriculture and physiology.* Taylor and Walton, London.

Lindemann, R.I. 1942. The trophic-dynamic aspect of ecology. *Ecology* 23: 399–418.

Lotka, A.J. 1922. The stability of the normal age distribution. *Proc. Nat. Acad. Sci.* 8:339–345.

Lotka, A.J. 1925. *Elements of physical biology.* Williams and Wilkins, Baltimore. (Reprinted as *Elements of mathematical biology* in 1956 by Dover, New York.) 460 pp.

Lotka, A.J. 1956. *Elements of mathematical biology.* Dover, New York. 465 pp.

Loucks, O.L. 1970. Evolution of diversity, efficiency, and community stability. *Amer. Zool.* 10:17–25.

Love, G.A., and R.M. Love (Eds.). 1970. *Ecological crisis, readings for survival.* Harcourt Brace Jovanovich, New York. 342 pp.

Lowry, W.P. 1969. *Weather and life: an introduction to biometeorology.* Academic Press, New York. 305 pp.

MacArthur, R.H. 1955. Fluctuations of animal populations, and a measure of community stability. *Ecology* 36:533–536.

MacArthur, R.H. 1957. On the relative abundance of bird species. *Proc. Nat. Acad. Sci.* 43:293–295.

MacArthur, R.H. 1958. Population ecology of some warblers of northeastern coniferous forests. *Ecology* 39:599–619.

MacArthur, R.H. 1959. On the breeding distribution pattern of North American migrant birds. *Auk* 76:318–325.

MacArthur, R.H. 1960a. On the relative abundance of species. *Amer. Natur.* 94:25–36.

MacArthur, R.H. 1960b. On the relation between reproductive value and optimal predation. *Proc. Nat. Acad. Sci.* 46:143–145.

MacArthur, R.H. 1961. Population effects of natural selection. *Amer. Natur.* 95:195–199.

MacArthur, R.H. 1962. Some generalized theorems of natural selection. *Proc. Nat. Acad. Sci.* 48: 1893–1897.

MacArthur, R.H. 1965. Patterns of species diversity. *Biol. Rev.* 40:510–533.

MacArthur, R.H. 1968. The theory of the niche. pp. 159–176. In R.C. Lewontin (Ed.), *Population biology and evolution.* Syracuse Univ. Press, Syracuse, N.Y. 205 pp.

MacArthur, R.H. 1970. Species packing and competitive equilibrium for many species. *Theoret. Pop. Biol.* 1:1–11.

MacArthur, R.H. 1971. Patterns of terrestrial bird communities. Chapter 5 (pp. 189–221). In D.S. Farner and J.R. King (Eds.), *Avian Biology.* Vol. I. Academic Press, New York. 586 pp.

MacArthur, R.H. 1972. *Geographical ecology: patterns in the distribution of species.* Harper & Row, New York. 269 pp.

MacArthur, R.H., and J.H. Connell. 1966. *The biology of populations.* Wiley, New York. 200 pp.

MacArthur, R.H., and R. Levins. 1964. Competition, habitat selection, and character displacement in a patchy environment. *Proc. Nat. Acad. Sci.* 51:1207–1210.

MacArthur, R.H., and R. Levins. 1967. The limiting similarity, convergence, and divergence of coexisting species. *Amer. Natur.* 101:377–385.

MacArthur, R.H., and J.W. MacArthur. 1961. On bird species diversity. *Ecology* 42:594–598.

MacArthur, R.H., and E.R. Pianka. 1966. On optimal use of a patchy environment. *Amer. Natur.* 100:603–609.

MacArthur, R.H., and E.O. Wilson. 1963. An equilibrium theory of insular zoogeography. *Evolution* 17:373–387.

MacArthur, R.H., and E.O. Wilson. 1967. *The theory of island biogeography.* Princeton Univ. Press, Princeton, N.J. 203 pp.

MacArthur, R.H., J.M. Diamond, and J.R. Karr. 1972. Density compensation in island faunas. *Ecology* 53:330–342.

MacArthur, R.H., J.W. MacArthur, and J. Preer. 1962. On bird species diversity. II. Prediction of bird census from habitat measurements. *Amer. Natur.* 96:167–174.

MacArthur, R.H., H. Recher, and M. Cody. 1966. On the relation between habitat selection and species diversity. *Amer. Natur.* 100:319–332.

MacFadyen, A. 1963. *Animal ecology.* Pitman, London. 344 pp.

McIntosh, R.P. 1967. The continuum concept of vegetation. *Bot. Rev.* 33:130–187.

McLaren, I.A. 1971. *Natural regulation of animal populations.* Atherton, New York. 195 pp.

McNab, B.K. 1963. Bioenergetics and the determination of home range size. *Amer. Natur.* 97:133–140.

McNaughton, S.J., and L.L. Wolf. 1970. Dominance and the niche in ecological systems. *Science* 167:131–139.

Maguire, B. 1967. A partial analysis of the niche. *Amer. Natur.* 101:515–523.

Maguire, B. 1973. Niche response structure and the analytic potentials of its relationship to the habitat. *Amer. Natur.* 107:213–246.

Mann, K.H. 1969. The dynamics of aquatic ecosystems. *Adv. Ecol. Res.* 6: 1–81.

Margalef, R. 1958a. Information theory in ecology. *Gen. Syst.* 3:36–71.

Margalef, R. 1958b. Temporal succession and spatial heterogeneity in phytoplankton. In Buzzati-Traverso (Ed.), *Perspectives in marine biology.* Univ. of California Press, Berkeley. 621 pp.

Margalef, R. 1963. On certain unifying principles in ecology. *Amer. Natur.* 97:357–374.

Margalef, R. 1968. *Perspectives in ecological theory.* Univ. of Chicago Press, Chicago. 111 pp.

Margalef, R. 1969. Diversity and stability: a practical proposal and a model of interdependence. *Brookhaven Symp. Biol.* 22:25–37.

Marine, G. 1969. *America the raped.* Discus Books, New York.

Martin, P., and P.J. Mehringer, Jr. 1965. Pleistocene pollen analysis and biogeography of the southwest. pp. 433–451. In Wright and Fry (Eds.), *The quaternary of the U.S.* Princeton Univ. Press, Princeton, N.J.

Marx, W. 1967. *The frail ocean.* Coward-McCann, New York.

May, R.M. 1973. *Stability and complexity in model ecosystems.* Princeton Univ. Press, Princeton, N.J.

May, R.M., and R.H. MacArthur. 1972. Niche overlap as a function of environmental variability. *Proc. Nat. Acad. Sci.* 69:1109–1113.

Maynard Smith, J. 1956. Fertility, mating behavior, and sexual selection in *Drosophila subobscura. J. Genet.* 54:261–279.

Maynard Smith, J. 1958. *The theory of evolution.* Penguin, Baltimore. 320 pp.

Maynard Smith, J. 1964. Group selection and kin selection: a rejoinder. *Nature* 201:1145–1147.

Maynard Smith, J. 1968. *Mathematical ideas in biology.* Cambridge Univ. Press, Cambridge. 152 pp.

Maynard Smith, J. 1971. The origin and maintenance of sex. pp. 163–175. In G.C. Williams (Ed.), *Group selection.* Aldine, Chicago. 210 pp.

Mayr, E. 1959. Where are we? *Cold Spring Harbor Symp. Quant. Biol.* 24: 1–14.

Mayr, E. 1961. Cause and effect in biology. *Science* 134:1501–1506.

Meadows, D.H., D.L. Meadows, J. Randers, and W.W. Behrens, III. 1972. *The limits to growth: a global challenge.* Universe, New York. 205 pp.

Medawar, P.B. 1957. *The uniqueness of the individual.* Methuen, London. 191 pp.

Mendel, G. 1865. Versuche über Pflanzenhybriden. *Verh. naturforsch. Verein Brunn* 4:3–17. (Translated and reprinted in W. Bateson. 1909. *Mendel's principles of heredity.* Cambridge Univ. Press, Cambridge.)

Menge, B.A. 1972. Foraging strategy of a starfish in relation to actual prey availability and environmental predictability. *Ecol. Monogr.* 42:25–50.

Merriam, C.H. 1890. Results of a biological survey of the San Francisco mountain region and the desert of the Little Colorado, Arizona. *North American Fauna* 3:1–113.

Mertz, D.B. 1970. Notes on methods used in life-history studies. pp. 4–17. In J.H. Connell, D.B. Mertz, and W.W. Murdoch (Eds.), *Readings in ecology and ecological genetics*. Harper & Row, New York. 397 pp.

Mertz, D.B. 1971a. Life history phenomena in increasing and decreasing populations. pp. 361–399. In E.C. Pielou and W.E. Waters (Eds.), *Statistical ecology*. Vol. II. *Sampling and modeling biological populations and population dynamics*. Pennsylvania State Univ. Press, University Park.

Mertz, D.B. 1971b. The mathematical demography of the California condor population. *Amer. Natur.* 105:437–453.

Mettler, L.E., and T.G. Gregg. 1969. *Population genetics and evolution*. Prentice-Hall, Englewood Cliffs, N.J. 212 pp.

Meyer, B.S., D.B. Anderson, and R.H. Bohning. 1960. *Introduction to plant physiology*. Van Nostrand, New York. 784 pp.

Millar, J.S. 1973. Evolution of litter size in the pika, *Ochotona princeps*. *Evolution* 27:134–143.

Miller, R.S. 1967. Pattern and process in competition. *Adv. Ecol. Res.* 4:1–74.

Milne, A. 1961. Definition of competition among animals. pp. 40–61. In F.L. Milthorpe (Ed.), *Mechanisms in biological competition*. Symp. Soc. Exp. Biol. No. 15. Cambridge Univ. Press, London.

Milthorpe, F.L. (Ed.). 1961. *Mechanisms in biological competition*. Symp. Soc. Exp. Biol. No. 15. Cambridge Univ. Press, London. 365 pp.

Mishan, E.J. 1970. *Technology and growth: the price we pay*. Praeger, New York.

Mohr, C.O. 1940. Comparative populations of game, fur and other mammals. *Amer. Midl. Natur.* 24:581–584.

Mohr, C.O. 1943. Cattle droppings as ecological units. *Ecol. Monogr.* 13:275–298.

Morris, D. 1967. *The naked ape*. McGraw-Hill, New York. 205 pp.

Morse, D.H. 1971. The insectivorous bird as an adaptive strategy. *Ann. Rev. Ecol. Syst.* 2:177–200.

Motomura, I. 1932. A statistical treatment of associations (in Japanese). *Japan. J. Zool.* 44:379–383.

Murdoch, W.W. 1966a. Community structure, population control, and competition — a critique. *Amer. Natur.* 100:219–226.

Murdoch, W.W. 1966b. Population stability and life history phenomena. *Amer. Natur.* 100:5–11.

Murdoch, W.W. 1969. Switching in general predators: experiments on predator specificity and stability of prey populations. *Ecol. Monogr.* 39:335–354.

Murdoch, W.W. 1970. Population regulation and population inertia. *Ecology* 51:497–502.

Murdoch, W.W. (Ed.). 1971. *Environment: resources, pollution and society.* Sinauer, Stamford, Conn. 440 pp.

Murdoch, W.W., F.C. Evans, and C.H. Peterson. 1972. Diversity and pattern in plants and insects. *Ecology* 53:819–829.

Murphy, G.I. 1968. Pattern in life history and the environment. *Amer. Natur.* 102:391–403.

National Academy of Science. 1969. *Eutrophication: causes, consequences and correctives.* Int. Symp. Eutrophication, Washington, D.C. 661 pp.

Neill, W.E. 1972. Effects of size-selective predation on community structure in laboratory aquatic microcosms. Ph.D. Dissertation, Univ. of Texas, Austin. 177 pp.

Newbigin, M.I. 1936. *Plant and animal geography.* Methuen, London. 298 pp.

Newell, N.D. 1949. Phyletic size increase: an important trend illustrated by fossil invertebrates. *Evolution* 3:103–124.

Neyman, J., T. Park, and E.L. Scott. 1956. Struggle for existence. The *Tribolium* model: biological and statistical aspects. pp. 41–79. In *Proc. 3rd Berkeley symp. on mathematical statistics and probability.* Vol. IV. Univ. California Press, Berkeley.

Nicholson, A.J. 1933. The balance of animal populations. *J. Anim. Ecol.* 2:132–178.

Nicholson, A.J. 1954. An outline of the dynamics of animal populations. *Aust. J. Zool.* 2:9–65.

Nicholson, A.J. 1957. The self-adjustment of populations to change. *Cold Spring Harbor Symp. Quant. Biol.* 22:153–173.

Odum, E.P. 1959. *Fundamentals of ecology* (2nd ed.). Saunders, Philadelphia. 564 pp.

Odum, E.P. 1963. *Ecology.* Holt, Rinehart and Winston, New York. 152 pp.

Odum, E.P. 1968. Energy flow in ecosystems: a historical review. *Amer. Zool.* 8:11–18.

Odum, E.P. 1969. The strategy of ecosystem development. *Science* 164:262–270.

Odum, E.P. 1971. *Fundamentals of ecology* (3rd ed.). Saunders, Philadelphia. 574 pp.

Odum, H.T. 1971. *Environment, power, and society.* Wiley, New York. 331 pp.

Oparin, A.I. 1957. *The origin of life on the earth* (3rd ed.). Oliver and Boyd, London. 495 pp.

Oosting, H.J. 1958. *The study of plant communities* (2nd ed.). Freeman, San Francisco. 440 pp.

Orians, G.H. 1962. Natural selection and ecological theory. *Amer. Natur.* 96: 257–263.

Orians, G.H. 1969a. The number of bird species in some tropical forests. *Ecology* 50:783–797.

Orians, G.H. 1969b. On the evolution of mating systems in birds and mammals. *Amer. Natur.* 103:589–603.

Orians, G.H., and H.S. Horn. 1969. Overlap in foods and foraging of four species of blackbirds in the potholes of central Washington. *Ecology* 50:930–938.

Orians, G.H., and M.F. Willson. 1964. Interspecific territories of birds. *Ecology* 45:736–745.

Osborn, F. 1948. *Our plundered planet.* Little, Brown, Boston.

Otte, D., and K. Williams. 1972. Environmentally induced color dimorphisms in grasshoppers, *Syrbula admirablis, Dichromorpha viridis,* and *Chortophaga viridifasciata. Ann. Entomol. Soc. Amer.* 65:1154–1161.

Paddock, W., and P. Paddock. 1967. *Famine — 1975!* Little, Brown, Boston.

Paine, R.T. 1966. Food web complexity and species diversity. *Amer. Natur.* 100:65–76.

Paine, R.T. 1971. The measurement and application of the calorie to ecological problems. *Ann. Rev. Ecol. Syst.* 2:145–164.

Park, T. 1948. Experimental studies of interspecific competition. I. Competition between populations of flour beetles *Tribolium confusum* Duval and *T. castaneum* Herbst. *Physiol. Zool.* 18:265–308.

Park, T. 1954. Experimental studies of interspecific competition. II. Temperature, humidity, and competition in two species of *Tribolium. Physiol. Zool.* 27:177–238.

Park, T. 1962. Beetles, competition, and populations. *Science* 138:1369–1375.

Park, T., P.H. Leslie, and D.B. Mertz. 1964. Genetic strains and competition in populations of *Tribolium. Physiol. Zool.* 37:97–162.

Parker, B.C., and B.L. Turner. 1961. "Operational niche" and "community-interaction values" as determined from *in vitro* studies of some soil algae. *Evolution* 15:228–238.

Patten, B.C. 1959. An introduction to the cybernetics of the ecosystem: the trophic-dynamic aspect. *Ecology* 40:221–231.

Patten, B.C. 1961. Competitive exclusion. *Science* 134:1599–1601.

Patten, B.C. 1962. Species diversity in net phytoplankton of Raritan Bay. *J. Marine Res.* 20:57–75.

Patten, B.C. (Ed.). 1971. *Systems analysis and simulation in ecology.* Vol. I. Academic Press, New York. 607 pp.

Patten, B.C. (Ed.). 1972. *Systems analysis and simulation in ecology.* Vol. II. Academic Press, New York. 592 pp.

Pearl, R. 1922. *The biology of death.* Lippincott, Philadelphia. 275 pp.

Pearl, R. 1927. The growth of populations. *Quart. Rev. Biol.* 2:532–548.

Pearl, R. 1928. *The rate of living.* Knopf, New York.

Pearl, R. 1930. *The biology of population growth.* Knopf, New York. 260 pp.

Perrins, C.M. 1964. Survival of young swifts in relation to brood-size. *Nature* 201:1147–1149.

Perrins, C.M. 1965. Population fluctuations and clutch size in the great tit, *Parus major* L. *J. Anim. Ecol.* 34:601–647.

Phillipson, J. 1966. *Ecological energetics*. Edward Arnold, London. 57 pp.

Pianka, E.R. 1966a. Latitudinal gradients in species diversity: a review of concepts. *Amer. Natur.* 100:33–46.

Pianka, E.R. 1966b. Convexity, desert lizards, and spatial heterogeneity. *Ecology* 47:1055–1059.

Pianka, E.R. 1969. Sympatry of desert lizards (*Ctenotus*) in western Australia. *Ecology* 50:1012–1030.

Pianka, E.R. 1970. On *r*- and *K*- selection. *Amer. Natur.* 104:592–597

Pianka, E.R. 1971a. Species diversity. pp. 401–406. In *Topics in the study of life: the bio source book*. Harper & Row, New York. 482 pp.

Pianka, E.R. 1971b. Ecology of the agamid lizard *Amphibolurus isolepis* in Western Australia. *Copeia* 1971:527–536.

Pianka, E.R. 1972. *r* and *K* selection or *b* and *d* selection? *Amer. Natur.* 106: 581–588.

Pianka, E.R. 1973. The structure of lizard communities. *Ann. Rev. Ecol. Syst.* 4:53–74.

Pielou, E.C. 1969. *An introduction to mathematical ecology*. Wiley-Interscience, New York. 286 pp.

Pielou, E.C. 1972. Niche width and niche overlap: a method for measuring them. *Ecology* 53:687–692.

Pimentel, D. 1968. Population regulation and genetic feedback. *Science* 159: 1432–1437.

Pitelka, F.A. 1964. The nutrient-recovery hypothesis for arctic microtine cycles. I. Introduction. pp. 55–56. In D.J. Crisp (Ed.), *Grazing in terrestrial and marine environments*. Brit. Ecol. Soc. Symposium.

Pittendrigh, C.S. 1961. Temporal organization in living systems. *Harvey Lecture Series* 56:93–125. Academic Press, New York.

Platt, J.R. 1964. Strong inference. *Science* 146:347–353.

Ponnamperuma, C. 1972. *The origins of life*. Dutton, New York. 215 pp.

Potter, V.R. 1971. *Bioethics, bridge to the future*. Prentice-Hall, Englewood Cliffs, N.J. 205 pp.

Poulson, T.L., and D.C. Culver. 1969. Diversity in terrestrial cave communities. *Ecology* 50:153–158.

Preston, F.W. 1948. The commonness and rarity of species. *Ecology* 29: 254–283.

Preston, F.W. 1960. Time and space and the variation of species. *Ecology* 41: 611–627.

Preston, F.W. 1962a. The canonical distribution of commonness and rarity. Part I. *Ecology* 43:185–215.

Preston, F.W. 1962b. The canonical distribution of commonness and rarity. Part II. *Ecology* 43:410–432.

Rapport, D.J. 1971. An optimization model of food selection. *Amer. Natur.* 105:575–587.

Ratcliffe, J. 1970. Changes attributable to pesticides in egg breakage frequency and egg shell thickness in some British birds. *J. Applied Ecology* 7: 67–115.

Raunkaier, C. 1934. *The life form of plants and statistical plant geography.* Clarendon, Oxford. 632 pp.

Recher, H.F. 1969. Bird species diversity and habitat diversity in Australia and North America. *Amer. Natur.* 103:75–80.

Reichle, D. (Ed.). 1970. *Analysis of temperate forest ecosystems.* Springer-Verlag, Heidelberg, Berlin. 304 pp.

Richards, P.W. 1952. *The tropical rain forest.* Cambridge Univ. Press, New York. 450 pp.

Ricklefs, R.E. 1966. The temporal component of diversity among species of birds. *Evolution* 20:235–242.

Ricklefs, R.E. 1973. *Ecology.* Chiron Press, Portland, Oregon. 861 pp.

Ricklefs, R.E., and G.W. Cox. 1972. The taxon cycle in the land bird fauna of the West Indies. *Amer. Natur.* 106:195–219.

Rosenzweig, M.L. 1968. Net primary productivity of terrestrial communities: prediction from climatological data. *Amer. Natur.* 102:67–74.

Rosenzweig, M.L. 1971. The paradox of enrichment: destabilization of exploitation ecosystems in ecological time. *Science* 171:385–387.

Rosenzweig, M.L. 1973a. Exploitation in three trophic levels. *Amer. Natur.* 107:275–294.

Rosenzweig, M.L. 1973b. Evolution of the predator isocline. *Evolution* 27: 84–94.

Rosenzweig, M.L., and R.H. MacArthur. 1963. Graphical representation and stability conditions of predator–prey interactions. *Amer. Natur.* 97:209–223.

Ross, H.H. 1957. Principles of natural coexistence indicated by leafhopper populations. *Evolution* 11:113–129.

Ross, H.H. 1958. Further comments on niches and natural coexistence. *Evolution* 12:112–113.

Roughgarden, J. 1971. Density-dependent natural selection. *Ecology* 52:453–468.

Roughgarden, J. 1972. Evolution of niche width. *Amer. Natur.* 106:683–718.

Royama, T. 1969. A model for the global variation of clutch size in birds. *Oikos* 20:562–567.

Royama, T. 1970. Factors governing the hunting behaviour and selection of food by the great tit (*Parus major* L.). *J. Anim. Ecol.* 39:619–668.

Rudd, R.L. 1964. *Pesticides and the living landscape.* Univ. of Wisconsin Press, Madison. 320 pp.

Ruttner, F. 1953. *Fundamentals of limnology.* Univ. of Toronto Press, Toronto Canada. 242 pp.

Salisbury, E.J. 1942. *The reproductive capacity of plants; studies in quantitative biology.* Bell and Sons, London. 244 pp.

Salt, G.W. 1967. Predation in an experimental protozoan population (*Woodruffia-Paramecium*). *Ecol. Monogr.* 37:113–144.

Salthe, S.N. 1972. *Evolutionary biology*. Holt, Rinehart and Winston, New York. 437 pp.

Savage, J.M. 1958. The concept of ecologic niche with reference to the theory of natural coexistence. *Evolution* 12:111–121.

Sawyer, J.S. (Ed.). 1966. *World climate from 8,000 to 0* B.C. *Proc. Int. Symp. on World Climate 8,000 to 0* B.C., Imperial College, London. Royal Meteorological Society, London. 229 pp.

Sax, K. 1955. *Standing room only*. Beacon, Boston. 209 pp.

Schaffer, W.M., and R.H. Tamarin. 1973. Changing reproductive rates and population cycles in lemmings and voles. *Evolution* 27:111–124.

Schaller, F. 1968. *Soil animals*. Univ. of Michigan Press, Ann Arbor. 144 pp.

Schmidt-Nielson, K. 1964. *Desert animals: physiological problems of heat and water*. Oxford Univ. Press, London. 277 pp.

Schoener, T.W. 1965. The evolution of bill size differences among sympatric congeneric species of birds. *Evolution* 19:189-213.

Schoener, T.W. 1967. The ecological significance of sexual dimorphism in size in the lizard *Anolis conspersus*. *Science* 155:474–477.

Schoener, T.W. 1968a. The *Anolis* lizards of Bimini: resource partitioning in a complex fauna. *Ecology* 49:704–726.

Schoener, T.W. 1968b. Sizes of feeding territories among birds. *Ecology* 49: 123–141.

Schoener, T.W. 1969a. Models of optimal size for solitary predators. *Amer. Natur.* 103:277–313.

Schoener, T.W. 1969b. Optimal size and specialization in constant and fluctuating environments: an energy-time approach. *Brookhaven Symp. Biol.* 22:103–114.

Schoener, T.W. 1970. Nonsynchronous spatial overlap of lizards in patchy habitats. *Ecology* 51:408–418.

Schoener, T.W. 1971. Theory of feeding strategies. *Ann. Rev. Ecol. Syst.* 2: 369–404.

Schoener, T.W. 1973. Population growth regulated by intraspecific competition for energy or time: some simple representations. *Theoret. Pop. Biol.* 4:56–84.

Schoener, T.W., and G.C. Gorman. 1968. Some niche differences in three Lesser Antillean lizards of the genus *Anolis*. *Ecology* 49:819–830.

Schoener, T.W., and D. Janzen. 1968. Notes on environmental determinants of tropical versus temperate insect size patterns. *Amer. Natur.* 102: 207–224.

Schultz, A.M. 1964. The nutrient-recovery hypothesis for arctic microtine cycles. pp. 57–68. In D.J. Crisp (Ed.), *Grazing in terrestrial and marine environments*. Brit. Ecol. Soc. Symposium.

Schultz, A. M. 1969. A study of an ecosystem: the arctic tundra. pp. 77–93. In G. Van Dyne (Ed.), *The ecosystem concept in natural resource management*. Academic Press, New York.

Selander, R. K. 1965. On mating systems and sexual selection. *Amer. Natur.* 99:129–141.

Selander, R. K. 1966. Sexual dimorphism and differential niche utilization in birds. *Condor* 68:113–151.

Selander, R. K. 1972. Sexual selection and dimorphism in birds. pp. 180–230. In B. G. Campbell (Ed.), *Sexual selection and the descent of man (1871–1971)*. Aldine-Atherton, Chicago.

Selander, R. K., and W. E. Johnson. 1972. Genetic variation among vertebrate species. *Proc XVII Int. Congr. Zool.*, 1972. (abridged version to appear in *Ann. Rev. Ecol. Syst.* 4.)

Shannon, C. E. 1948. The mathematical theory of communication. pp. 3–91. In Shannon and Weaver (Eds.), *The mathematical theory of communication*. Univ. Illinios Press, Urbana. 117 pp.

Shelford, V. E. 1913a. *Animal communities in temperate America*. Univ. of Chicago Press, Chicago. 368 pp.

Shelford, V. E. 1913b. The reactions of certain animals to gradients of evaporating power and air. A study in experimental ecology. *Biol. Bull.* 25: 79–120.

Shelford, V. E. 1963. *The ecology of North America*. Univ. of Illinois Press, Urbana. 610 pp.

Shepard, P., and D. McKinley (Eds.). 1969. *The subversive science: essays towards an ecology of man*. Houghton-Mifflin, Boston. 453 pp.

Sheppard, P. M. 1959. *Natural selection and heredity*. Hutchinson Univ. Library, London. 212 pp.

Shimwell, D. W. 1971. *Description and classification of vegetation*. Univ. Washington Press, Seattle. 264 pp.

Shugart, H.H., and B.C. Patten. 1972. Niche quantification and the concept of niche pattern. pp. 284–327. In B. Patten (Ed.), *Systems analysis and simulation in ecology*. Vol. II. Academic Press, New York. 592 pp.

Simberloff, D. S., and E. O. Wilson. 1970. Experimental zoogeography of islands. A two-year record of colonization. *Ecology* 51:934–937.

Simpson, E. H. 1949. Measurement of diversity. *Nature* 163:688.

Simpson, G. G. 1969. Species density of North American recent mammals. *Syst. Zool.* 13:57–73.

Skutch, A. F. 1949. Do tropical birds rear as many young as they can nourish? *Ibis* 91:430–455.

Slobodkin, L. B. 1960. Ecological energy relationships at the population level. *Amer. Natur.* 94:213–236.

Slobodkin, L.B. 1962. Energy in animal ecology. *Adv. Ecol. Res.* 1:69–101.

Slobodkin, L. B. 1962. *Growth and regulation of animal populations*. Holt, Rinehart and Winston, New York. 184 pp.

Slobodkin, L.B. 1968. How to be a predator. *Amer. Zool.* 8:43–51.

Smith, A.D. 1940. A discussion of the application of a climatological diagram, the hythergraph, to the distribution of natural vegetation types. *Ecology* 21:184–191.

Smith, C.C. 1968. The adaptive nature of social organization in the genus of tree squirrels *Tamiasciurus*. *Ecol. Monogr.* 38:31–63.

Smith, C.C. 1970. The coevolution of pine squirrels (*Tamiasciurus*) and conifers. *Ecol. Monogr.* 40:349–371.

Smith, F.E. 1952. Experimental methods in population dynamics. A critique. *Ecology* 33:441–450.

Smith, F.E. 1954. Quantitative aspects of population growth. pp. 277–294. In E. Boell (Ed.), *Dynamics of growth processes*. Princeton Univ. Press, Princeton, N.J. 307 pp.

Smith, F.E. 1961. Density dependence in the Australian thrips. *Ecology* 42: 403–407.

Smith, F.E. 1963a. Population dynamics in *Daphnia magna* and a new model for population growth. *Ecology* 44:651–663.

Smith, F.E. 1963b. Density dependence. *Ecology* 44:220.

Smith, F.E. 1970a. Analysis of ecosystems. pp. 7–18. In D. Reichle (Ed.), *Analysis of temperate forest ecosystems*. Springer, Berlin.

Smith, F.E. 1970b. Effects of enrichment in mathematical models. pp. 631–645. In *Eutrophication: causes, consequences, correctives*. National Acad. Sciences, Washington, D.C.

Smith, F.E. 1972. Spatial heterogeneity, stability, and diversity in ecosystems. pp. 309–335. In E.S. Deevey (Ed.), *Growth by intussusception: Ecological essays in honor of G. Evelyn Hutchinson. Trans. Conn. Acad. Arts. Sci.* 44:1–443.

Smith, H.W. 1952. *Man and his gods*. Little, Brown, Boston. 501 pp.

Smith, N. 1968. The advantage of being parasitized. *Nature* 219:690–694.

Smith, R. 1966. *Ecology and field biology*. Harper & Row, New York. 686 pp.

Smouse, P.E. 1971. The evolutionary advantages of sexual dimorphism. *Theoret. Pop. Biol.* 2:469–481.

Sokal, R.R. 1970. Senescence and genetic load: evidence from *Tribolium*. *Science* 167:1733–1734.

Solomon, M.E. 1949. The natural control of animal populations. *J. Anim. Ecol.* 18:1–32.

Solomon, M.E. 1972. *Population dynamics*. Edward Arnold, London.

Somero, G.N. 1969. Enzymic mechanisms of temperature compensation. *Amer. Natur.* 103:517–530.

Soulé, M. 1971. The variation problem: the gene flow-variation hypothesis. *Taxon* 20:37–50.

Soulé, M., and B.R. Stewart. 1970. The "niche-variation" hypothesis: a test and alternatives. *Amer. Natur.* 104:85–97.

Southwood, T. R. E. 1966. *Ecological methods with particular reference to the study of insect populations*. Methuen, London. 391 pp.

Spicer, E. H. 1952. *Human problems in technological change*. Wiley, New York.

Spinage, C. A. 1972. African ungulate life tables. *Ecology* 53:645–652.

Stahl, E. 1888. Pflanzen und Schnecken. Biolosche Studie über die Schutzmittel der Pflanzen gegen Schneckenfrass. *Jena Z. Med. Naturw.* 22:557–684.

Stewart, R. E., and J. W. Aldrich. 1951. Removal and repopulation of breeding birds in a spruce-fir community. *Auk* 68:471–482.

Tamarin, R.H., and C.J. Krebs. 1969. *Microtus* population biology. II. Genetic changes at the transferrin locus in fluctuating populations of two vole species. *Evolution* 23:183–211.

Taylor, G. 1920. *Australian meteorology*. Clarendon Press, Oxford. 312 pp.

Teal, J. M. 1962. Energy flow in the salt marsh ecosystem of Georgia. *Ecology* 43:614–624.

Terborgh, J. 1971. Distribution on environmental gradients: theory and a preliminary interpretation of distributional patterns in the avifauna of the Cordillera Vilcabamba, Peru. *Ecology* 52:23–40.

Terborgh, J., and J. M. Diamond. 1970. Niche overlap in feeding assemblages of New Guinea birds. *Wilson Bull.* 82:29–52.

Terborgh, J., and J.S. Weske. 1969. Colonization of secondary habitats by Peruvian birds. *Ecology* 50:765–782.

Thornthwaite, C.W. 1948. An approach toward a rational classification of climate. *Geogr. Rev.* 38:55–94.

Tinbergen, N. 1957. The functions of territory. *Bird Study* 4:14–27.

Tinkle, D. W. 1967. The life and demography of the side-blotched lizard, *Uta stansburiana. Misc. Publ. Mus. Zool., Univ. Mich.* No. 132. 182 pp.

Tinkle, D. W. 1969. The concept of reproductive effort and its relation to the evolution of life histories of lizards. *Amer. Natur.* 103:501–516.

Tinkle, D. W., H. M. Wilbur, and S. G. Tilley. 1970. Evolutionary strategies in lizard reproduction. *Evolution* 24:55–74.

Tosi, J. A. 1964. Climatic control of terrestrial ecosystems: a report on the Holdridge model. *Econ. Geogr.* 40:173–181.

Tramer, E. J. 1969. Bird species diversity: components of Shannon's formula. *Ecology* 50:927–929.

Trewartha, G. T. 1943. *An introduction to weather and climate*. McGraw-Hill, New York. 545 pp.

Trivers, R. L. 1971. The evolution of reciprocal altruism. *Quart. Rev. Biol.* 46: 35–57.

Trivers, R.L. 1972. Parental investment and sexual selection. pp. 136–179. In B. G. Campbell (Ed.), *Sexual selection and the descent of man* (*1871–1971*). Aldine-Atherton, Chicago.

Trivers, R. L., and D. E. Willard. 1973. Natural selection of parental ability to vary the sex ratio of offspring. *Science* 179:90–92.

Tullock, G. 1970. The coal tit as a careful shopper. *Amer. Natur.* 104:77–80.

Turk, A., J. Turk, and J.T. Wittes. 1972. *Ecology pollution environment.* Saunders, Philadelphia. 217 pp.

Turner, F.B., G.A. Hoddenbach, P.A. Medica, and J.R. Lannom. 1970. The demography of the lizard *Uta stansburiana* (Baird and Girard), in southern Nevada. *J. Anim. Ecol.* 39:505–519.

Turner, F.B., R.I. Jennrich, and J.D. Weintraub. 1969. Home ranges and body size of lizards. *Ecology* 50:1076–1081.

Udvardy, M.D.F. 1959. Notes on the ecological concepts of habitat, biotope, and niche. *Ecology* 40:725–728.

Udvardy, M.D.F. 1969. *Dynamic zoogeography with special reference to land animals.* Van Nostrand Reinhold, New York. 445 pp.

United Nations. 1968. *Demographic Year Book.* U.N., New York.

United States Department of Agriculture. 1941. *Climate and man.* Washington, D.C. 1248 pp.

Utida, S. 1957. Population fluctuation, an experimental and theoretical approach. *Cold Spring Harbor Symp. Quant. Biol.* 22:139–151.

Van Dyne, G.M. (Ed.). 1966. *The ecosystem concept in natural resource management.* Academic Press, New York. 383 pp.

Van Valen, L. 1965. Morphological variation and width of the ecological niche. *Amer. Natur.* 94:377–390.

Van Valen, L. 1971. Group selection and the evolution of dispersal. *Evolution* 25:591–598.

Van Valen, L., and P.R. Grant. 1970. Variation and niche width reexamined. *Amer. Natur.* 104:589–590.

Vandermeer, J.H. 1968. Reproductive value in a population of arbitrary age distribution. *Amer. Natur.* 102:586–589.

Vandermeer, J.H. 1970. The community matrix and the number of species in a community. *Amer. Natur.* 104:73–83.

Vandermeer, J.H. 1972a. On the covariance of the community matrix. *Ecology* 53:187–189.

Vandermeer, J.H. 1972b. Niche theory. *Ann. Rev. Ecol. Syst.* 3:107–132.

Vaurie, C. 1951. Adaptive differences between two sympatric species of nuthatches (*Sitta*). *Proc. Int. Ornithol. Congr.* 19:163–166.

Verner, J. 1964. Evolution of polygamy in the long-billed marsh wren. *Evolution* 18:252–261.

Verner, J. 1965. Breeding biology of the long-billed marsh wren. *Condor* 67: 6–30.

Verner, J., and G.H. Engelsen. 1970. Territories, multiple nest building, and polygyny in the long-billed marsh wren. *Auk* 87:557–567.

Verner, J., and M.F. Willson. 1966. The influence of habitats on mating systems of North American passerine birds. *Ecology* 47:143–147.

Volterra, V. 1926a. Fluctuations in the abundance of a species considered mathematically. *Nature* 118:558–560.

Volterra, V. 1926b. Variazioni e fluttuazioni del numero d'individui in specie animali conviventi. *Mem. Acad. Lincei* 2:31–113.

Volterra, V. 1931. Variation and fluctuations of the number of individuals in animal species living together. Appendix (pp. 409–448). In R.N. Chapman (1939), *Animal ecology*. McGraw-Hill, New York.

Wagner, R.H. 1971. *Environment and man*. Norton, New York. 491 pp.

Waksman, S.A. 1952. *Soil microbiology*. Wiley, New York. 356 pp.

Wald, G. 1964. The origins of life. *Proc. Nat. Acad. Sci.* 52:595–611.

Wallace, A.R. 1876. *The geographical distribution of animals* (2 volumes). Hafner, New York. 503 pp. and 607 pp. (Reprinted in 1962).

Wallace, B. 1973. Misinformation, fitness, and kin selection. *Amer. Natur.* 107:1–7.

Walter, H. 1939. Grassland, Savanne und Busch der Arideren teile Afrikas in ihrer ökologischen Bedingtheit. *Jahrbucher für wissenschaftliche Botanik* 87:750–860.

Wangersky, P.J., and W.J. Cunningham. 1956. On time lags in equations of growth. *Proc. Nat. Acad. Sci.* 42:699–702.

Warburg, M. 1965. The evolutionary significance of the ecological niche. *Oikos* 16:205–213.

Waterman, T.H. 1968. Systems theory and biology — view of a biologist. In M.D. Mesarovic (Ed.), *Systems theory and biology*. Proc. 3rd Syst. Symp. Case Inst. Tech. Springer-Verlag, New York. 408 pp.

Watt, K.E.F. 1965. Community stability and the strategy of biological control. *Canad. Entomol.* 97:887–895.

Watt, K.E.F. (Ed.). 1966. *Systems analysis in ecology*. Academic Press, New York. 276 pp.

Watt, K.E.F. 1968. *Ecology and resource management*. McGraw-Hill, New York. 450 pp.

Watt, K.E.F. 1973. *Principles of environmental science*. McGraw-Hill, New York. 319 pp.

Watts, D. 1971. *Principles of biogeography*. McGraw-Hill, New York. 402 pp.

Weatherley, A.H. 1963. Notions of niche and competition among animals, with special reference to freshwater fish. *Nature* 197:14–17.

Weaver, J.E., and F.E. Clements. 1938. *Plant ecology* (2nd ed.). McGraw-Hill, New York. 601 pp.

Weedon, J.S., and J.B. Falls. 1959. Differential responses of male ovenbirds to recorded songs of neighboring and more distant individuals. *Auk* 76: 343–351.

Welch, P.S. 1952. *Limnology* (2nd ed.). McGraw-Hill, New York. 538 pp.

Wellington, W.G. 1957. Individual differences as a factor in population dynamics: the development of a problem. *Canad. J. Zool.* 35:293–323.

Wellington, W.G. 1960. Qualitative changes in natural populations during changes in abundance. *Canad. J. Zool.* 38:289–314.

Weyl, P.K. 1970. *Oceanography: an introduction to the marine environment.* Wiley, New York. 535 pp.

Whiteside, M.C., and R.B. Hainsworth. 1967. Species diversity in chydorid (*Cladocera*) communities. *Ecology* 48:664–667.

Whittaker, R.H. 1953. A consideration of climax theory: the climax as a population and pattern. *Ecol. Monogr.* 23:41–78.

Whittaker, R.H. 1962. Classification of natural communities. *Bot. Rev.* 28: 1–239.

Whittaker, R.H. 1965. Dominance and diversity in land plant communities. *Science* 147:250–260.

Whittaker, R.H. 1967. Gradient analysis of vegetation. *Biol. Rev.* 42:207–264.

Whittaker, R.H. 1969. Evolution of diversity in plant communities. *Brookhaven Symp. Biol.* 22:178–196.

Whittaker, R.H. 1970. *Communities and ecosystems.* Macmillan, New York. 162 pp.

Whittaker, R.H. 1972. Evolution and measurement of species diversity. *Taxon* 21:213–251.

Whittaker, R.H., and P.P. Feeny. 1971. Allelochemics: chemical interactions between species. *Science* 171:757–770.

Whittaker, R.H., and G.M. Woodwell. 1971. Evolution of a natural communities. pp. 137–159. In J.A. Wiens (Ed.), *Ecosystem structure and function.* Proc. 31st Ann. Biol. Coll., Oregon State Univ. Press.

Whittaker, R.H., S.A. Levin, and R.B. Root. 1973. Niche, habitat, and ecotope. *Amer. Natur.* 107:321–338.

Whittaker, R.H., R.B. Walker, and A.R. Kruckeberg. 1954. The ecology of serpentine soils. *Ecology* 35:258–288.

Wiegert, R.C. 1968. Thermodynamic considerations in animal nutrition. *Amer. Zool.* 8:71–81.

Wiens, J.A. 1966. Group selection and Wynne-Edward's hypothesis. *Amer. Sci.* 54:273–287.

Wilbur, H.M. 1972. Competition, predation, and the structure of the *Ambystoma-Rana sylvatica* community. *Ecology* 53:3–21.

Williams, C.B. 1944. Some applications of the logarithmic series and the index of diversity to ecological problems. *J. Ecol.* 32:1–44.

Williams, C.B. 1953. The relative abundance of different species in a wild animal population. *J. Anim. Ecol.* 22:14–31.

Williams, C.B. 1964. *Patterns in the balance of nature.* Academic Press, New York. 324 pp.

Williams, G.C. 1957. Pleiotropy, natural selection, and the evolution of senescence. *Evolution* 11:398–411.

Williams, G.C. 1966a. *Adaptation and natural selection.* Princeton Univ. Press, Princeton, N.J. 307 pp.

Williams, G.C. 1966b. Natural selection, the costs of reproduction, and a refinement of Lack's principle. *Amer. Natur.* 100:687–690.

Williams, G.C. 1971. *Group selection.* Aldine-Atherton, Chicago. 210 pp.

Williamson, M. 1971. *The analysis of biological populations.* Edward Arnold, London. 180 pp.

Willson, M.F. 1969. Avian niche size and morphological variation. *Amer. Natur.* 103:531–542.

Willson, M.F. 1971. Life history consequences of death rates. *The Biologist* 53:49–56.

Willson, M.F., and E.R. Pianka. 1963. Sexual selection, sex ratio, and mating system. *Amer. Natur.* 97:405–407.

Wilson, E.O. 1961. The nature of the taxon cycle in the Melanesian ant fauna. *Amer. Natur.* 95:169–193.

Wilson, E.O. 1969. The species equilibrium. *Brookhaven Symp. Biol.* 22:38–47.

Wilson, E.O. 1971. *The insect societies.* Belknap Press, Cambridge, Mass. 548 pp.

Wilson, E.O., and W.H. Bossert. 1971. *A primer of population biology.* Sinauer, Stamford, Conn. 192 pp.

Wilson, J.T. 1971. Continental drift. pp. 88–92. In *Topics in animal behavior, ecology, and evolution.* Harper & Row, New York. 184 pp.

Wilson, J.T. (Ed.). 1973. *Continents adrift. A collection of articles from Scientific American.* Freeman, San Francisco. 172 pp.

Wiseman, J.D.H. 1966. Evidence for recent climatic changes in cores from the ocean bed. In J.S. Sawyer (Ed.), *World climate from 8,000 to 0* B.C. Imperial College, London, 1966. Royal Meteorological Society, London.

Wolf, L.L., and F.R. Hainsworth. 1971. Time and energy budgets of territorial hummingbirds. *Ecology* 52:980–988.

Wolf, L.L., F.R. Hainsworth, and F.G. Stiles. 1972. Energetics of foraging: rate and efficiency of nectar extraction by hummingbirds. *Science* 176:1351–1352.

Woodwell, G.M. and H. Smith (Eds.). 1969. *Diversity and stability in ecological systems. Brookhaven Symp. Biol.* No. 22. Upton, N.Y. 264 pp.

Woodwell, G.M., and R.H. Whittaker. 1968. Primary production in terrestrial communities. *Amer. Zool.* 8:19–30.

Wright, H.E., and D. Frey (Eds). 1965. *The quaternary of the United States.* Princeton Univ. Press, Princeton, N.J.

Wright, S. 1931. Evolution in Mendelian populations. *Genetics* 16:97–159.

Wurster, C.F. 1968. DDT reduces photosynthesis by marine phytoplankton. *Science* 159:1474–1475.

Wurster, C.F. 1969. Chlorinated hydrocarbon insecticides and the world ecosystem. *Biol. Cons.* 1:123–129.

Wurster, C.F., and D.B. Wingate. 1968. DDT residues and declining reproduction in the Bermuda petrel. *Science* 159:979–981.

Wynne-Edwards, V.C. 1955. Low reproductive rates in birds, especially seabirds. *Acta XI Int. Orn. Congr., Basel* 1954:540–547.

Wynne-Edwards, V.C. 1962. *Animal dispersion in relation to social behaviour*. Oliver and Boyd, Edinburgh. 653 pp.

Wynne-Edwards, V.C. 1964. Group selection and kin selection. *Nature* 201: 1145–1147.

Wynne-Edwards, V.C. 1965a. Self-regulating systems in populations of animals. *Science* 147:1543–1548.

Wynne-Edwards, V.C. 1965b. Social organization as a population regulator. *Symp. Zool. Soc. London* 14:173–178.

Zeuthen, E. 1953. Oxygen uptake as related to body size in organisms. *Quart. Rev. Biol.* 28:1–12.

Zweifel, R.G., and C.H. Lowe. 1966. The ecology of a population of *Xantusia vigilis*, the desert night lizard. *Amer. Mus. Novitates* 2247:1–57.

Index

75 76 9 8 7 6 5 4 3